Foundations
of Nuclear Physics

Foundations

OF NUCLEAR PHYSICS

Facsimiles of THIRTEEN FUNDAMENTAL STUDIES AS THEY

WERE ORIGINALLY REPORTED IN THE SCIENTIFIC JOURNALS

With A BIBLIOGRAPHY *compiled by* Robert T. Beyer

Assistant Professor of Physics, Brown University

DOVER PUBLICATIONS, INC.

New York

Designed by James T. Parker

Manufactured in
the United States of America

The Publisher wishes to express his appreciation to the various journals for their kind permission to reproduce the papers herein.

C O N T E N T S

THE COLLECTION of papers in this volume has been prepared with the purpose of bringing within the covers of a single volume the original accounts of fundamental studies in nuclear physics as they were reported by the investigators themselves.

The discovery of the atomic bomb has led to the publication of many texts in which the present knowledge of the nucleus is set forth in detail. The aim of this volume is somewhat different—namely, to give the reader an historical view of nuclear physics, and to bring him into closer contact with the ideas of the leaders of this research.

A few words should be said about the method of selection. These papers represent the original reports of what the editor considers the most momentous hypotheses, discoveries, and inventions in nuclear physics. As criteria of their importance, careful notice has been taken of the frequency with which these papers have been cited by subsequent investigators and of the amount of new research which has been stimulated by them.

The selection of theoretical papers has been particularly difficult because advances in that branch have been made, for the most part, in small steps and by a great number of workers.

The editor realizes that there will be some disagreement on the present selection, particularly concerning papers which have been omitted. He believes this to be a natural consequence of the rapidity with which nuclear physics has developed, and a condition of editorship in general. He points to the limited size of this volume as his defense.

A bibliography of journal articles has been appended which is reasonably complete up to the spring of 1947. It should be pointed out that necessities of type-setting required the use of italics rather than bold-face for the volume numbers of the various journals. Also, the language of the original article is the same as that used in the name of the journal unless otherwise indicated.

ROBERT T. BEYER

Brown University
July 1947

PAPERS

The Positive Electron

CARL D. ANDERSON, *California Institute of Technology, Pasadena, California*

(Received February 28, 1933)

Out of a group of 1300 photographs of cosmic-ray tracks in a vertical Wilson chamber 15 tracks were of positive particles which could not have a mass as great as that of the proton. From an examination of the energy-loss and ionization produced it is concluded that the charge is less than twice, and is probably exactly equal to, that of the proton. If these particles carry unit positive charge the curvatures and ionizations produced require the mass to be less than twenty times the electron mass. These particles will be called positrons. Because they occur in groups associated with other tracks it is concluded that they must be secondary particles ejected from atomic nuclei.

Editor

O N August 2, 1932, during the course of photographing cosmic-ray tracks produced in a vertical Wilson chamber (magnetic field of 15,000 gauss) designed in the summer of 1930 by Professor R. A. Millikan and the writer, the tracks shown in Fig. 1 were obtained, which seemed to be interpretable only on the basis of the existence in this case of a particle carrying a positive charge but having a mass of the same order of magnitude as that normally possessed by a free negative electron. Later study of the photograph by a whole group of men of the Norman Bridge Laboratory only tended to strengthen this view. The reason that this interpretation seemed so inevitable is that the track appearing on the upper half of the figure cannot possibly have a mass as large as that of a proton for as soon as the mass is fixed the energy is at once fixed by the curvature. The energy of a proton of that curvature comes out 300,000 volts, but a proton of that energy according to well established and universally accepted determinations[1] has a total range of about 5 mm in air while that portion of the range actually visible in this case exceeds 5 cm without a noticeable change in curvature. The only escape from this conclusion would be to assume that at exactly the same instant (and the sharpness of the tracks determines that instant to within about a fiftieth of a second) two independent electrons happened to produce two tracks so placed as to give the impression of a single particle shooting through the lead plate. This assumption was dismissed on a probability basis, since a sharp track of this order of curvature under the experimental conditions prevailing occurred in the chamber only once in some 500 exposures, and since there was practically no chance at all that two such tracks should line up in this way. We also discarded as completely untenable the assumption of an electron of 20 million volts entering the lead on one side and coming out with an energy of 60 million volts on the other side. A fourth possibility is that a photon, entering the lead from above, knocked out of the nucleus of a lead atom two particles, one of which shot upward and the other downward. But in this case the upward moving one would be a positive of small mass so that either of the two possibilities leads to the existence of the positive electron.

In the course of the next few weeks other photographs were obtained which could be interpreted logically only on the positive-electron basis, and a brief report was then published[2] with due reserve in interpretation in view of the importance and striking nature of the announcement.

MAGNITUDE OF CHARGE AND MASS

It is possible with the present experimental data only to assign rather wide limits to the

---

[1] Rutherford, Chadwick and Ellis, *Radiations from Radioactive Substances*, p. 294. Assuming $R \propto v^3$ and using data there given the range of a 300,000 volt proton in air S.T.P. is about 5 mm.

[2] C. D. Anderson, Science **76**, 238 (1932).

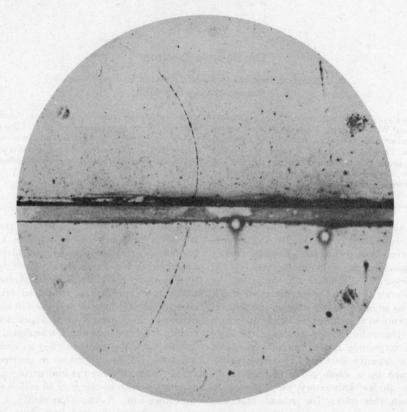

FIG. 1. A 63 million volt positron ($H\rho = 2.1 \times 10^5$ gauss-cm) passing through a 6 mm lead plate and emerging as a 23 million volt positron ($H\rho = 7.5 \times 10^4$ gauss-cm). The length of this latter path is at least ten times greater than the possible length of a proton path of this curvature.

magnitude of the charge and mass of the particle. The specific ionization was not in these cases measured, but it appears very probable, from a knowledge of the experimental conditions and by comparison with many other photographs of high- and low-speed electrons taken under the same conditions, that the charge cannot differ in magnitude from that of an electron by an amount as great as a factor of two. Furthermore, if the photograph is taken to represent a positive particle penetrating the 6 mm lead plate, then the energy lost, calculated for unit charge, is approximately 38 million electron-volts, this value being practically independent of the proper mass of the particle as long as it is not too many times larger than that of a free negative electron.

This value of 63 million volts per cm energy-loss for the positive particle it was considered legitimate to compare with the measured mean of approximately 35 million volts[3] for negative electrons of 200–300 million volts energy since the rate of energy-loss for particles of small mass is expected to change only very slowly over an energy range extending from several million to several hundred million volts. Allowance being made for experimental uncertainties, an upper limit to the rate of loss of energy for the positive particle can then be set at less than four times that for an electron, thus fixing, by the usual relation between rate of ionization and

[3] C. D. Anderson, Phys. Rev. **43**, 381A (1933).

charge, an upper limit to the charge less than twice that of the negative electron. It is concluded, therefore, that the magnitude of the charge of the positive electron which we shall henceforth contract to positron is very probably equal to that of a free negative electron which from symmetry considerations would naturally then be called a negatron.

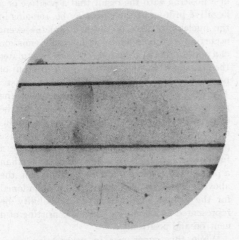

FIG. 2. A positron of 20 million volts energy ($H\rho = 7.1 \times 10^4$ gauss-cm) and a negatron of 30 million volts energy ($H\rho = 10.2 \times 10^4$ gauss-cm) projected from a plate of lead. The range of the positive particle precludes the possibility of ascribing it to a proton of the observed curvature.

It is pointed out that the effective depth of the chamber in the line of sight which is the same as the direction of the magnetic lines of force was 1 cm and its effective diameter at right angles to that line 14 cm, thus insuring that the particle crossed the chamber practically normal to the lines of force. The change in direction due to scattering in the lead,[3] in this case about 8° measured in the plane of the chamber, is a probable value for a particle of this energy though less than the most probable value.

The magnitude of the proper mass cannot as yet be given further than to fix an upper limit to it about twenty times that of the electron mass. If Fig. 1 represents a particle of unit charge passing through the lead plate then the curvatures, on the basis of the information at hand on ionization, give too low a value for the energy-loss unless the mass is taken less than

twenty times that of the negative electron mass. Further determinations of $H\rho$ for relatively low energy particles before and after they cross a known amount of matter, together with a study of ballistic effects such as close encounters with electrons, involving large energy transfers, will enable closer limits to be assigned to the mass.

To date, out of a group of 1300 photographs of cosmic-ray tracks 15 of these show positive particles penetrating the lead, none of which can be ascribed to particles with a mass as large as that of a proton, thus establishing the existence of positive particles of unit charge and of mass small compared to that of a proton. In many other cases due either to the short section of track available for measurement or to the high energy of the particle it is not possible to differentiate with certainty between protons and positrons. A comparison of the six or seven hundred positive-ray tracks which we have taken is, however, still consistent with the view that the positive particle which is knocked out of the nucleus by the incoming primary cosmic ray is in many cases a proton.

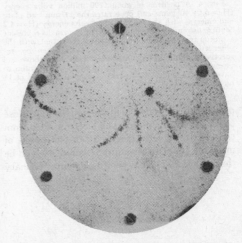

FIG. 3. A group of six particles projected from a region in the wall of the chamber. The track at the left of the central group of four tracks is a negatron of about 18 million volts energy ($H\rho = 6.2 \times 10^4$ gauss-cm) and that at the right a positron of about 20 million volts energy ($H\rho = 7.0 \times 10^4$ gauss-cm). Identification of the two tracks in the center is not possible. A negatron of about 15 million volts is shown at the left. This group represents early tracks which were broadened by the diffusion of the ions. The uniformity of this broadening for all the tracks shows that the particles entered the chamber at the same time.

From the fact that positrons occur in groups associated with other tracks it is concluded that they must be secondary particles ejected from an atomic nucleus. If we retain the view that a nucleus consists of protons and neutrons (and α-

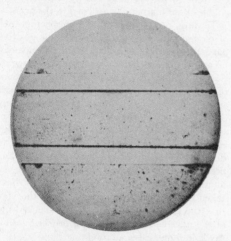

FIG. 4. A positron of about 200 million volts energy ($H\rho = 6.6 \times 10^5$ gauss-cm) penetrates the 11 mm lead plate and emerges with about 125 million volts energy ($H\rho = 4.2 \times 10^5$ gauss-cm). The assumption that the tracks represent a proton traversing the lead plate is inconsistent with the observed curvatures. The energies would then be, respectively, about 20 million and 8 million volts above and below the lead, energies too low to permit the proton to have a range sufficient to penetrate a plate of lead of 11 mm thickness.

particles) and that a neutron represents a close combination of a proton and electron, then from the electromagnetic theory as to the origin of mass the simplest assumption would seem to be that an encounter between the incoming primary ray and a proton may take place in such a way as to expand the diameter of the proton to the same value as that possessed by the negatron. This process would release an energy of a billion electron-volts appearing as a secondary photon. As a second possibility the primary ray may disintegrate a neutron (or more than one) in the nucleus by the ejection either of a negatron or a positron with the result that a positive or a negative proton, as the case may be, remains in the nucleus in place of the neutron, the event occurring in this instance without the emission of a photon. This alternative, however, postulates the existence in the nucleus of a proton of negative charge, no evidence for which exists. The greater symmetry, however, between the positive and negative charges revealed by the discovery of the positron should prove a stimulus to search for evidence of the existence of negative protons. If the neutron should prove to be a fundamental particle of a new kind rather than a proton and negatron in close combination, the above hypotheses will have to be abandoned for the proton will then in all probability be represented as a complex particle consisting of a neutron and positron.

While this paper was in preparation press reports have announced that P. M. S. Blackett and G. Occhialini in an extensive study of cosmic-ray tracks have also obtained evidence for the existence of light positive particles confirming our earlier report.

I wish to express my great indebtedness to Professor R. A. Millikan for suggesting this research and for many helpful discussions during its progress. The able assistance of Mr. Seth H. Neddermeyer is also appreciated.

692

The Existence of a Neutron.

By J. CHADWICK, F.R.S.

(Received May 10, 1932.)

§ 1. It was shown by Bothe and Becker* that some light elements when bombarded by α-particles of polonium emit radiations which appear to be of the γ-ray type. The element beryllium gave a particularly marked effect of this kind, and later observations by Bothe, by Mme. Curie-Joliot† and by Webster‡ showed that the radiation excited in beryllium possessed a penetrating power distinctly greater than that of any γ-radiation yet found from the radioactive elements. In Webster's experiments the intensity of the radiation was measured both by means of the Geiger-Müller tube counter and in a high pressure ionisation chamber. He found that the beryllium radiation had an absorption coefficient in lead of about $0 \cdot 22$ cm.$^{-1}$ as measured under his experimental conditions. Making the necessary corrections for these conditions, and using the results of Gray and Tarrant to estimate the relative contributions of scattering, photoelectric absorption, and nuclear absorption in the absorption of such penetrating radiation, Webster concluded that the radiation had a quantum energy of about 7×10^6 electron volts. Similarly he found that the radiation from boron bombarded by α-particles of polonium consisted in part of a radiation rather more penetrating than that from beryllium, and he estimated the quantum energy of this component as about 10×10^6 electron volts. These conclusions agree quite well with the supposition that the radiations arise by the capture of the α-particle into the beryllium (or boron) nucleus and the emission of the surplus energy as a quantum of radiation.

The radiations showed, however, certain peculiarities, and at my request the beryllium radiation was passed into an expansion chamber and several photographs were taken. No unexpected phenomena were observed though, as will be seen later, similar experiments have now revealed some rather striking events. The failure of these early experiments was partly due to the weakness of the available source of polonium, and partly to the experimental arrangement, which, as it now appears, was not very suitable.

* ' Z. Physik,' vol. 66, p. 289 (1930).

† I. Curie, ' C. R. Acad. Sci. Paris,' vol. 193, p. 1412 (1931).

‡ ' Proc. Roy. Soc.,' A, vol. 136, p. 428 (1932).

Quite recently, Mme. Curie-Joliot and M. Joliot* made the very striking observation that these radiations from beryllium and from boron were able to eject protons with considerable velocities from matter containing hydrogen. In their experiments the radiation from beryllium was passed through a thin window into an ionisation vessel containing air at room pressure. When paraffin wax, or other matter containing hydrogen, was placed in front of the window, the ionisation in the vessel was increased, in some cases as much as doubled. The effect appeared to be due to the ejection of protons, and from further experiment they showed that the protons had ranges in air up to about 26 cm., corresponding to a velocity of nearly 3×10^9 cm. per second. They suggested that energy was transferred from the beryllium radiation to the proton by a process similar to the Compton effect with electrons, and they estimated that the beryllium radiation had a quantum energy of about 50×10^6 electron volts. The range of the protons ejected by the boron radiation was estimated to be about 8 cm. in air, giving on a Compton process an energy of about 35×10^6 electron volts for the effective quantum.†

There are two grave difficulties in such an explanation of this phenomenon. Firstly, it is now well established that the frequency of scattering of high energy quanta by electrons is given with fair accuracy by the Klein-Nishina formula, and this formula should also apply to the scattering of quanta by a proton. The observed frequency of the proton scattering is, however, many thousand times greater than that predicted by this formula. Secondly, it is difficult to account for the production of a quantum of 50×10^6 electron volts from the interaction of a beryllium nucleus and an α-particle of kinetic energy of 5×10^6 electron volts. The process which will give the greatest amount of energy available for radiation is the capture of the α-particle by the beryllium nucleus, Be^9, and its incorporation in the nuclear structure to form a carbon nucleus C^{13}. The mass defect of the C^{13} nucleus is known both from data supplied by measurements of the artificial disintegration of boron B^{10} and from observations of the band spectrum of carbon; it is about 10×10^6 electron volts. The mass defect of Be^9 is not known, but the assumption that it is zero will give a maximum value for the possible change of energy in the reaction $Be^9 + \alpha \rightarrow C^{13} +$ quantum. On this assumption it follows that the energy of the quantum emitted in such a reaction cannot be greater than about 14×10^6 electron volts. It must, of course, be admitted that this argument

---

* Curie and Joliot, ' C. R. Acad. Sci. Paris,' vol. 194, p. 273 (1932).

† Many of the arguments of the subsequent discussion apply equally to both radiations, and the term " beryllium radiation " may often be taken to include the boron radiation.

from mass defects is based on the hypothesis that the nuclei are made as far as possible of α-particles; that the Be^9 nucleus consists of 2 α-particles + 1 proton + 1 electron and the C^{13} nucleus of 3 α-particles + 1 proton + 1 electron. So far as the lighter nuclei are concerned, this assumption is supported by the evidence from experiments on artificial disintegration, but there is no general proof.

Accordingly, I made further experiments to examine the properties of the radiation excited in beryllium. It was found that the radiation ejects particles not only from hydrogen but from all other light elements which were examined. The experimental results were very difficult to explain on the hypothesis that the beryllium radiation was a quantum radiation, but followed immediately if it were supposed that the radiation consisted of particles of mass nearly equal to that of a proton and with no net charge, or neutrons. A short statement of some of these observations was published in 'Nature.'[*] This paper contains a fuller description of the experiments, which suggest the existence of neutrons and from which some of the properties of these particles can be inferred. In the succeeding paper Dr. Feather will give an account of some observations by means of the expansion chamber of the collisions between the beryllium radiation and nitrogen nuclei, and this is followed by an account by Mr. Dee of experiments to observe the collisions with electrons.

§ 2. *Observations of Recoil Atoms.*—The properties of the beryllium radiation were first examined by means of the valve counter used in the work[†] on the artificial disintegration by α-particles and described fully there. Briefly, it consists of a small ionisation chamber connected to a valve amplifier. The sudden production of ions in the chamber by the entry of an ionising particle is detected by means of an oscillograph connected in the output circuit of the amplifier. The deflections of the oscillograph were recorded photographically on a film of bromide paper.

The source of polonium was prepared from a solution of radium $(D+E+F)$[‡] by deposition on a disc of silver. The disc had a diameter of 1 cm. and was placed close to a disc of pure beryllium of 2 cm. diameter, and both were enclosed in a small vessel which could be evacuated, fig. 1. The first ionisation chamber used had an opening of 13 mm. covered with aluminium foil of 4·5 cm. air equivalent, and a depth of 15 mm. This chamber had a very low natural effect, giving on the average only about 7 deflections per hour.

[*] 'Nature,' vol. 129, p. 312 (1932).

[†] Chadwick, Constable and Pollard, 'Proc. Roy. Soc.,' A, vol. 130, p. 463 (1931).

[‡] The radium D was obtained from old radon tubes generously presented by Dr. C. F. Burnam and Dr. F. West, of the Kelly Hospital, Baltimore.

When the source vessel was placed in front of the ionisation chamber, the number of deflections immediately increased. For a distance of 3 cm. between the beryllium and the counter the number of deflections was nearly 4 per minute. Since the number of deflections remained sensibly the same when thick metal sheets, even as much as 2 cm. of lead, were interposed between the source vessel and the counter, it was clear that these deflections were due to a penetrating radiation emitted from the beryllium. It will be shown later that the deflections were due to atoms of nitrogen set in motion by the impact of the beryllium radiation.

When a sheet of paraffin wax about 2 mm. thick was interposed in the path of the radiation just in front of the counter, the number of deflections recorded by the oscillograph increased markedly. This increase was due to particles

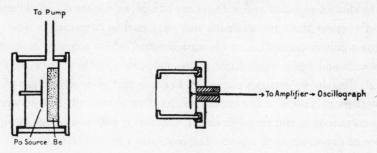

FIG. 1.

ejected from the paraffin wax so as to pass into the counter. By placing absorbing screens of aluminium between the wax and the counter the absorption curve shown in fig. 2, curve A, was obtained. From this curve it appears that the particles have a maximum range of just over 40 cm. of air, assuming that an Al foil of 1·64 mg. per square centimetre is equivalent to 1 cm. of air. By comparing the sizes of the deflections (proportional to the number of ions produced in the chamber) due to these particles with those due to protons of about the same range it was obvious that the particles were protons. From the range-velocity curve for protons we deduce therefore that the maximum velocity imparted to a proton by the beryllium radiation is about $3 \cdot 3 \times 10^9$ cm. per second, corresponding to an energy of about $5 \cdot 7 \times 10^6$ electron volts.

The effect of exposing other elements to the beryllium radiation was then investigated. An ionisation chamber was used with an opening covered with a gold foil of 0·5 mm. air equivalent. The element to be examined was fixed on a clean brass plate and placed very close to the counter opening. In this way lithium, beryllium, boron, carbon and nitrogen, as paracyanogen, were tested. In each case the number of deflections observed in the counter increased when the element was bombarded by the beryllium radiation. The ranges of the particles ejected from these elements were quite short, of the order of some millimetres in air. The deflections produced by them were of different sizes, but many of them were large compared with the deflection produced even by a slow proton. The particles therefore have a large ionising power and are probably in each case recoil atoms of the elements. Gases were investigated by filling the ionisation chamber with the required gas by circulation for several minutes. Hydrogen, helium, nitrogen, oxygen, and argon were examined in this way. Again, in each case deflections were observed which were attributed to the production of recoil atoms in the different gases. For a given position of the beryllium source relative to the counter, the number of recoil atoms was roughly the same for each gas. This point will be referred to later. It appears then that the beryllium radiation can impart energy to the atoms of matter through which it passes and that the chance of an energy transfer does not vary widely from one element to another.

It has been shown that protons are ejected from paraffin wax with energies up to a maximum of about $5 \cdot 7 \times 10^6$ electron volts. If the ejection be ascribed to a Compton recoil from a quantum of radiation, then the energy of the quantum must be about 55×10^6 electron volts, for the maximum energy which can be given to a mass m by a quantum $h\nu$ is $\dfrac{2}{2 + mc^2/h\nu} \cdot h\nu$.

The energies of the recoil atoms produced by this radiation by the same process in other elements can be readily calculated. For example, the nitrogen recoil atoms should have energies up to a maximum of 450,000 electron volts. Taking the energy necessary to form a pair of ions in air as 35 electron volts, the recoil atoms of nitrogen should produce not more than about 13,000 pairs of ions. Many of the deflections observed with nitrogen, however, corresponded to far more ions than this; some of the recoil atoms produced from 30,000 to 40,000 ion pairs. In the case of the other elements a similar discrepancy was noted between the observed energies and ranges of the recoil atoms and the values calculated on the assumption that the atoms were set in motion by recoil from a quantum of 55×10^6 electron volts. The energies of the recoil atoms were estimated from the number of ions produced in the counter, as given by the size of the oscillograph deflections. A sufficiently good measurement of the ranges could be made either by varying the distance between the element and the counter or by interposing thin screens of gold between the element and the counter.

The nitrogen recoil atoms were also examined, in collaboration with Dr. N. Feather, by means of the expansion chamber. The source vessel was placed immediately above an expansion chamber of the Shimizu type, so that a large proportion of the beryllium radiation traversed the chamber. A large number of recoil tracks was observed in the course of a few hours. Their range, estimated by eye, was sometimes as much as 5 or 6 mm. in the chamber, or, correcting for the expansion, about 3 mm. in standard air. These visual estimates were confirmed by a preliminary series of experiments by Dr. Feather with a large automatic expansion chamber, in which photographs of the recoil tracks in nitrogen were obtained. Now the ranges of recoil atoms of nitrogen of different velocities have been measured by Blackett and Lees. Using their results we find that the nitrogen recoil atoms produced by the beryllium radiation may have a velocity of at least 4×10^8 cm. per second, corresponding to an energy of about $1 \cdot 2 \times 10^6$ electron volts. In order that the nitrogen nucleus should acquire such an energy in a collision with a quantum of radiation, it is necessary to assume that the energy of the quantum should be about 90×10^6 electron volts, if energy and momentum are conserved in the collision. It has been shown that a quantum of 55×10^6 electron volts is sufficient to explain the hydrogen collisions. In general, the experimental results show that if the recoil atoms are to be explained by collision with a quantum, we must assume a larger and larger energy for the quantum as the mass of the struck atom increases.

§ 3. *The Neutron Hypothesis.*—It is evident that we must either relinquish the application of the conservation of energy and momentum in these collisions or adopt another hypothesis about the nature of the radiation. If we suppose that the radiation is not a quantum radiation, but consists of particles of mass very nearly equal to that of the proton, all the difficulties connected with the collisions disappear, both with regard to their frequency and to the energy transfer to different masses. In order to explain the great penetrating power of the radiation we must further assume that the particle has no net charge. We may suppose it to consist of a proton and an electron in close combination, the " neutron " discussed by Rutherford* in his Bakerian Lecture of 1920.

When such neutrons pass through matter they suffer occasionally close

---

* Rutherford, ' Proc. Roy. Soc.,' A, vol. 97, p. 374 (1920). Experiments to detect the formation of neutrons in a hydrogen discharge tube were made by J. L. Glasson, ' Phil. Mag.,' vol. 42, p. 596 (1921), and by J. K. Roberts, ' Proc. Roy. Soc.,' A, vol. 102, p. 72 (1922). Since 1920 many experiments in search of these neutrons have been made in this laboratory.

698   J. Chadwick.

collisions with the atomic nuclei and so give rise to the recoil atoms which are
observed. Since the mass of the neutron is equal to that of the proton, the
recoil atoms produced when the neutrons pass through matter containing
hydrogen will have all velocities up to a maximum which is the same as the
maximum velocity of the neutrons. The experiments showed that the maximum velocity of the protons ejected from paraffin wax was about $3 \cdot 3 \times 10^9$
cm. per second. This is therefore the maximum velocity of the neutrons
emitted from beryllium bombarded by α-particles of polonium. From this
we can now calculate the maximum energy which can be given by a colliding
neutron to other atoms, and we find that the results are in fair agreement with
the energies observed in the experiments. For example, a nitrogen atom will
acquire in a head-on collision with the neutron of mass 1 and velocity $3 \cdot 3 \times 10^9$
cm. per second a velocity of $4 \cdot 4 \times 10^8$ cm. per second, corresponding to an
energy of $1 \cdot 4 \times 10^6$ electron volts, a range of about $3 \cdot 3$ mm. in air, and a
production of ions of about 40,000 pairs. Similarly, an argon atom may acquire
an energy of $0 \cdot 54 \times 10^6$ electron volts, and produce about 15,000 ion pairs.
Both these values are in good accord with experiment.*

It is possible to prove that the mass of the neutron is roughly equal to that
of the proton, by combining the evidence from the hydrogen collisions with
that from the nitrogen collisions. In the succeeding paper, Feather records
experiments in which about 100 tracks of nitrogen recoil atoms have been
photographed in the expansion chamber. The measurement of the tracks
shows that the maximum range of the recoil atoms is $3 \cdot 5$ mm. in air at 15° C.
and 760 mm. pressure, corresponding to a velocity of $4 \cdot 7 \times 10^8$ cm. per second
according to Blackett and Lees. If M, V be the mass and velocity of the
neutron then the maximum velocity given to a hydrogen atom is

$$u_p = \frac{2M}{M+1} \cdot V,$$

and the maximum velocity given to a nitrogen atom is

$$u_n = \frac{2M}{M+14} \cdot V,$$

whence

$$\frac{M+14}{M+1} = \frac{u_p}{u_n} = \frac{3 \cdot 3 \times 10^9}{4 \cdot 7 \times 10^8},$$

---

* It was noted that a few of the nitrogen recoil atoms produced about 50 to 60,000 ion
pairs. These probably correspond to the cases of disintegration found by Feather and
described in his paper.

and
$$M = 1 \cdot 15.$$

The total error in the estimation of the velocity of the nitrogen recoil atom may easily be about 10 per cent., and it is legitimate to conclude that the mass of the neutron is very nearly the same as the mass of the proton.

We have now to consider the production of the neutrons from beryllium by the bombardment of the α-particles. We must suppose that an α-particle is captured by a Be^9 nucleus with the formation of a carbon C^{12} nucleus and the emission of a neutron. The process is analogous to the well-known artificial disintegrations, but a neutron is emitted instead of a proton. The energy relations of this process cannot be exactly deduced, for the masses of the Be^9 nucleus and the neutron are not known accurately. It is, however, easy to show that such a process fits the experimental facts. We have

$$Be^9 + He^4 + \text{kinetic energy of } \alpha$$
$$= C^{12} + n^1 + \text{kinetic energy of } C^{12} + \text{kinetic energy of } n^1.$$

If we assume that the beryllium nucleus consists of two α-particles and a neutron, then its mass cannot be greater than the sum of the masses of these particles, for the binding energy corresponds to a defect of mass. The energy equation becomes

$$(8 \cdot 00212 + n^1) + 4 \cdot 00106 + \text{K.E. of } \alpha > 12 \cdot 0003 + n^1$$
$$+ \text{K.E. of } C^{12} + \text{K.E. of } n^1$$

or

$$\text{K.E. of } n^1 < \text{K.E. of } \alpha + 0 \cdot 003 - \text{K.E. of } C^{12}.$$

Since the kinetic energy of the α-particle of polonium is $5 \cdot 25 \times 10^6$ electron volts, it follows that the energy of emission of the neutron cannot be greater than about 8×10^6 electron volts. The velocity of the neutron must therefore be less than $3 \cdot 9 \times 10^9$ cm. per second. We have seen that the actual maximum velocity of the neutron is about $3 \cdot 3 \times 10^9$ cm. per second, so that the proposed disintegration process is compatible with observation.

A further test of the neutron hypothesis was obtained by examining the radiation emitted from beryllium in the opposite direction to the bombarding α-particles. The source vessel, fig. 1, was reversed so that a sheet of paraffin wax in front of the counter was exposed to the " backward " radiation from the beryllium. The maximum range of the protons ejected from the wax was determined as before, by counting the numbers of protons observed through different thicknesses of aluminium interposed between the wax and the counter.

J. Chadwick.

The absorption curve obtained is shown in curve B, fig. 2. The maximum range of the protons was about 22 cm. in air, corresponding to a velocity of about $2 \cdot 74 \times 10^9$ cm. per second. Since the polonium source was only about 2 mm. away from the beryllium, this velocity should be compared with that of the neutrons emitted not at 180° but at an angle not much greater than 90°

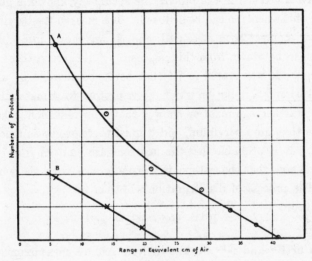

Fig. 2.

to the direction of the incident α-particles. A simple calculation shows that the velocity of the neutron emitted at 90° when an α-particle of full range is captured by a beryllium nucleus should be $2 \cdot 77 \times 10^9$ cm. per second, taking the velocity of the neutron emitted at 0° in the same process as $3 \cdot 3 \times 10^9$ cm. per second. The velocity found in the above experiment should be less than this, for the angle of emission is slightly greater than 90°. The agreement with calculation is as good as can be expected from such measurements.

§ 4. *The Nature of the Neutron.*—It has been shown that the origin of the radiation from beryllium bombarded by α-particles and the behaviour of the radiation, so far as its interaction with atomic nuclei is concerned, receive a simple explanation on the assumption that the radiation consists of particles of mass nearly equal to that of the proton which have no charge. The simplest hypothesis one can make about the nature of the particle is to suppose that it consists of a proton and an electron in close combination, giving a net charge 0 and a mass which should be slightly less than the mass of the hydrogen atom. This hypothesis is supported by an examination of the evidence which can be obtained about the mass of the neutron.

As we have seen, a rough estimate of the mass of the neutron was obtained from measurements of its collisions with hydrogen and nitrogen atoms, but such measurements cannot be made with sufficient accuracy for the present purpose. We must turn to a consideration of the energy relations in a process in which a neutron is liberated from an atomic nucleus; if the masses of the atomic nuclei concerned in the process are accurately known, a good estimate of the mass of the neutron can be deduced. The mass of the beryllium nucleus has, however, not yet been measured, and, as was shown in § 3, only general conclusions can be drawn from this reaction. Fortunately, there remains the case of boron. It was stated in § 1 that boron bombarded by α-particles of polonium also emits a radiation which ejects protons from materials containing hydrogen. Further examination showed that this radiation behaves in all respects like that from beryllium, and it must therefore be assumed to consist of neutrons. It is probable that the neutrons are emitted from the isotope B^{11}, for we know that the isotope B^{10} disintegrates with the emission of a proton.* The process of disintegration will then be

$$B^{11} + He^4 \rightarrow N^{14} + n^1.$$

The masses of B^{11} and N^{14} are known from Aston's measurements, and the further data required for the deduction of the mass of the neutron can be obtained by experiment.

In the source vessel of fig. 1 the beryllium was replaced by a target of powdered boron, deposited on a graphite plate. The range of the protons ejected by the boron radiation was measured in the same way as with the beryllium radiation. The effects observed were much smaller than with beryllium, and it was difficult to measure the range of the protons accurately. The maximum range was about 16 cm. in air, corresponding to a velocity of $2 \cdot 5 \times 10^9$ cm. per second. This then is the maximum velocity of the neutron liberated from boron by an α-particle of polonium of velocity $1 \cdot 59 \times 10^9$ cm. per second. Assuming that momentum is conserved in the collision, the velocity of the recoiling N^{14} nucleus can be calculated, and we then know the kinetic energies of all the particles concerned in the disintegration process. The energy equation of the process is

Mass of B^{11} + mass of He^4 + K.E. of He^4

   = mass of N^{14} + mass of n^1 + K.E. of N^{14} + K.E. of n^1.

* Chadwick, Constable and Pollard, *loc. cit.*

702   J. Chadwick.

The masses are $B^{11} = 11 \cdot 00825 \pm 0 \cdot 0016$; $He^4 = 4 \cdot 00106 \pm 0 \cdot 0006$; $N^{14} = 14 \cdot 0042 \pm 0 \cdot 0028$. The kinetic energies in mass units are α-particle $= 0 \cdot 00565$; neutron $= 0 \cdot 0035$; and nitrogen nucleus $= 0 \cdot 00061$. We find therefore that the mass of the neutron is $1 \cdot 0067$. The errors quoted for the mass measurements are those given by Aston. They are the maximum errors which can be allowed in his measurements, and the probable error may be taken as about one-quarter of these.* Allowing for the errors in the mass measurements it appears that the mass of the neutron cannot be less than $1 \cdot 003$, and that it probably lies between $1 \cdot 005$ and $1 \cdot 008$.

Such a value for the mass of the neutron is to be expected if the neutron consists of a proton and an electron, and it lends strong support to this view. Since the sum of the masses of the proton and electron is $1 \cdot 0078$, the binding energy, or mass defect, of the neutron is about 1 to 2 million electron volts. This is quite a reasonable value. We may suppose that the proton and electron form a small dipole, or we may take the more attractive picture of a proton embedded in an electron. On either view, we may expect the " radius " of the neutron to be a few times 10^{-13} cm.

§ 5. *The Passage of the Neutron through Matter.*—The electrical field of a neutron of this kind will clearly be extremely small except at very small distances of the order of 10^{-12} cm. In its passage through matter the neutron will not be deflected unless it suffers an intimate collision with a nucleus. The potential of a neutron in the field of a nucleus may be represented roughly by fig. 3. The radius of the collision area for sensible deflection of the neutron

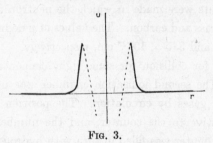

FIG. 3.

will be little greater than the radius of the nucleus. Further, the neutron should be able to penetrate the nucleus easily, and it may be that the scattering of the neutrons will be largely due to the internal field of the nucleus, or, in other words, that the scattered neutrons are mainly those which have penetrated

* The mass of B^{11} relative to B^{10} has been checked by optical methods by Jenkins and McKellar (' Phys. Rev.,' vol. 39, p. 549 (1932)). Their value agrees with Aston's to 1 part in 10^5. This suggests that great confidence may be put in Aston's measurements.

the potential barrier. On these views we should expect the collisions of a neutron with a nucleus to occur very seldom, and that the scattering will be roughly equal in all directions, at least as compared with the Coulomb scattering of a charged particle.

These conclusions were confirmed in the following way. The source vessel, with Be target, was placed rather more than 1 inch from the face of a closed counter filled with air, fig. 1. The number of deflections, or the number of nitrogen recoil atoms produced in the chamber, was observed for a certain time. The number observed was 190 per hour, after allowing for the natural effect. A block of lead 1 inch thick was then introduced between the source vessel and the counter. The number of deflections fell to 166 per hour. Since the number of recoil atoms produced must be proportional to the number of neutrons passing through the counter, these observations show that 13 per cent. of the neutrons had been absorbed or scattered in passing through 1 inch of lead.

Suppose that a neutron which passes within a distance p from the centre of the lead nucleus is scattered and removed from the beam. Then the fraction removed from the beam in passing through a thickness t of lead will be $\pi p^2 nt$, where n is the number of lead atoms per unit volume. Hence $\pi p^2 nt = 0 \cdot 13$, and $p = 7 \times 10^{-13}$ cm. This value for the collision radius with lead seems perhaps rather small, but it is not unreasonable. We may compare it with the radii of the radioactive nuclei calculated from the disintegration constants by Gamow and Houtermans,* viz., about 7×10^{-13} cm.

Similar experiments were made in which the neutron radiation was passed through blocks of brass and carbon. The values of p deduced in the same way were 6×10^{-13} cm. and $3 \cdot 5 \times 10^{-13}$ cm. respectively.

The target areas for collision for some light elements were compared by another method. The second ionisation chamber was used, which could be filled with different gases by circulation. The position of the source vessel was kept fixed relative to the counter, and the number of deflections was observed when the counter was filled in turn with hydrogen, nitrogen, oxygen, and argon. Since the number of neutrons passing through the counter was the same in each case, the number of deflections should be proportional to the target area for collision, neglecting the effect of the material of the counter, and allowing for the fact that argon is monatomic. It was found that nitrogen, oxygen, and argon gave about the same number of deflections ; the target areas of nitrogen and oxygen are thus roughly equal, and the target area of argon is

* ' Z. Physik,' vol. 52, p. 453 (1928).

nearly twice that of these. With hydrogen the measurements were very difficult, for many of the deflections were very small owing to the low ionising power of the proton and the low density of the gas. It seems probable from the results that the target area of hydrogen is about two-thirds that of nitrogen or oxygen, but it may be rather greater than this.

There is as yet little information about the angular distribution of the scattered neutrons. In some experiments kindly made for me by Dr. Gray and Mr. Lea, the scattering by lead was compared in the backward and forward directions, using the ionisation in a high pressure chamber to measure the neutrons. They found that the amount of scattering was about that to be expected from the measurements quoted above, and that the intensity per unit solid angle was about the same between 30° to 90° in the forward direction as between 90° to 150° in the backward direction. The scattering by lead is therefore not markedly anisotropic.

Two types of collision may prove to be of peculiar interest, the collision of a neutron with a proton and the collision with an electron. A detailed study of these collisions with an elementary particle is of special interest, for it should provide information about the structure and field of the neutron, whereas the other collisions will depend mainly on the structure of the atomic nuclei. Some preliminary experiments by Mr. Lea, using the pressure chamber to measure the scattering of neutrons by paraffin wax and by liquid hydrogen, suggest that the collision with a proton is more frequent than with other light atoms. This is not in accord with the experiments described above, but the results are at present indecisive. These collisions can be more directly investigated by means of the expansion chamber or by counting methods, and it is hoped to do so shortly.

The collision of a neutron with an electron has been examined in two ways, by the expansion chamber and by the counter. An account of the expansion chamber experiments is given by Mr. Dee in the third paper of this series. Mr. Dee has looked for the general ionisation produced by a large number of neutrons in passing through the expansion chamber, and also for the short electron tracks which should be the result of a very close collision between a neutron and an electron. His results show that collisions with electrons are extremely rare compared even with those with nitrogen nuclei, and he estimates that a neutron can produce on the average not more than 1 ion pair in passing through 3 metres of air.

In the counter experiments a beam of neutrons was passed through a block of brass, 1 inch thick, and the maximum range of the protons ejected from paraffin wax by the emergent beam was measured. From this range the maximum velocity of the neutrons after travelling through the brass is obtained and it can be compared with the maximum velocity in the incident beam. No change in the velocity of the neutrons due to their passage through the brass could be detected. The accuracy of the experiment is not high, for the estimation of the end of the range of the protons was rather difficult. The results show that the loss of energy of a neutron in passing through 1 inch of brass is not more than about $0 \cdot 4 \times 10^6$ electron volts. A path of 1 inch in brass corresponds as regards electron collisions to a path of nearly 2×10^4 cm. of air, so that this result would suggest that a neutron loses less than 20 volts per centimetre path in air in electron collisions. This experiment thus lends general support to those with the expansion chamber, though it is of far inferior accuracy. We conclude that the transfer of energy from the neutron to electrons is of very rare occurrence. This is not unexpected. Bohr* has shown on quite general ideas that collisions of a neutron with an electron should be very few compared with nuclear collisions. Massey,† on plausible assumptions about the field of the neutron, has made a detailed calculation of the loss of energy to electrons, and finds also that it should be small, not more than 1 ion pair per metre in air.

General Remarks.

It is of interest to examine whether other elements, besides beryllium and boron, emit neutrons when bombarded by α-particles. So far as experiments have been made, no case comparable with these two has been found. Some evidence was obtained of the emission of neutrons from fluorine and magnesium, but the effects were very small, rather less than 1 per cent. of the effect obtained from beryllium under the same conditions. There is also the possibility that some elements may emit neutrons spontaneously, *e.g.*, potassium, which is known to emit a nuclear β-radiation accompanied by a more penetrating radiation. Again no evidence was found of the presence of neutrons, and it seems fairly certain that the penetrating type is, as has been assumed, a γ-radiation.

Although there is certain evidence for the emission of neutrons only in two cases of nuclear transformations, we must nevertheless suppose that the neutron is a common constituent of atomic nuclei. We may then proceed to build up nuclei out of α-particles, neutrons and protons, and we are able to

---

* Bohr, Copenhagen discussions, unpublished.

† Massey, 'Nature,' vol. 129, p. 469, corrected p. 691 (1932).

706 J. Chadwick.

avoid the presence of uncombined electrons in a nucleus. This has certain advantages for, as is well known, the electrons in a nucleus have lost some of the properties which they have outside, *e.g.*, their spin and magnetic moment. If the α-particle, the neutron, and the proton are the only units of nuclear structure, we can proceed to calculate the mass defect or binding energy of a nucleus as the difference between the mass of the nucleus and the sum of the masses of the constituent particles. It is, however, by no means certain that the α-particle and the neutron are the only complex particles in the nuclear structure, and therefore the mass defects calculated in this way may not be the true binding energies of the nuclei. In this connection it may be noted that the examples of disintegration discussed by Dr. Feather in the next paper are not all of one type, and he suggests that in some cases a particle of mass 2 and charge 1, the hydrogen isotope recently reported by Urey, Brickwedde and Murphy, may be emitted. It is indeed possible that this particle also occurs as a unit of nuclear structure.

It has so far been assumed that the neutron is a complex particle consisting of a proton and an electron. This is the simplest assumption and it is supported by the evidence that the mass of the neutron is about 1·006, just a little less than the sum of the masses of a proton and an electron. Such a neutron would appear to be the first step in the combination of the elementary particles towards the formation of a nucleus. It is obvious that this neutron may help us to visualise the building up of more complex structures, but the discussion of these matters will not be pursued further for such speculations, though not idle, are not at the moment very fruitful. It is, of course, possible to suppose that the neutron may be an elementary particle. This view has little to recommend it at present, except the possibility of explaining the statistics of such nuclei as N^{14}.

There remains to discuss the transformations which take place when an α-particle is captured by a beryllium nucleus, Be^9. The evidence given here indicates that the main type of transformation is the formation of a C^{12} nucleus and the emission of a neutron. The experiments of Curie-Joliot and Joliot,[*] of Auger,[†] and of Dee show quite definitely that there is some radiation emitted by beryllium which is able to eject fast electrons in passing through matter. I have made experiments using the Geiger point counter to investigate this radiation and the results suggest that the electrons are produced by a

* 'C. R. Acad. Sci. Paris,' vol. 194, p. 708 and p. 876 (1932).
† 'C. R. Acad. Sci. Paris,' vol. 194, p. 877 (1932).

γ-radiation. There are two distinct processes which may give rise to such a radiation. In the first place, we may suppose that the transformation of Be^9 to C^{12} takes place sometimes with the formation of an excited C^{12} nucleus which goes to the ground state with the emission of γ-radiation. This is similar to the transformations which are supposed to occur in some cases of disintegration with proton emission, *e.g.*, B^{10}, F^{19}, Al^{27} ; the majority of transformations occur with the formation of an excited nucleus, only in about one-quarter is the final state of the residual nucleus reached in one step. We should then have two groups of neutrons of different energies and a γ-radiation of quantum energy equal to the difference in energy of the neutron groups. The quantum energy of this radiation must be less than the maximum energy of the neutrons emitted, about $5 \cdot 7 \times 10^6$ electron volts. In the second place, we may suppose that occasionally the beryllium nucleus changes to a C^{13} nucleus and that all the surplus energy is emitted as radiation. In this case the quantum energy of the radiation may be about 10×10^6 electron volts.

It is of interest to note that Webster has observed a soft radiation from beryllium bombarded by polonium α-particles, of energy about 5×10^5 electron volts. This radiation may well be ascribed to the first of the two processes just discussed, and its intensity is of the right order. On the other hand, some of the electrons observed by Curie-Joliot and Joliot had energies of the order of 2 to 10×10^6 volts, and Auger recorded one example of an electron of energy about $6 \cdot 5 \times 10^6$ volts. These electrons may be due to a hard γ-radiation produced by the second type of transformation.*

It may be remarked that no electrons of greater energy than the above appear to be present. This is confirmed by an experiment† made in this laboratory by Dr. Occhialini. Two tube counters were placed in a horizontal plane and the number of coincidences recorded by them was observed by means of the method devised by Rossi. The beryllium source was then brought up in the plane of the counters so that the radiation passed through both counters in turn. No increase in the number of coincidences could be detected. It follows that there are few, if any, β-rays produced with energies sufficient to pass through the walls of both counters, a total of 4 mm. brass ; that is, with energies greater than about 6×10^6 volts. This experiment further shows that the neutrons very rarely produce coincidences in tube counters under the usual conditions of experiment.

---

* Although the presence of fast electrons can be easily explained in this way, the possibility that some may be due to secondary effects of the neutrons must not be lost sight of.

† *Cf.* also Rasetti, ' Naturwiss.,' vol. 20, p. 252 (1932).

In conclusion, I may restate briefly the case for supposing that the radiation the effects of which have been examined in this paper consists of neutral particles rather than of radiation quanta. Firstly, there is no evidence from electron collisions of the presence of a radiation of such a quantum energy as is necessary to account for the nuclear collisions. Secondly, the quantum hypothesis can be sustained only by relinquishing the conservation of energy and momentum. On the other hand, the neutron hypothesis gives an immediate and simple explanation of the experimental facts ; it is consistent in itself and it throws new light on the problem of nuclear structure.

Summary.

The properties of the penetrating radiation emitted from beryllium (and boron) when bombarded by the α-particles of polonium have been examined. It is concluded that the radiation consists, not of quanta as hitherto supposed, but of neutrons, particles of mass 1, and charge 0. Evidence is given to show that the mass of the neutron is probably between $1 \cdot 005$ and $1 \cdot 008$. This suggests that the neutron consists of a proton and an electron in close combination, the binding energy being about 1 to 2×10^6 electron volts. From experiments on the passage of the neutrons through matter the frequency of their collisions with atomic nuclei and with electrons is discussed.

I wish to express my thanks to Mr. H. Nutt for his help in carrying out the experiments.

Experiments with High Velocity Positive Ions.   II.—*The Disintegration of Elements by High Velocity Protons.*

By J. D. Cockcroft, Ph.D., Fellow of St. John's College, Cambridge, and E. T. S. Walton, Ph.D.

(Communicated by Lord Rutherford, O.M., F.R.S.—Received June 15, 1932.)

[PLATE 12.]

1. *Introduction.*

In a previous paper* we have described a method of producing high velocity positive ions having energies up to 700,000 electron volts. We first used this method to determine the range of high-speed protons in air and hydrogen and the results obtained will be described in a subsequent paper. In the present communication we describe experiments which show that protons having energies above 150,000 volts are capable of disintegrating a considerable number of elements.

Experiments in artificial disintegration have in the past been carried out with streams of α-particles as the bombarding particles; the resulting transmutations have in general been accompanied by the emission of a proton and in some cases γ-radiation.† The present experiments show that under the bombardment of protons, α-particles are emitted from many elements; the disintegration process is thus in a sense the reverse process to the α-particle transformation.

---

* 'Proc. Roy. Soc.,' A, vol. 136, p. 619 (1932) denoted as (I) hereafter.
† Rutherford, Chadwick and Ellis, "Radioactive Substances."

2. *The Experimental Method.*

Positive ions of hydrogen obtained from a hydrogen canal ray tube are accelerated by voltages up to 600 kilovolts in the experimental tube described in (I) and emerge through a 3-inch diameter brass tube into a chamber well shielded by lead and screened from electrostatic fields. To this brass tube is attached by a flat joint and plasticene seal the apparatus shown in fig. 1. A target, A, of the metal to be investigated is placed at an angle of 45 degrees to the direction of the proton stream. Opposite the centre of the target is a side tube across which is sealed at B either a zinc sulphide screen or a mica window.

Stream of fast protons

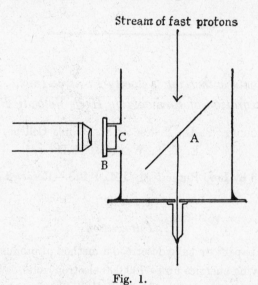

Fig. 1.

In our first experiments we used a round target of lithium 5 cm. in diameter and sealed the side tube with a zinc sulphide screen, the sensitive surface being towards the target. The distance from the centre of the target to the screen was 5 cm. A sheet of mica, C, of stopping power 1·4 cm. was placed between the screen and target and was more than adequate to prevent any scattered protons reaching the screen, since our range determinations* and the experiments of Blackett† have shown that the maximum range of protons accelerated by 600 kilovolts is of the order of 10 mm. in air. The screen is observed with a microscope having a numerical aperture of 0·6, the area of screen covered being 12 sq. mm. This arrangement with the fluorescent surface inside the vacuum is generally used in the preliminary investigations

* In course of publication.

† 'Proc. Roy. Soc.' A, vol. 134, p. 658 (1931).

of elements and when it is necessary to detect the presence of particles of short range.

The current to the target is measured by a galvanometer and controlled by varying the speed of the motor used for driving the alternator exciting the discharge tube (see Paper I). Currents of up to 5 microamperes can be obtained. Since metals bombarded by high-speed positive ions emit large numbers of secondary electrons for each incident ion, it is necessary to prevent the emission of these electrons if an accurate determination of the number of incident ions is required. This has been effected by applying a magnetic field of the order of 700 gauss to the target. Since it is well known that the majority of the secondary electrons have energies below 20 volts, such a field should be adequate to prevent secondary electron emission being a serious source of error.

An accurate determination of the exact composition of the beam of ions has not yet been made, but deflection experiments with a magnetic field in a subsidiary apparatus have shown that approximately half the current is carried by protons and half by H_2^+ ions. The number of neutral atoms appears to be small.

The accelerating voltage used in the experiments is controlled by varying the field of the alternator exciting the main high tension transformer. The secondary voltage of this transformer is measured by the method described in an earlier paper,* which rectifies the current passing through a condenser. A microammeter on the control table allows a continuous reading of this voltage to be obtained. The value of the steady potential produced by the rectifier system varies between 3 and 3·5 times the maximum of the transformer voltage according to the brightness of the rectifier filaments. The actual value of the voltage is determined by using a sphere gap consisting of two 75-cm. diameter aluminium spheres, one of which is earthed. In each experiment the multiplication factor of the rectifier system is determined for several voltages and intermediate points obtained by interpolation. The accuracy of the determination of the voltage by the sphere gaps has been checked by measuring the deflection of the protons in a magnetic field. It has been found that corrections of the order of 15 per cent. may be required as a result of the proximity of neighbouring objects or unfavourable arrangements of the connecting leads. The voltages given in this paper have all been corrected by reference to the magnetic deflection experiments.

* 'Proc. Roy. Soc.' A, vol. 129, p. 477 (1930).

3. *The Disintegration of Lithium.*

When the current passing to the target was of the order of 1 microampere and the accelerating potential was increased to 125 kilovolts, a number of bright scintillations were observed on the screen, the numbers being proportional to the current collected and of the order of 5 per minute per microampere at 125 kilovolts.

No scintillations were observed when the proton current was cut off by shutting off the discharge tube excitation or by interposing a brass flap between the beam and the target. Since the scintillations were very similar in appearance and brightness to α-particle scintillations, the apparatus was now changed to allow a determination of their range to be made. For this purpose a mica window having a stopping power of 2 cm. was sealed to the side tube in place of the fluorescent screen, which was now placed outside the window. It was then possible to insert mica screens of known stopping power between the window and the screen. In this way it became apparent that the scintillations were produced by particles having a well-defined range of about 8 cm. Variations of voltage between 250 and 500 kilovolts did not appear to alter the range appreciably.

In order to check this conclusion, the particles were now passed into a Shimizu expansion chamber, through a mica window in the side of the chamber having a stopping power of 3·6 cm. When the accelerating voltage was applied to the tube a number of discrete tracks were at once observed in the chamber whose lengths agreed closely with the first range determinations. From the appearance of the tracks and the brightness of the scintillations it seemed now fairly clear that we were observing α-particles ejected from the lithium nuclei under the proton bombardment, and that the lithium isotope of mass 7 was breaking up into two α-particles.

In order to obtain a further proof of the nature of the particles the experiments were repeated with an ionisation chamber, amplifier and oscillograph of the type described by Wynn Williams and Ward.* The mica window on the side tube was reduced to a thickness corresponding to a stopping power of 1·2 mm. with an area of about 1 sq. cm., the mica being supported on a grid structure. The lithium target was at the same time reduced in size to a circle of 1 cm. diameter in order to reduce the angular spread of the particles entering the counter. The ionisation chamber was of the parallel plane type having a total depth of 3 mm. and was sealed by an aluminium window having a stopping

* 'Proc. Roy. Soc.,' A, vol. 132, p. 391 (1931).

power of 5 mm. The degree of resolution of the amplifier and oscillograph was such that it was possible to record accurately up to 2000 particles per minute. With the full potential applied to the apparatus but with no proton current, the number of spurious deflections in the oscillograph was of the order of 2 per minute, whilst with an accelerating potential of 500 kilovolts and a current of 0·3 microamperes the number of particles entering the ionisation chamber per minute was of the order of 700.

In figs. 8, 9, 10 and 11, Plate 12, are shown the oscillograph records obtained as additional mica absorbers are inserted. It will be seen that the size of the deflections increases as additional mica is inserted, whilst the numbers fall off rapidly when the total absorber thickness is increased beyond 7 cm. In fig. 2 is plotted the number of particles entering the chamber per minute per micro-ampere for increasing absorber thickness and for accelerating potentials of 270 kilovolts and 450 kilovolts. The stopping power of the mica screens of windows has been checked and the final range determination made by a com-parison with the α-particles from thorium C. We find that the range is 8·4 cm. Preliminary observations showed that between the lowest and highest voltages used, the range remained approximately constant. It is, however, of great interest to test whether the whole of the energy of the proton is communicated to the α-particles, and it is intended at a later date to examine this point more carefully. The general shape of the range curve, together with the evidence from the size of the oscillograph deflections, suggests that the great majority of the particles have initially a uniform velocity, but further investigation will be required with lower total absorption to exclude the possibility of the existence of particles of short range.

As is well known, the size of the oscillograph kicks are a measure of the ionisation produced by the particles. At the beginning of the range the size of the kicks observed was very uniform, whilst the average size varied with the range of the particle corresponding to the ionisation given by the Bragg curve. Fig. 3 shows the variation of the ionisation of the most numerous particles with range.

The sizes of the deflections were now compared with the deflections produced in the same ionisation chamber by α-particles from a polonium source, these deflections being recorded in fig. 12, Plate 12, for comparison. It has been shown in this way that the maximum deflection for the two types of particle is the same. This result, together with the uniformity of the ionisation pro-duced by the particles, is sufficient to exclude the possibility of some of the particles being protons, since the maximum ionisation produced by a

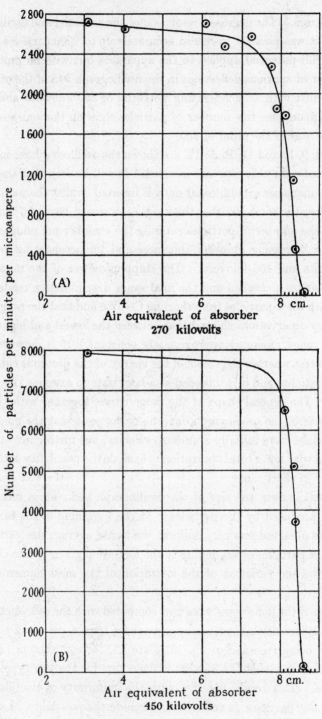

FIG. 2.

proton is less than 40 per cent. of the maximum ionisation produced by an α-particle.

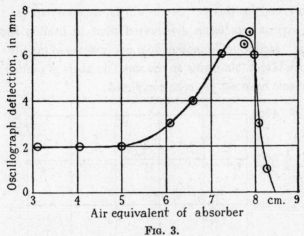

FIG. 3.

The variation of the numbers of particles with accelerating voltage was determined from the oscillograph records between 200 kilovolts and 500 kilovolts, the change in numbers being clear from the records, figs. 13, 14, 15, Plate 12. For voltages between 70 kilovolts and 250 kilovolts, the numbers of particles entering the ionisation chamber were counted by a single stage thyratron counter of the type described by Wynn Williams and Ward.* The results are plotted in fig. 4. The numbers increase roughly exponentially with the voltage at the lower voltages and linearly with voltage above 300 kilovolts.†

It is of great interest to estimate the number of particles produced by the bombardment of a thick layer of lithium by a fixed number of protons. In making this estimate we have assumed that the particles are emitted uniformly in all directions and that the molecular ions produce no effect. With these assumptions the number of disintegrations for a voltage of 250 kilovolts is 1 per 10^9 protons, and for a voltage of 500 kilovolts is 10 per 10^9 protons.

In considering the variation in numbers of particles with voltage it has, of course, to be borne in mind that with a thick target the effects are due to

* 'Proc. Roy. Soc.,' A, vol. 131, p. 191 (1931).

† All the measurements in a single run, in which more than 2000 particles were counted are included in the figure. The spread of the points in the centre part of the curve is probably due to variations in the vacuum and therefore in the voltage applied during the experiment. In other runs no evidence was obtained for such a variation.

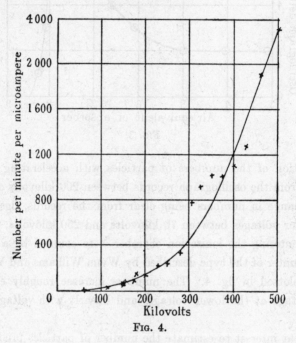

protons of all energies from the maximum to zero energy. It will be very important to determine the probability of disintegration for protons of one definite energy, and for this purpose it will be necessary to use thin targets. Preliminary experiments using evaporated films of lithium show that the probability or " excitation " function does not increase so rapidly with voltage as for the thick target, but owing to the small numbers of particles obtainable these experiments have not yet been completed.

Fig. 4.

4. *The Interpretation of Results.*

We have already stated that the obvious interpretation of our results is to assume that the lithium isotope of mass 7 captures a proton and that the resulting nucleus of mass 8 breaks up into two α-particles. If momentum is conserved in the process, then each of the α-particles must take up equal amounts of energy, and from the observed range of the α-particles we conclude that an energy of 17·2 million volts would be liberated in this disintegration process. The mass of the Li$_7$ nucleus from Costa's determination is 7·0104 with a probable error of 0·003. The decrease of mass in the disintegration process is therefore $7 \cdot 0104 + 1 \cdot 0072 - 8 \cdot 0022 = 0 \cdot 0154 \pm 0 \cdot 003$. This is equivalent to an energy liberation of $(14 \cdot 3 \pm 2 \cdot 7) \times 10^6$ volts. We conclude, therefore, that the observed energies of the α-particles are consistent with our hypothesis. An additional test can, however, be applied. If momentum is conserved in the disintegration, the two α-particles must be ejected in practically opposite directions and, therefore, if we arrange two zinc sulphide screens opposite to a small target of lithium as shown in the arrangement of fig. 5, we should observe a large proportion of coincidences in the time of appearance of the scintillations on the two screens. The lithium used in the experiments was evaporated on to a thin film of mica having an area of 1 sq. mm. and a stopping power of 1·1 cm., so that α-particles ejected from the lithium would pass easily through the mica and reach the screen on the opposite side of the lithium layer.

The two screens were observed through microscopes each covering an area of 7 sq. mm. and a tape recording machine was used to record the scintillations,

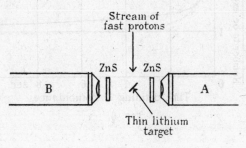

Fig. 5.

a buzzer being installed in the observation chamber to prevent the noise of the recording keys being audible to the observers. Five hundred and sixty-five scintillations were observed in microscope A and 288 scintillations in microscope B, the former being nearer the target. Analysis of the records showed that the results are consistent with the assumption that about 25 per cent. of the scintillations recorded in B have a corresponding scintillation in A. If we calculate the chance of a scintillation being recorded by B within x seconds of the record of a scintillation in A, assuming a perfectly random distribution of scintillations, and compare this with the observed record, the curve shown in fig. 6 is obtained. It will be seen that as the interval x is made less, the ratio of the observed to the random coincidences increases. We also plot for comparison the theoretical curve (shown by broken line) which would be obtained if there were 25 per cent. of coincidences. It will be seen that the two curves are in good accord. The number of coincidences observed is about that to be expected on our theory of the disintegration process, when we take into account the geometry of the experimental arrangement and the efficiency of the zinc

238   J. D. Cockcroft and E. T. S. Walton.

sulphide screens. It is clear that there is strong evidence supporting the
hypothesis that the α-particles are emitted in pairs. A more complete investigation will be made later, using larger areas for the counting device, when it is
to be expected that the fraction of coincidences should increase.

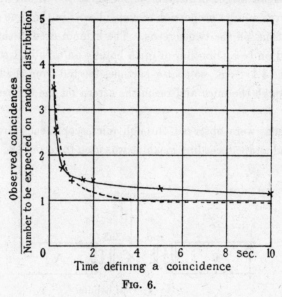

FIG. 6.

5. *Comparison with the Gamow Theory.*

In a paper which was largely responsible for stimulating the present investigation, Gamow† has calculated the probability W_1^* of a particle of charge Ze,
mass m and energy E, entering a nucleus of charge $Z'e$. Gamow's formula is

$$W_1^* = e^{\frac{-4\pi \sqrt{(2m)}}{h} \cdot \frac{ZZ'e^2}{\sqrt{(E)}} \cdot J_k},$$

where J_k is a function varying slowly with E and Z. Using this formula, we
have calculated W_1^*, the probability of a proton entering a lithium nucleus,
for 600, 300 and 100 kilovolts, and find the values $0\cdot187$. $2\cdot75 \times 10^{-2}$ and
$1\cdot78 \times 10^{-4}$. Using these figures, our observed variation of proton range with
velocity for a thick target, and assuming a target area of 10^{-25} cm.², the
number of protons N required to produce one disintegration may be calculated.
For 600 kilovolts we find N to be of the order of 10^6, and for 300 kilovolts of
the order of 2×10^7.

The order of magnitude of the numbers observed is thus smaller than the

† 'Z. Physik,' vol. 52, p. 510 (1928).

number predicted by the Gamow theory, but a closer comparison must be deferred until the results for a thin target are available.

6. *The Disintegration of other Elements.*

Preliminary investigations have been made to determine whether any evidence of disintegration under proton bombardment could be obtained for the following elements : Be, B, C, O, F, Na, Al, K, Ca, Fe, Co, Ni, Cu, Ag, Pb, U. Using the fluorescent screen as a detector we have observed some bright scintillations from all these elements, the numbers varying markedly from element to element, the relative orders of magnitude being indicated by fig. 7 for 300 kilovolts. The results of the scintillation method have been confirmed by the electrical counter for Ca, K, Ni, Fe and Co, and the size of the oscillograph kicks suggests that the majority of the particles ejected are α-particles.

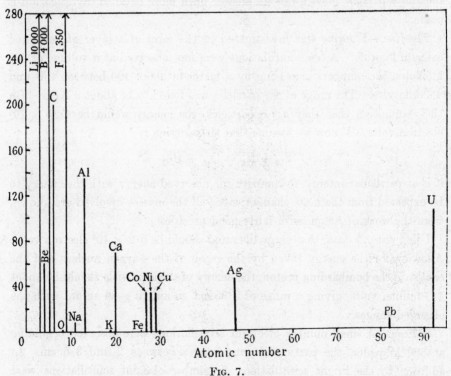

Fig. 7.

The numbers of particles counted have up to the present not been sufficient to enable these figures to be taken as anything other than an order of magnitude. In particular, the possibility must be borne in mind that some of the particles observed may be due to impurities. It may, however, be of some interest to describe briefly the general character of the effects observed in some of the more interesting cases.

Beryllium.—Two types of scintillation were observed with beryllium, a few bright scintillations having the appearance of α-particle scintillations together with a much greater number of faint scintillations appearing at about 500 kilovolts, the numbers increasing rapidly with voltage. We were not able to observe the faint scintillations outside the vacuum chamber, so that they are presumably due to particles of short range.

Boron.—Next to lithium, boron gave the greatest number of scintillations, most of the particles having a range of about 3·5 cm. Scintillations were first observed at voltages of the order of 115 kilovolts, the numbers increasing by more than 100 between this voltage and 375 kilovolts. The interesting problem as to whether the boron splits up into three α-particles or into Be_8 plus an α-particle must await an answer until more detailed investigation is made.

Fluorine.—Fluorine was investigated in the form of a layer of powdered calcium fluoride. A few scintillations were first observed at a voltage of 200 kilovolts, the numbers increasing by a factor of about 100 between this and 450 kilovolts. The range of the particles was found to be about 2·8 cm. On the assumption that they are α-particles, the energy would be $4·15 \times 10^6$ electron volts. If now we assume that the reaction is

$$F_{19} + H_1 = O_{16} + He_4$$

it is of particular interest to compare the observed energy with the energy to be expected from the mass changes, since all the masses involved are known, from the work of Aston, with fairly good precision.

Using Aston's data, the energy liberated should be $5·2 \times 10^6$ electron volts. Allowing for the energy taken by the recoil of the oxygen nucleus and the energy of the bombarding proton, the energy of the α-particle should be about 4·3 million volts, giving a range of 2·95 cm. in air, in good accord with the observed ranges.

Sodium.—A small number of bright scintillations were observed beginning at 300 kilovolts, the particles having ranges between 2 and 3·5 cm. In addition to the bright scintillations, a number of faint scintillations were observed similar to those seen in the case of beryllium. The faint scintillations are again presumably due to particles of short range since they could not be observed outside the tube. The probable α-particle transition would be

$$Na_{23} + H_1 = Ne_{20} + He_4.$$

Cockcroft and Walton.   *Proc. Roy. Soc., A, vol.* 137, *Pl.* 12.

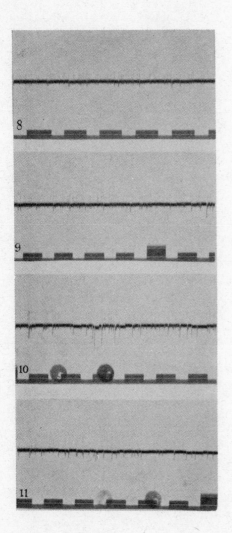
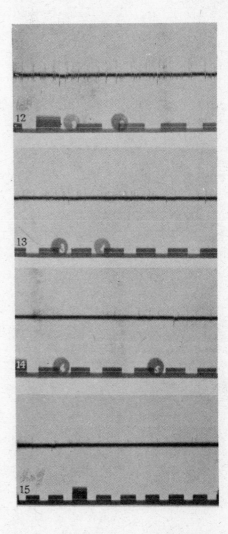

FIG. 8.—270 Kv. 4·0 cm. absorber.

FIG. 12.—Polonium α-particles 2 cm. absorber.

FIG. 9.—270 Kv. 5·0 cm. absorber.

FIG. 13.—250 Kv. 3·1 cm. absorber.

FIG. 10.—270 Kv. 6·6 cm. absorber.

FIG. 14.—210 Kv. 3·1 cm. absorber.

FIG. 11.—270 Kv. 7·9 cm. absorber.

FIG. 15.—175 Kv. 3·1 cm. absorber.

Potassium.—Potassium is of special interest on account of its radioactivity. The very small effects observed may easily be due to an impurity. The most likely reaction to occur

$$K_{39} + H_1 = A_{36} + He_4,$$

would probably have a negative energy balance.

Iron, Nickel, Cobalt, Copper.—These elements follow each other in the periodic table, so that the small result obtained for iron compared with that for the following three elements is of special interest. The effect for iron is of the same order as that for potassium, and again may be due to impurity. For these elements most of the particles had a range of about 2·5 cm., but a few particles were present having a slightly longer range.

Uranium.—Using potentials of up to 600 kilovolts and strong proton currents, the number of scintillations observed was about four times the natural radio-active effect, and the artificially produced particles appeared to have a longer range than the natural ones. The numbers obtained did not appear to vary markedly with voltage.

We hope in the near future to investigate the above and other elements in much greater detail and in particular to determine whether any of the effects described are due to impurities. There seems to be little doubt, however, that most of the effects are due to transformations giving rise to an α-particle emission. In view of the very small probability of a proton of 500 kilovolts energy penetrating the potential barrier of the heavier nuclei by any process other than a resonance process, it would appear most likely that such processes are responsible for the effects observed with the heavier elements.

We have seen that the three elements, lithium, boron and fluorine give the largest emission of particles, the emission varying similarly with rise of voltage. These elements are all of the $4n + 3$ type, and presumably the nuclei are made up of α-particles with the addition of three protons and two electrons. It is natural to suppose that the addition of a captured proton leads to the formation of a new α-particle inside the nucleus. In the case of lithium, it seems probable that the capture of the proton, the formation of the α-particle and the dis-integration of the resulting nucleus into two α-particles must at this stage be regarded as a single process, the excess energy appearing in the form of kinetic energy of the expelled α-particles.* Until further and more accurate data are available it is not desirable to discuss at this stage the possible bearing of

---

* Such a view does not preclude the possibility that sometimes part of the energy may appear in another form, for example, as γ-radiation.

these new observations on the problems of astrophysics and on the question of the abundance of the elements.

In conclusion, we wish to express our thanks to Lord Rutherford for his constant encouragement and advice. We are indebted to Dr. Wynn Williams for considerable assistance with the electrical recording apparatus, and to members of the research staff of Metropolitan-Vickers Electrical Company for their assistance in supplying much of the apparatus used in this work. One of us (E.T.S.W.) has been in receipt of a senior research award from the Department of Scientific and Industrial Research.

254   ACADÉMIE DES SCIENCES.

PHYSIQUE NUCLÉAIRE. — *Un nouveau type de radioactivité.*
Note de Mme IRÈNE CURIE et M. F. JOLIOT, présentée par M. Jean Perrin.

Nous avons montré récemment par la méthode de Wilson ([1]) que certains éléments légers (glucinium, bore, aluminium) émettent des électrons positifs quand on les bombarde avec des rayons α du polonium. Selon notre interprétation l'émission des électrons positifs de Be serait due à la *matérialisation interne* du rayonnement γ tandis que les électrons positifs émis par B et Al seraient des *électrons de transmutation* accompagnant l'émission des neutrons.

En cherchant à préciser le mécanisme de ces émissions nous avons découvert le phénomène suivant :

L'émission des électrons positifs par certains éléments légers irradiés par les rayons α du polonium subsiste pendant des temps plus ou moins longs, pouvant atteindre plus d'une demi-heure dans le cas du bore, après l'enlèvement de la source de rayons α.

Nous plaçons une feuille d'aluminium à 1^{mm} d'une source de polonium. L'aluminium ayant été irradié pendant 10 minutes environ, nous le plaçons au-dessus d'un compteur de Geiger Müller portant un orifice fermé par un écran de $7/100^e$ de millimètre d'aluminium. Nous observons que la feuille émet un rayonnement dont l'intensité décroît exponentiellement en fonction du temps avec une période de 3 minutes 15 secondes. On obtient un résultat analogue avec le bore et le magnésium mais les périodes de décroissance sont *différentes*, 14 minutes pour le bore et 2 minutes 30 secondes pour le magnésium.

L'intensité du rayonnement (immédiatement après l'exposition aux rayons α) augmente avec le temps d'irradiation jusqu'à une valeur limite. On a alors des intensités initiales du même ordre pour B, Mg, Al d'environ 150 impulsions par minute dans le compteur en utilisant une source de polonium de 60 millicuries.

Avec les éléments H, Li, C, Be, N, O, F, Na, Ca, Ni, Ag, aucun effet n'a été observé ([2]). Pour certain de ces éléments le phénomène ne se produit probablement pas, pour d'autres la période de décroissance est peut-être trop courte.

([1]) *Comptes rendus*, 196, 1933, p. 1885.; *J. de Phys. et Rad.*, 4, 1933, p. 494.
([2]) Ce phénomène ne peut donc pas être dû à une contamination par la source de polonium.

Les expériences faites par la méthode de Wilson ou par la méthode de la trochoïde introduite par Thibaud ont montré que le rayonnement émis par le bore et par l'aluminium est constitué par des électrons positifs. Il est probable qu'il en est de même pour le rayonnement du magnésium.

En introduisant des écrans de cuivre entre le compteur et la feuille irradiée on trouve que la majeure partie du rayonnement est absorbée dans

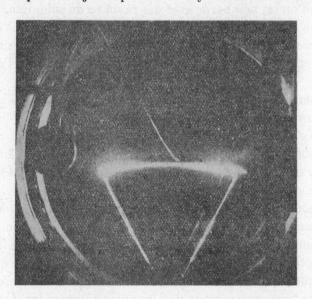

$0,88$ g/cm² pour Al, $0,26$ g/cm² pour B et Mg, ce qui correspond, en admettant les mêmes lois d'absorption que pour les électrons négatifs, à une énergie de $2,2 \times 10^6 e$ V pour Al et $0,7 \times 10^6 e$ V pour B et Mg.

Lorsqu'on réduit l'énergie des rayons α irradiant l'aluminium, le nombre des électrons positifs diminue, mais la période de décroissance ne semble pas modifiée. Quand l'énergie des rayons α est réduite de $10^6 e$ V, on n'observe presque plus de ces électrons.

Ces expériences montrent l'existence d'un nouveau type de radioactivité avec émission d'électrons positifs. Nous pensons que le processus d'émission serait le suivant pour l'aluminium :

$$^{27}_{13}\text{Al} + ^{4}_{2}\text{He} = ^{30}_{15}\text{P} + ^{1}_{0}n.$$

L'isotope $^{30}_{15}$P du phosphore serait radioactif avec une période de $3^m 15^s$ et émettrait des électrons positifs suivant la réaction

$$^{30}_{15}\text{P} = ^{30}_{14}\text{Si} + \overset{+}{\varepsilon}.$$

Une réaction analogue pourrait être envisagée pour le bore et le magnésium, les noyaux instables étant $^{13}_{7}N$ et $^{27}_{14}Si$. Les isotopes $^{13}_{7}N$, $^{27}_{14}Si$, $^{30}_{15}P$ ne peuvent exister que des temps assez courts, c'est pourquoi on ne les observerait pas dans la nature.

Nous considérons comme peu vraisemblable l'explication suivant laquelle

$$^{27}_{13}Al + {}^{4}_{2}He = {}^{30}_{14}Si + {}^{1}_{1}H, \qquad {}^{30}_{14}Si = {}^{30}_{14}Si + \overset{+}{\varepsilon} + \overset{-}{\varepsilon},$$

l'isotope $^{30}_{14}Si$ étant excité et pouvant se désactiver au cours du temps, l'énergie se matérialiserait en donnant une paire d'électrons. On n'observe pas d'émission d'électrons négatifs et il est théoriquement très improbable que la différence d'énergie entre les électrons soit suffisante pour que les négatifs ne soient pas observés ([1]). D'autre part ce processus supposerait une durée de l'état excité extraordinairement longue avec un coefficient de matérialisation interne unité.

En définitive il a été possible pour la première fois de créer à l'aide d'une cause extérieure la radioactivité de certains noyaux atomiques pouvant subsister un temps mesurable en l'absence de la cause excitatrice.

Des radioactivités durables, analogues à celles que nous avons observé, peuvent sans doute exister dans le cas de bombardement par d'autres particules. Un même atome radioactif pourrait sans doute être créé par plusieurs réactions nucléaires. Par exemple le noyau $^{13}_{7}N$ qui est radioactif selon notre hypothèse, pourrait être obtenu par l'action d'un deuton sur le carbone, après émission d'un neutron.

CHIMIE PHYSIQUE. — *Étude cinétique de la réaction iodure de potassium-eau oxygénée en solution acide.* Note de Mme **P. Rumpf**, présentée par M. G. Urbain.

L'eau oxygénée et l'iodure de potassium réagissent en solution acide en donnant naissance à de l'iode. La cinétique de cette réaction a été l'objet de nombreux travaux; les auteurs ont généralement dosé l'iode mis en liberté; plus récemment, Abel ([2]), puis Liebhafsky ([3]) ont mesuré le dégagement d'oxygène au fur et à mesure de la décomposition de l'eau oxygénée. Cependant la question reste obscure; il a semblé intéressant de l'aborder

---

([1]) Nedelsky et Oppenheimer, *Phys. Rev.*, 44, 1933, p. 948.

([2]) *Zeit. phys. Chem.*, 96, 1920, p. 1-180, et 136, 1928, p. 161-182.

([3]) *J. amer. chem. Soc.*, 54, 1932, p. 1792-1806 et 3499-3508.

ERRATA.

———

(Séance du 15 janvier 1934.)

Note de Mme *Irène Curie* et M. *F. Joliot*, Un nouveau type de radioactivité :
La figure représentée page 255 doit être supprimée.

———————◦◦◦◦———————

Possible Production of Elements of Atomic Number Higher than 92

By Prof. E. Fermi, Royal University of Rome

UNTIL recently it was generally admitted that an atom resulting from artificial disintegration should normally correspond to a stable isotope. M. and Mme. Joliot first found evidence that it is not necessarily so ; in some cases the product atom may be radioactive with a measurable mean life, and go over to a stable form only after emission of a positron.

The number of elements which can be activated either by the impact of an α-particle (Joliot) or a proton (Cockcroft, Gilbert, Walton) or a deuton (Crane, Lauritsen, Henderson, Livingston, Lawrence) is necessarily limited by the fact that only light elements can be disintegrated, owing to the Coulomb repulsion.

This limitation is not effective in the case of neutron bombardment. The high efficiency of these particles in producing disintegrations compensates fairly for the weakness of available neutron sources as compared with α-particle or proton sources. As a matter of fact, it has been shown[1] that a large number of elements (47 out of 68 examined until now) of any atomic weight could be activated, using neutron sources consisting of a small glass tube filled with beryllium powder and radon up to 800 millicuries. This source gives a yield of about one million neutrons per second.

All the elements activated by this method with intensity large enough for a magnetic analysis of the sign of the charge of the emitted particles were found to give out only negative electrons. This is theoretically understandable, as the absorption of the bombarding neutron produces an excess in the number of neutrons present inside the nucleus ; a stable state is therefore reached generally through transformation of a neutron into a proton, which is connected to the emission of a β-particle.

In several cases it was possible to carry out a chemical separation of the β-active element, following the usual technique of adding to the irradiated substance small amounts of the neighbouring elements. These elements are then separated by chemical analysis and separately checked for the β-activity with a Geiger-Müller counter. The activity always followed completely a certain element, with which the active element could thus be identified.

In three cases (aluminium, chlorine, cobalt) the active element formed by bombarding the element of atomic number Z has atomic number $Z - 2$. In four cases (phosphorus, sulphur, iron, zinc) the atomic number of the active product is $Z - 1$. In two cases (bromine, iodine) the active element is an isotope of the bombarded element.

This evidence seems to show that three main processes are possible : (a) capture of a neutron with instantaneous emission of an α-particle ; (b) capture of the neutron with emission of a proton ; (c) capture of the neutron with emission of a γ-quantum, to get rid of the surplus energy. From a theoretical point of view, the probability of processes (a) and (b) depends very largely on the energy of the emitted α- or H-particle ; the more so the higher the atomic weight of the element. The probability of process (c) can be evaluated only very roughly in the present state of nuclear theory ; nevertheless, it would appear to be smaller than the observed value by a factor 100 or 1,000.

It seemed worth while to direct particular attention to the heavy radioactive elements thorium and uranium, as the general instability of nuclei in this range of atomic weight might give rise to successive transformations. For this reason an investigation of these elements was undertaken by the writer in collaboration with F. Rasetti and O. D'Agostino.

Experiment showed that both elements, previously freed of ordinary active impurities, can be strongly activated by neutron bombardment. The initial induced activity corresponded in our experiments to about 1,000 impulses per minute in a Geiger counter made of aluminium foil of 0·2 mm. thickness. The curves of decay of these activities show that the phenomenon is rather complex. A rough survey of thorium activity showed in this element at least two periods.

Better investigated is the case of uranium ; the existence of periods of about 10 sec., 40 sec., 13 min., plus at least two more periods from 40 minutes to one day is well established. The large uncertainty in the decay curves due to the statistical fluctuations makes it very difficult to establish whether these periods represent successive or alternative processes of disintegration.

Attempts have been made to identify chemically the β-active element with the period of 13 min. The general scheme of this research consisted in adding to the irradiated substance (uranium nitrate in concentrated solution, purified of its decay products) such an amount of an ordinary β-active element as to give some hundred impulses per minute on the counter. Should it be possible to prove that the induced activity, recognisable by its characteristic period, can be chemically separated from the added activity, it is reasonable to assume that the two activities are not due to isotopes.

The following reaction enables one to separate the 13 min.-product from most of the heaviest elements. The irradiated uranium solution is diluted in 50 per cent nitric acid ; a small amount of a manganese salt is added and then the manganese is precipitated as dioxide (MnO_2) from the boiling solution by addition of sodium chlorate. The manganese dioxide precipitate carries a large percentage of the activity.

This reaction proves at once that the 13 min.-activity is not isotopic with uranium. For testing the possibility that it might be due to an element 90 (thorium) or 91 (palladium), we repeated the reaction at least ten times, adding

an amount of uranium $X_1 + X_2$ corresponding to about 2,000 impulses per minute ; also some cerium and lanthanum were added in order to sustain uranium X. In these conditions the manganese reaction carried only the 13 min.-activity ; no trace of the 2,000 impulses of uranium X_1 (period 24 days) was found in the precipitate ; and none of uranium X_2, although the operation had been performed in less than two minutes from the precipitation of the manganese dioxide, so that several hundreds of impulses of uranium X_2 (period 75 sec.) would have been easily recognisable.

Similar evidence was obtained for excluding atomic numbers 88 (radium) and 89 (actinium). For this, mesothorium-1 and -2 were used, adding barium and lanthanum ; the evidence was completely negative, as in the former case. The eventual precipitation of uranium-X_1 and mesothorium-1, which do not emit β-rays penetrating enough to be detectable in our counters, would have been revealed by the subsequent formation respectively of uranium-X_2 and mesothorium-2.

Lastly, we added to the irradiated uranium solution some inactive lead and bismuth, and proved that the conditions of the manganese dioxide reaction could be regulated in such a way as to obtain the precipitation of manganese dioxide with the 13 min.-activity, without carrying down lead and bismuth.

In this way it appears that we have excluded the possibility that the 13 min.-activity is due to isotopes of uranium (92), palladium (91), thorium (90), actinium (89), radium (88), bismuth (83), lead (82). Its behaviour excludes also ekacaesium (87) and emanation (86).

This negative evidence about the identity of the 13 min.-activity from a large number of heavy elements suggests the possibility that the atomic number of the element may be greater than 92. If it were an element 93, it would be chemically homologous with manganese and rhenium. This hypothesis is supported to some extent also by the observed fact that the 13 min.-activity is carried down by a precipitate of rhenium sulphide insoluble in hydrochloric acid. However, as several elements are easily precipitated in this form, this evidence cannot be considered as very strong.

The possibility of an atomic number 94 or 95 is not easy to distinguish from the former, as the chemical properties are probably rather similar. Valuable information on the processes involved could be gathered by an investigation of the possible emission of heavy particles. A careful search for such heavy particles has not yet been carried out, as they require for their observation that the active product should be in the form of a very thin layer. It seems therefore at present premature to form any definite hypothesis on the chain of disintegrations involved.

¹ E. Fermi, *Ricerca Scientifica*, 1, 5, 283 ; 6, 330. NATURE, **133**, 757, May 19, 1934. E. Amaldi, O. D'Agostino, E. Fermi, F. Rasetti, E. Segrè, *Ricerca Scientifica*, 8, 452 ; 1934.

Versuch einer Theorie der β-Strahlen. I[1]).

Von E. Fermi in Rom.

Mit 3 Abbildungen. (Eingegangen am 16. Januar 1934.)

Eine quantitative Theorie des β-Zerfalls wird vorgeschlagen, in welcher man die Existenz des Neutrinos annimmt, und die Emission der Elektronen und Neutrinos aus einem Kern beim β-Zerfall mit einer ähnlichen Methode behandelt, wie die Emission eines Lichtquants aus einem angeregten Atom in der Strahlungstheorie. Formeln für die Lebensdauer und für die Form des emittierten kontinuierlichen β-Strahlenspektrums werden abgeleitet und mit der Erfahrung verglichen.

1. Grundannahmen der Theorie.

Bei dem Versuch, eine Theorie der Kernelektronen sowie der β-Emission aufzubauen, begegnet man bekanntlich zwei Schwierigkeiten. Die erste ist durch das kontinuierliche β-Strahlenspektrum bedingt. Falls der Erhaltungssatz der Energie gültig bleiben soll, muß man annehmen, daß ein Bruchteil der beim β-Zerfall frei werdenden Energie unseren bisherigen Beobachtungsmöglichkeiten entgeht. Nach dem Vorschlag von W. Pauli kann man z. B. annehmen, daß beim β-Zerfall nicht nur ein Elektron, sondern auch ein neues Teilchen, das sogenannte „Neutrino" (Masse von der Größenordnung oder kleiner als die Elektronenmasse; keine elektrische Ladung) emittiert wird. In der vorliegenden Theorie werden wir die Hypothese des Neutrinos zugrunde legen.

Eine weitere Schwierigkeit für die Theorie der Kernelektronen besteht darin, daß die jetzigen relativistischen Theorien der leichten Teilchen (Elektronen oder Neutrinos) nicht imstande sind, in einwandfreier Weise zu erklären, wie solche Teilchen in Bahnen von Kerndimensionen gebunden werden können.

Es scheint deswegen zweckmäßiger, mit Heisenberg[2]) anzunehmen, daß ein Kern nur aus schweren Teilchen, Protonen und Neutronen, besteht. Um trotzdem die Möglichkeit der β-Emission zu verstehen, wollen wir versuchen, eine Theorie der Emission leichter Teilchen aus einem Kern in Analogie zur Theorie der Emission eines Lichtquants aus einem angeregten Atom beim gewöhnlichen Strahlungsprozeß aufzubauen. In der Strahlungstheorie ist die totale Anzahl der Lichtquanten keine Konstante: Lichtquanten entstehen, wenn sie von einem Atom emittiert werden, und verschwinden, wenn sie absorbiert werden. In Analogie hierzu wollen wir der β-Strahlentheorie folgende Annahmen zugrunde legen:

---

[1]) Vgl. die vorläufige Mitteilung: La Ricerca Scientifica 2, Heft 12, 1933. —
[2]) W. Heisenberg, ZS. f. Phys. 77, 1, 1932.

11 *

a) Die totale Anzahl der Elektronen, sowie der Neutrinos, ist nicht notwendigerweise konstant. Elektronen (oder Neutrinos) können entstehen und verschwinden. Diese Möglichkeit hat jedoch keine Analogie zum Entstehen oder Verschwinden eines Paares aus einem Elektron und einem Positron; falls man das Positron als Diracsches „Loch" interpretiert, kann man in der Tat diesen letzten Prozeß einfach als einen Quantensprung eines Elektrons zwischen einem Zustand mit negativer Energie und einem Zustand mit positiver Energie mit Erhaltung der totalen (unendlich großen) Anzahl der Elektronen auffassen.

b) Die schweren Teilchen, Neutronen und Protonen, können wie bei Heisenberg als zwei innere Quantenzustände des schweren Teilchens betrachtet werden. Wir formulieren dies durch die Einführung einer inneren Koordinate ϱ des schweren Teilchens, welche nur zwei Werte annehmen kann: $\varrho = 1$, falls das Teilchen ein Neutron ist; $\varrho = -1$, falls das Teilchen ein Proton ist.

c) Die Hamilton-Funktion des aus schweren und leichten Teilchen bestehenden Systems muß so gewählt werden, daß jedem Übergang von Neutron zu Proton das Entstehen eines Elektrons und eines Neutrinos zugeordnet ist. Dem umgekehrten Prozeß, Verwandlung eines Protons in ein Neutron, soll dagegen das Verschwinden eines Elektrons und eines Neutrinos zugeordnet sein. Man bemerke, daß hierdurch die Erhaltung der Ladung gesichert ist.

2. Die in der Theorie auftretenden Operatoren.

Ein mathematischer Formalismus der Theorie in Einklang mit diesen drei Forderungen kann am leichtesten mit Hilfe der Dirac-Jordan-Kleinschen Methode[1]) der „zweiten Quantelung" aufgebaut werden. Wir werden also die Wahrscheinlichkeitsamplituden ψ und φ der Elektronen und der Neutrinos sowie die komplex konjugierten Größen ψ^* und φ^* als Operatoren auffassen; für die Beschreibung der schweren Teilchen werden wir dagegen die übliche Darstellung im Konfigurationsraum benutzen, wobei natürlich auch ϱ als Koordinate mitgezählt werden muß.

Wir führen zuerst zwei Operatoren Q und Q^* ein, welche auf die Funktionen der zweiwertigen Variablen ϱ als die linearen Substitutionen

$$Q = \begin{vmatrix} 0 & 1 \\ 0 & 0 \end{vmatrix}; \quad Q^* = \begin{vmatrix} 0 & 0 \\ 1 & 0 \end{vmatrix} \tag{1}$$

---

[1]) Vgl. z. B. P. Jordan u. O. Klein, ZS. f. Phys. **45**, 751, 1927; W. Heisenberg, Ann. d. Phys. **10**, 888, 1931.

wirken. Man sieht ohne weiteres, daß Q einem Übergang von Proton zu Neutron entspricht und Q^* einem Übergang von Neutron zu Proton.

Die Bedeutung der als Operatoren aufgefaßten Wahrscheinlichkeitsamplituden ψ und φ ist bekanntlich die folgende: Sei

$$\psi_1 \psi_2 \cdots \psi_s \cdots$$

ein System individueller Quantenzustände für die Elektronen. Man setze weiter

$$\psi = \sum_s \psi_s a_s; \quad \psi^* = \sum_s \psi_s^* a_s^*. \tag{2}$$

Die Amplituden a_s und die komplex konjugierten Größen a_s^* sind Operatoren, welche auf die Funktionen der Besetzungszahlen $N_1, N_2, \ldots, N_s, \ldots$ der individuellen Quantenzustände wirken. Im Falle des Pauli-Prinzips ist jedes der N_s nur der beiden Werte 0 und 1 fähig. Die Operatoren a_s und a_s^* sind dann folgendermaßen definiert:

$$\left. \begin{aligned} a_s \Psi(N_1 N_2 \ldots N_s \ldots) &= (-1)^{N_1 + N_2 + \cdots + N_s - 1}(1 - N_s)\Psi(N_1 N_2 \ldots 1 - N_s \ldots) \\ a_s^* \Psi(N_1 N_2 \ldots N_s \ldots) &= (-1)^{N_1 + N_2 + \cdots + N_s - 1} N_s \, \Psi(N_1 N_2 \ldots 1 - N_s \ldots) \end{aligned} \right\} \tag{3}$$

Der Operator a_s^* entspricht der Erzeugung und der Operator a_s dem Verschwinden eines Elektrons im Quantenzustand s.

Entsprechend zu (2) setze man für die Neutrinos

$$\varphi = \sum_\sigma \varphi_\sigma b_\sigma \quad \varphi^* = \sum_\sigma \varphi_\sigma^* b_\sigma^*. \tag{4}$$

Die komplex-konjugierten Größen b_σ und b_σ^* sind Operatoren, die auf die Funktionen der Besetzungszahlen $M_1, M_2, \ldots, M_\sigma, \ldots$ der individuellen Quantenzustände $\varphi_1, \varphi_2, \ldots, \varphi_\sigma, \ldots$ der Neutrinos wirken. Nimmt man an, daß auch für die Neutrinos das Pauli-Prinzip gilt, so sind die Zahlen M_σ nur der beiden Werte 0 und 1 fähig. Es ist ferner:

$$\left. \begin{aligned} b_\sigma \Phi(M_1 M_2 \ldots M_\sigma \ldots) &= (-1)^{M_1 + M_2 + \cdots + M_\sigma - 1}(1 - M_\sigma)\Phi(M_1 M_2 \ldots 1 - M_\sigma \ldots) \\ b_\sigma^* \Phi(M_1 M_2 \ldots M_\sigma \ldots) &= (-1)^{M_1 + M_2 + \cdots + M_\sigma - 1} M_\sigma \, \Phi(M_1 M_2 \ldots 1 - M_\sigma \ldots) \end{aligned} \right\} \tag{5}$$

Die Operatoren b_σ und b_σ^* entsprechen dem Verschwinden bzw. dem Entstehen eines Neutrinos im Quantenzustand σ.

3. Aufstellung der Hamilton-Funktion.

Die Energie des gesamten, aus schweren und leichten Teilchen bestehenden, Systems ist die Summe der Energien H_{schwer} der schweren Teilchen $+ \, H_{\text{leicht}}$ der leichten Teilchen $+$ der Wechselwirkungsenergie H zwischen schweren und leichten Teilchen.

Das erste Glied schreiben wir, indem wir vorläufig nur ein einziges schweres Teilchen betrachten, in der Form

$$H_{\text{schwer}} = \frac{1+\varrho}{2} N + \frac{1-\varrho}{2} P, \qquad (6)$$

wo N und P die Energieoperatoren des Neutrons bzw. des Protons darstellen. Für $\varrho = 1$ (Neutron) reduziert sich in der Tat (6) auf N; für $\varrho = -1$ (Proton) reduziert sich (6) auf P.

Die Energie H_{leicht} der leichten Teilchen nimmt die einfachste Form an, wenn man als Quantenzustände $\psi_1 \psi_2 \ldots \psi_s \ldots$ und $\varphi_1 \varphi_2 \ldots \varphi_\sigma \ldots$ stationäre Zustände für die Elektronen bzw. die Neutrinos nimmt. Für die Elektronen soll man dabei etwa die stationären Zustände im Coulomb-Feld des Kerns, unter Berücksichtigung der Elektronenabschirmung, wählen. Für die Neutrinos kann man einfach ebene de Broglie-Wellen annehmen, da wohl die auf die Neutrinos wirkenden Kräfte keine wesentliche Rolle spielen. Seien $H_1 H_2 \ldots H_s \ldots$ und $K_1 K_2 \ldots K_\sigma \ldots$ die Energien der stationären Zustände der Elektronen und der Neutrinos; dann haben wir:

$$H_{\text{leicht}} = \sum_s H_s N_s + \sum_\sigma K_\sigma M_\sigma. \qquad (7)$$

Es bleibt nur noch die Wechselwirkungsenergie zu schreiben. Diese besteht erstens aus der Coulomb-Energie zwischen Proton und Elektronen; bei schweren Kernen spielt jedoch die Anziehung durch ein einziges Proton nur eine untergeordnete Rolle[1]) und trägt in keinem Falle zum Prozeß des β-Zerfalls bei. Wir wollen also dies Glied der Einfachheit halber nicht berücksichtigen. Wir müssen hingegen zur Hamilton-Funktion ein Glied addieren, das die Bedingung c) von Ziffer 1 erfüllt.

Ein Glied, das notwendigerweise die Verwandlung eines Protons in ein Neutron mit dem Verschwinden eines Elektrons und eines Neutrinos koppelt, hat nun nach Ziffer 2 die Form

$$Q \, a_s \, b_\sigma. \qquad (8)$$

Der komplex konjugierte Operator

$$Q^* \, a_s^* \, b_\sigma^* \qquad (8')$$

koppelt dagegen die umgekehrten Prozesse (Verwandlung eines Neutrons in ein Proton und Entstehen eines Elektrons und eines Neutrinos).

Ein Wechselwirkungsglied, das die Bedingung c) erfüllt, kann also in der folgenden Form geschrieben werden:

$$H = Q \sum_{s\sigma} c_{s\sigma} a_s b_\sigma + Q^* \sum_{s\sigma} c_{s\sigma}^* a_s^* b_\sigma^*, \qquad (9)$$

---

[1]) Die Coulombsche Wirkung der zahlreichen übrigen Protonen muß natürlich als statisches Feld in Betracht gezogen werden.

49

Versuch einer Theorie der β-Strahlen. I. 165

wo c_{s_0} und $c_{s_0}^{*}$ Größen darstellen, die von den Koordinaten, Impulsen usw. des schweren Teilchens abhängen können.

Zur näheren Bestimmung von H ist man auf Einfachheitskriterien angewiesen. Eine wesentliche Einschränkung in der Freiheit der Wahl von H ist durch die Erhaltung des Impulses sowie durch die Bedingung gesetzt, daß bei einer Drehung oder einer Translation der Raumkoordinaten (9) invariant bleiben muß.

Sehen wir zunächst von den Relativitätskorrektionen und der Spinwirkung ab, so ist wohl die eintachst mögliche Wahl von (9) die folgende:

$$H = g\left\{Q\,\psi\,(x)\,\varphi\,(x) + Q^{*}\,\psi^{*}\,(x)\,\varphi^{*}\,(x)\right\}, \qquad (10)$$

wo g eine Konstante mit den Dimensionen $L^5 M\,T^{-2}$ darstellt; x repräsentiert die Koordinaten des schweren Teilchens; ψ, φ, ψ^{*}, φ^{*} sind durch (2) und (4) gegeben und sind an dem Orte x, y, z des schweren Teilchens zu nehmen.

(10) stellt keineswegs die einzig mögliche Wahl von H dar. Jeder skalare Ausdruck, wie etwa

$$L\,(p)\,\psi\,(x)\,M\,(p)\,\varphi\,(x)\,N\,(p) + \text{kompl. konjug.},$$

wo $L\,(p)$, $M\,(p)$, $N\,(p)$ passende Funktionen des Impulses des schweren Teilchens darstellen, würde ebensogut möglich sein. Da jedoch die Folgerungen aus (10) bisher mit der Erfahrung in Einklang zu sein scheinen, ist es wohl besser, sich vorläufig auf die einfachste Wahl zu beschränken.

Wesentlich ist es jedoch, den Ausdruck (10) derart zu verallgemeinern, daß man mindestens die leichten Teilchen relativistisch behandeln kann. Auch bei dieser Verallgemeinerung ist natürlich eine gewisse Willkür nicht auszuschließen. Die einfachste Lösung des Problems dürfte die folgende sein:

Relativistisch treten an Stelle von ψ und φ je vier Diracsche Funktionen $\psi_1\,\psi_2\,\psi_3\,\psi_4$ und $\varphi_1\,\varphi_2\,\varphi_3\,\varphi_4$. Wir betrachten nun die 16 unabhängigen bilinearen Kombinationen aus $\psi_1\,\psi_2\,\psi_3\,\psi_4$ und $\varphi_1\,\varphi_2\,\varphi_3\,\varphi_4$. Bei einer Lorentz-Transformation der Koordinaten erfahren diese 16 Größen eine lineare Transformation, eine Darstellung der Ordnung 16 der Lorentz-Gruppe. Diese Darstellung spaltet sich in verschiedene einfachere Darstellungen; im besonderen transformieren sich die vier bilinearen Kombinationen:

$$\left.\begin{aligned}
A_0 &= -\,\psi_1\varphi_2 + \psi_2\varphi_1 + \psi_3\varphi_4 - \psi_4\varphi_3,\\
A_1 &= \psi_1\varphi_3 - \psi_2\varphi_4 - \psi_3\varphi_1 + \psi_4\varphi_2,\\
A_2 &= i\,\psi_1\varphi_3 + i\,\psi_2\varphi_4 - i\,\psi_3\varphi_1 - i\,\psi_4\varphi_2,\\
A_3 &= -\,\psi_1\varphi_4 - \psi_2\varphi_3 + \psi_3\varphi_2 + \psi_4\varphi_1,
\end{aligned}\right\} \qquad (11)$$

166   E. Fermi,

wie die Komponenten eines polaren Vierervektors, also wie die Komponenten des elektromagnetischen Viererpotentials. Es liegt nun nahe, die
Größen

$$g\,(Q\,A_i + Q^*\,A_i^*)$$

in der Hamilton-Funktion des schweren Teilchens in einer Stellung auf-
zunehmen, die der Stellung der Komponenten des Viererpotentials entspricht.

Hier begegnen wir einer Schwierigkeit, welche davon herrührt, daß
die relativistische Wellengleichung für die schweren Teilchen unbekannt
ist. Falls die Geschwindigkeit des schweren Teilchens klein gegenüber c
ist, kann man sich jedoch auf den zu $e\,V$ (V = skalares Potential) analogen
Term beschränken und schreiben:

$$H = g\,[Q\,(-\,\psi_1\,\varphi_2 + \psi_2\,\varphi_1 + \psi_3\,\varphi_4 - \psi_4\,\varphi_3)$$
$$+ Q^*\,(-\,\psi_1^*\,\varphi_2^* + \psi_2^*\,\varphi_1^* + \psi_3^*\,\varphi_4^* - \psi_4^*\,\varphi_3^*)].\quad (12)$$

Zu diesem Glied sollen noch andere Glieder von der Größenordnung v/c
addiert werden. Da die Geschwindigkeiten der Neutronen und Protonen
in den Kernen gewöhnlich klein gegenüber der Lichtgeschwindigkeit sind,
wollen wir diese Glieder vorläufig vernachlässigen (vgl. hierzu Ziffer 9).

(12) kann in symbolischer Schreibweise folgendermaßen abgekürzt
werden:

$$H = g\,[Q\,\widetilde{\psi}^*\,\delta\,\varphi + Q^*\,\widetilde{\psi}\,\delta\,\varphi^*],\quad (13)$$

wo ψ und φ als vertikale Matrixspalten zu schreiben sind; das Zeichen $\sim$
verwandelt eine Matrix in die konjugiert transponierte; und es ist

$$\delta = \begin{vmatrix} 0 & -1 & 0 & 0 \\ 1 & 0 & 0 & 0 \\ 0 & 0 & 0 & 1 \\ 0 & 0 & -1 & 0 \end{vmatrix}\quad (14)$$

Mit diesen Bezeichnungen bekommt man durch Vergleich mit (9)

$$c_{s\sigma} = g\,\widetilde{\psi}_s^*\,\delta\,\varphi_\sigma; \quad c_{s\sigma}^* = g\,\widetilde{\psi}_s\,\delta\,\varphi_\sigma^*,\quad (15)$$

wo ψ_s und φ_s die normierten vierkomponentigen Eigenfunktionen der Zustände s (des Elektrons) und σ (des Neutrinos) darstellen. ψ und φ sind
in (15) an der Stelle des schweren Teilchens, also als Funktionen von x, y, z,
zu nehmen.

4. Die Störungsmatrix.

Die Theorie des β-Zerfalls kann mit Hilfe der aufgestellten Hamilton-
Funktion in voller Analogie zur Strahlungstheorie durchgeführt werden.
In dieser letzteren besteht die Hamilton-Funktion bekanntlich aus der

51

Versuch einer Theorie der β-Strahlen. I.   167

Summe: Energie des Atoms + Energie des reinen Strahlungsfeldes + Kopplungsenergie. Dies letzte Glied wird als Störung der beiden anderen aufgefaßt. In Analogie hierzu werden wir in unserem Falle die Summe

$$H_{\text{schwer}} + H_{\text{le cht}} \tag{16}$$

als ungestörte Hamilton-Funktion betrachten; hinzu kommt die durch das Kopplungsglied (13) dargestellte Störung.

Die Quantenzustände des ungestörten Systems können folgendermaßen numeriert werden:

$$(\varrho, n, N_1 N_2 \ldots N_s \ldots M_1 M_2 \ldots M_\sigma \ldots), \tag{17}$$

wo die erste Zahl ϱ einen der beiden Werte ± 1 annimmt und angibt, ob das schwere Teilchen ein Neutron oder ein Proton ist. Die zweite Zahl n numeriert den Quantenzustand des Neutrons oder des Protons. Für $\varrho = 1$ (Neutron) sei die entsprechende Eigenfunktion

$$u_n(x), \tag{18}$$

wo x die Koordinaten des schweren Teilchens, bis auf ϱ, darstellt. Für $\varrho = -1$ (Proton) sei die Eigenfunktion

$$v_n(x). \tag{19}$$

Die übrigen Zahlen $N_1 N_2 \ldots N_s \ldots M_1 M_2 \ldots M_\sigma \ldots$ sind nur der beiden Werte 0 und 1 fähig und geben an, ob der betreffende Zustand des Elektrons oder des Neutrinos besetzt ist.

Faßt man nun die allgemeine Form (9) der Störungsenergie ins Auge, so sieht man, daß sie von Null verschiedene Elemente nur für solche Übergänge hat, bei denen entweder das schwere Teilchen von einem Neutron in einen Protonenzustand übergeht und zugleich ein Elektron und ein Neutrino entstehen, oder umgekehrt.

Mit Hilfe von (1), (3), (5), (9), (18), (19) findet man ohne weiteres das betreffende Matrixelement

$$H^{1\, n\, N_1\, N_2\, \ldots\, O_s\, \ldots\, M_1 M_2\, \ldots\, O_\sigma\, \ldots}_{-1\, m\, N_1\, N_2\, \ldots\, 1_s\, \ldots\, M_1 M_2\, \ldots\, 1_\sigma\, \ldots} = \pm \int v_m^* c_{s\sigma}^* u_n \, d\tau, \tag{20}$$

wo die Integration über den Konfigurationsraum des schweren Teilchens (bis auf die Koordinate ϱ) erstreckt werden muß. Das $\pm$-Zeichen bedeutet genauer

$$(-1)^{N_1 + N_2 + \cdots + N_{s-1} + M_1 + M_2 + \cdots + M_{\sigma-1}}$$

und wird übrigens aus den folgenden Rechnungen herausfallen. Dem entgegengesetzten Übergang entspricht ein komplex konjugiertes Matrixelement.

Führt man für $c_{s\sigma}^{*}$ den Wert (15) ein, so erhält man

$$H_{-1\,m\,1_s\,1_\sigma}^{1\,n\,0_s\,0_\sigma} = \pm\, g \int v_m^* u_n\, \widetilde{\psi}_s\, \delta\, \varphi_\sigma^*\, \mathrm{d}\tau, \qquad (21)$$

wo der Kürze wegen im ersten Glied alle gleichbleibenden Indizes fort-gelassen worden sind.

5. Theorie des β-Zerfalls.

Ein β-Zerfall besteht in einem Prozeß, bei welchem ein Kernneutron sich in ein Proton verwandelt und gleichzeitig mit dem geschilderten Me-chanismus ein Elektron, das als β-Strahl beobachtet wird und ein Neu-trino emittiert werden. Um die Wahrscheinlichkeit dieses Prozesses zu berechnen, wollen wir annehmen, daß zur Zeit $t=0$ ein Neutron in einem Kernzustand mit Eigenfunktion $u_n\,(x)$ vorhanden ist und $N_s = M_\sigma = 0$, d. h. der Elektronenzustand s und der Neutrinozustand σ leer sind. Dann ist für $t=0$ die Wahrscheinlichkeitsamplitude des Zustands $(1, n, 0_s, 0_\sigma)$

$$a_{1\,n\,0_s\,0_\sigma} = 1 \qquad (22)$$

und die des Zustandes $(-1, m, 1_s, 1_\sigma)$, wo das Neutron in ein Proton mit der Eigenfunktion $v_m\,(x)$ unter Emission eines Elektrons und eines Neu-trinos übergegangen ist, gleich Null.

Mit Anwendung der gewöhnlichen Störungsformeln hat man nun für eine Zeit, die kurz genug ist, damit (22) noch angenähert gültig ist:

$$\dot{a}_{-1\,m\,1_s\,1_\sigma} = -\,\frac{2\,\pi\,i}{h}\, H_{-1\,m\,1_s\,1_\sigma}^{1\,n\,0_s\,0_\sigma}\, e^{\frac{2\,\pi\,i}{h}(-\,W\,+\,H_s\,+\,K_0)\,t}, \qquad (23)$$

wo W die Energiedifferenz des Neutronen- und des Protonenzustandes darstellt.

Aus (23) erhält man (da für $t=0$, $a_{-1\,m\,1_s\,1_\sigma} = 0$)

$$a_{-1\,m\,1_s\,1_\sigma} = -\,H_{-1\,m\,1_s\,1_\sigma}^{1\,n\,0_s\,0_\sigma}\, \frac{e^{\frac{2\,\pi\,i}{h}(-\,W\,+\,H_s\,+\,K_\sigma)\,t} - 1}{-\,W\,+\,H_s\,+\,K_\sigma}. \qquad (24)$$

Die Wahrscheinlichkeit des betrachteten Übergangs ist also zur Zeit t

$$\left|a_{-1\,m\,1_s\,1_\sigma}\right|^2 = 4\,\left|H_{-1\,m\,1_s\,1_\sigma}^{1\,n\,0_s\,0_\sigma}\right|^2\, \frac{\sin^2\dfrac{\pi\,t}{h}(-\,W\,+\,H_s\,+\,K_\sigma)}{(-\,W\,+\,H_s\,+\,K_\sigma)^2}. \qquad (25)$$

Um die Lebensdauer des Neutronenzustands u_n zu berechnen, hat man den Ausdruck (25) über alle freien Elektronen- und Neutrinozustände zu summieren.

53

Versuch einer Theorie der β-Strahlen. I.   169

Eine wesentliche Vereinfachung in der Ausführung der Summe erhält man durch die Bemerkung, daß die de Broglie-Wellenlänge für Elektronen und Neutrinos mit Energien von einigen Millionen Volt wesentlich größer ist als die Kerndimensionen. In erster Näherung kann man also die Eigenfunktionen ψ_s und φ_σ innerhalb des Kerns als Konstante betrachten. (21) wird dann:

$$H^{1\,n\,0_s\,0_\sigma}_{-1\,m\,1_s\,1_\sigma} = \pm\, g\, \widetilde{\psi}_s\, \delta\, \varphi_\sigma^*\, \int v_m^*\, u_n\, \mathrm{d}\tau, \tag{26}$$

wobei hier und im folgenden ψ_s und φ_σ an der Stelle des Kerns zu nehmen sind (vgl. Ziffer 8). Aus (26) hat man

$$\left| H^{1\,n\,0_s\,0_\sigma}_{-1\,m\,1_s\,1_\sigma} \right|^2 = g^2 \left| \int v_m^*\, u_n\, \mathrm{d}\tau \right|^2 \widetilde{\psi}_s\, \delta\, \varphi_\sigma^*\, \widetilde{\varphi}_\sigma^*\, \widetilde{\delta}\, \psi_s. \tag{27}$$

Die Zustände σ des Neutrinos sind durch ihren Impuls p_σ und die Spinrichtung bestimmt. Falls wir zu Normierungszwecken in einem Volumen Ω quantisieren, dessen Dimensionen wir nachher ins Unendliche wachsen lassen werden, so sind die normierten Neutrinoeigenfunktionen ebene Dirac-Wellen, mit der Dichte $1/\Omega$. Eine einfache Algebra erlaubt dann in (27) den Mittelwert über alle Richtungen von p_σ und alle Spinrichtungen des Neutrinos zu nehmen. (Zu betrachten sind dabei nur die positiven Eigenwerte; die negativen sind mit einem der Diracschen Löchertheorie analogen Kunstgriff zu beseitigen.) Man findet:

$$\overline{\left| H^{1\,n\,0_s\,0_\sigma}_{-1\,m\,1_s\,1_\sigma} \right|^2} = \frac{g^2}{4\,\Omega} \left| \int v_m^*\, u_n\, \mathrm{d}\tau \right|^2 \left(\widetilde{\psi}_s\, \psi_s - \frac{\mu\, c^2}{K_\sigma}\, \widetilde{\psi}_s \beta\, \psi_s \right), \tag{28}$$

wo μ die Ruhemasse des Neutrinos und β die Diracsche Matrix

$$\beta = \begin{vmatrix} 1 & 0 & 0 & 0 \\ 0 & 1 & 0 & 0 \\ 0 & 0 & -1 & 0 \\ 0 & 0 & 0 & -1 \end{vmatrix} \tag{29}$$

darstellt. Beachtet man nun:

daß die Anzahl der Neutrinozustände positiver Energie mit Impuls zwischen p_σ und $p_\sigma + \mathrm{d}p_\sigma$ $\dfrac{8\,\pi\,\Omega}{h^3}\, p_\sigma^2\, \mathrm{d}p_\sigma$ ist;

daß $\dfrac{\partial K_\sigma}{\partial p_\sigma} = v_\sigma$, wo v_σ die Geschwindigkeit des Neutrinos im Zustand σ darstellt;

daß (25) ein scharfes Maximum in der Nähe des Wertes von p_σ hat, für den die Variation der ungestörten Energie verschwindet, d. h.

$$- W + H_s + K_\sigma = 0, \tag{30}$$

170   E. Fermi,

so kann man die Summe von (25) über σ in bekannter Weise[1]) ausführen, und man findet:

$$t \cdot \frac{8\,\pi^3\,g^2}{h^4} \left| \int v_m^* u_n \, \mathrm{d}\,\tau \right|^2 \frac{p_o^3}{v_\sigma} \left(\widetilde{\psi}_s\,\psi_s - \frac{\mu\,c^2}{K_\sigma}\,\widetilde{\psi}_s\,\beta\,\psi_s \right), \qquad (31)$$

wo p_o hier den Wert des Neutrinoimpulses bedeutet, für den (30) gültig ist.

6. Bestimmungsstücke der Übergangswahrscheinlichkeit.

(31) gibt die Wahrscheinlichkeit dafür an, daß während der Zeit t ein β-Zerfall mit Übergang des Elektrons in den Zustand s stattfindet. Wie es sein soll, ist diese Wahrscheinlichkeit proportional der Zeit t (t ist als klein in bezug auf die Lebensdauer angenommen worden); der Koeffizient von t gibt die Übergangswahrscheinlichkeit für den geschilderten Prozeß an. Sie ist:

$$P_s = \frac{8\,\pi^3\,g^2}{h^4} \left| \int v_m^* u_n \, \mathrm{d}\,\tau \right|^2 \frac{p_o^3}{v_\sigma} \left(\widetilde{\psi}_s\,\psi_s - \frac{\mu\,c^2}{K_\sigma}\,\widetilde{\psi}_s\,\beta\,\psi_s \right). \qquad (32)$$

Man bemerke:

a) Für die freien Neutrinozustände ist immer $K_\sigma > \mu c^2$. Damit (30) befriedigt werden kann, ist also notwendig, daß

$$H_s \lesseqgtr W - \mu c^2. \qquad (33)$$

Dem $=$-Zeichen entspricht die obere Grenze des kontinuierlichen β-Strahlspektrums.

b) Da für die freien Elektronenzustände $H_s > mc^2$ ist, bekommt man die folgende, für die Möglichkeit des β-Zerfalls notwendige Bedingung

$$W \gtreqless (m + \mu)\,c^2. \qquad (34)$$

Ein besetzter Neutronenzustand n im Kerne muß also hoch genug über einem unbesetzten Protonenzustand m liegen, damit der β-Prozeß vor sich gehen kann.

c) Nach (32) hängt P_s von den Eigenfunktionen u_n, v_m des schweren Teilchens im Kerne durch das Matrixelement

$$Q_{m\,n}^* = \int v_m^* u_n \, \mathrm{d}\,\tau \qquad (35)$$

ab. Dies Matrixelement spielt in der β-Strahltheorie eine ähnliche Rolle wie das Matrixelement des elektrischen Moments eines Atoms in der Strahlungstheorie. Das Matrixelement (35) hat normalerweise die Größenordnung 1; durch besondere Symmetrieeigenschaften von u_n und v_m kann es jedoch oft vorkommen, daß $Q_{m\,n}^*$ verschwindet. In solchen Fällen sprechen

---

[1]) Für die genaue Beschreibung der Methode, solche Summen auszuführen. vgl. irgendeinen Aufsatz über Strahlungstheorie; etwa: E. Fermi, Rev. Mod. Phys. **4**, 87, 1932.

wir von *verbotenen β-Übergängen*. Man muß natürlich nicht erwarten, daß die verbotenen Übergänge überhaupt nicht vorkommen, da (32) nur eine Näherungsformel ist. Wir werden in Ziffer 9 etwas über diesen Typ von Übergängen sprechen.

7. Die Masse des Neutrinos.

Durch die Übergangswahrscheinlichkeit (32) ist die Form des kontinuierlichen β-Spektrums bestimmt. Wir wollen zuerst diskutieren, wie diese Form von der Ruhemasse μ des Neutrinos abhängt, um von einem Vergleich mit den empirischen Kurven diese Konstante zu bestimmen. Die Masse μ ist in dem Faktor p_o^2/v_σ enthalten. Die Abhängigkeit der Form der Energieverteilungskurve von μ ist am meisten ausgeprägt in der Nähe des Endpunktes

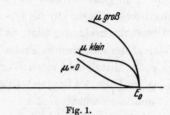

Fig. 1.

der Verteilungskurve. Ist E_0 die Grenzenergie der β-Strahlen, so sieht man ohne Schwierigkeit, daß die Verteilungskurve für Energien E in der Nähe von E_0 bis auf einen von E unabhängigen Faktor sich wie

$$\frac{p_\sigma^2}{v_\sigma} = \frac{1}{c^3} \left(\mu c^2 + E_0 - E\right) \sqrt{(E_0 - E)^2 + 2\mu c^2 (E_0 - E)} \qquad (36)$$

verhält.

In der Fig. 1 ist das Ende der Verteilungskurve für $\mu = 0$ und für einen kleinen und einen großen Wert von μ gezeichnet. Die größte Ähnlichkeit mit den empirischen Kurven zeigt die theoretische Kurve für $\mu = 0$.

Wir kommen also zu dem Schluß, daß die Ruhemasse des Neutrinos entweder Null oder jedenfalls sehr klein in bezug auf die Masse des Elektrons ist[1]. In den folgenden Rechnungen werden wir die einfachste Hypothese $\mu = 0$ einführen. Es wird dann (30)

$$v_\sigma = c; \quad K_\sigma = p_\sigma c; \quad p_\sigma = \frac{K_\sigma}{c} = \frac{W - H_s}{c}. \qquad (37)$$

Die Ungleichungen (33), (34) werden jetzt:

$$H_s \leqq W; \quad W \geqq m c^2. \qquad (38)$$

Und die Übergangswahrscheinlichkeit (32) nimmt die Form an:

$$P_s = \frac{8\pi^3 g^2}{c^3 h^4} \left| \int v_m^* u_n \, d\tau \right|^2 \widetilde{\psi}_s \psi_s (W - H_s)^2. \qquad (39)$$

---

[1]) In einer kürzlich erschienenen Notiz kommt F. Perrin, C. R. **197**, 1625, 1933, mit qualitativen Überlegungen zu demselben Schluß.

8. Lebensdauer und Form der Verteilungskurve für „erlaubte" Übergänge.

Aus (39) kann man eine Formel ableiten, welche angibt, wieviel β-Übergänge in der Zeiteinheit stattfinden, für welche das β-Teilchen einen Impuls zwischen $mc\eta$ und $mc\,(\eta + d\eta)$ erhält. Dazu muß man eine Formel für die Summe von $\widetilde{\psi}_s\psi_s$ am Orte des Kerns über alle im kontinuierlichen Spektrum liegenden Quantenzustände des betreffenden Intervalls ableiten.

Dabei sei bemerkt, daß die relativistischen Eigenfunktionen im Coulomb-Feld für die Zustände mit $j = {}^1/_2$ ($^2s_{1/2}$ und $^2p_{1/2}$) für $r = 0$ unendlich groß werden. Nun gehorcht aber die Kernanziehung für die Elektronen dem Coulombschen Gesetz nur bis $r > \varrho$, wo ϱ hier den Kernradius bedeutet. Eine Überschlagsrechnung zeigt, daß, wenn man plausible Annahmen über den Verlauf des elektrischen Feldes innerhalb des Kerns macht, der Wert von $\widetilde{\psi}_s\psi_s$ im Mittelpunkt einen Wert hat, der sehr nahe dem Werte liegt, den $\widetilde{\psi}_s\psi_s$ im Falle des Coulomb-Gesetzes in der Entfernung ϱ vom Mittelpunkt annehmen würde.

Durch Heranziehung der bekannten Formeln[1]) für die relativistischen Eigenfunktionen des Kontinuums im wasserstoffähnlichen Falle findet man also nach einer ziemlich langwierigen Rechnung

$$\sum_{d\eta}\widetilde{\psi}_s\psi_s = d\eta\cdot\frac{32\,\pi\,m^3\,c^3}{h^3[\Gamma(3+2S)]^2}\left(\frac{4\pi mc\varrho}{h}\right)^{2S}\eta^{2+2S}e^{\pi\gamma\frac{\sqrt{1+\eta^2}}{\eta}}\left|\Gamma\left(1+S+i\gamma\frac{\sqrt{1+\eta^2}}{\eta}\right)\right|^2, (40)$$

wo

$$\gamma = Z/137; \quad S = \sqrt{1-\gamma^2}-1. \tag{41}$$

Die Übergangswahrscheinlichkeit in einen Elektronenzustand mit einem Impuls des Intervalls $mcd\eta$ wird dann nach (39):

$$P(\eta)\,d\eta = d\eta\cdot g^2\,\frac{256\,\pi^4}{[\Gamma(3+2S)]^2}\,\frac{m^5c^4}{h^7}\left(\frac{4\pi mc\varrho}{h}\right)^{2S}\left|\int v_m^*\,u_n\,d\tau\right|^2$$

$$\cdot\eta^{2+2S}e^{\pi\gamma\frac{\sqrt{1+\eta^2}}{\eta}}\left|\Gamma\left(1+S+i\gamma\frac{\sqrt{1+\eta^2}}{\eta}\right)\right|^2\left(\sqrt{1+\eta_0^2}-\sqrt{1+\eta^2}\right)^2, (42)$$

wo η_0 den in Einheiten mc gemessenen maximalen Impuls der emittierten β-Strahlen darstellt.

Die numerische Auswertung von (42) kann man etwa für $\gamma = 0,6$, d. h. $Z = 82,2$ machen, da ja die Atomnummern der radioaktiven Stoffe nicht weit von diesem Wert liegen. Für $\gamma = 0,6$ ist nach (41) $S = -0,2$. Man findet weiter, daß für $\eta < 10$ die folgende Formel angenähert gilt:

$$\eta^{1,6}\,e^{0,6\,\pi\frac{\sqrt{1+\eta^2}}{\eta}}\left|\Gamma\left(0,8+0,6\,i\frac{\sqrt{1+\eta^2}}{\eta}\right)\right|^2 \simeq 4,5\,\eta + 1,6\,\eta^2. \tag{43}$$

---

[1]) R. H. Hulme, Proc. Roy. Soc. London (A) **133**, 381, 1931.

57

Versuch einer Theorie der β-Strahlen. I.   173

Formel (42) wird damit, wenn man $\varrho = 9 \cdot 10^{-13}$ setzt:

$$P(\eta)\,\mathrm{d}\eta = 1{,}75 \cdot 10^{95} g^2 \left| \int v_m^* u_n \,\mathrm{d}\tau \right|^2 (\eta + 0{,}355\,\eta^2)\left(\sqrt{1+\eta_0^2} - \sqrt{1+\eta^2}\right)^2. \quad (44)$$

Die reziproke Lebensdauer erhält man aus (44) durch Integration von $\eta = 0$ bis $\eta = \eta_0$; man findet:

$$\frac{1}{\tau} = 1{,}75 \cdot 10^{95} g^2 \left| \int v_m^* u_n \,\mathrm{d}\tau \right|^2 F(\eta_0), \quad (45)$$

wo

$$F(\eta_0) = \frac{2}{3}\left(\sqrt{1+\eta_0^2} - 1\right)$$
$$+ \frac{\eta_0^4}{12} - \frac{\eta_0^2}{3} + 0{,}355\left[-\frac{\eta_0}{4} - \frac{\eta_0^3}{12} + \frac{\eta_0^5}{30} + \frac{\sqrt{1+\eta_0^2}}{4}\log\left(\eta_0 + \sqrt{1+\eta_0^2}\right)\right]. \quad (46)$$

Für kleine Argumente verhält sich $F(\eta_0)$ wie $\eta_0^6/24$; für größere Argumente sind die Werte von F in der folgenden Tabelle zusammengestellt.

Tabelle 1.

η_0	$F(\eta_0)$	η_0	$F(\eta_0)$	η_0	$F(\eta_0)$	η_0	$F(\eta_0)$
0	$\eta_0^6/24$	2	1,2	4	29	6	185
1	0,03	3	7,5	5	80	7	380

9. Die verbotenen Übergänge.

Bevor wir zum Vergleich mit der Erfahrung übergehen, wollen wir noch einige Eigenschaften der verbotenen β-Übergänge diskutieren.

Wie schon bemerkt, ist ein Übergang verboten, wenn das zugehörige Matrixelement (35) verschwindet. Falls nun die Darstellung des Kerns mit individuellen Quantenzuständen der Neutronen und der Protonen eine gute Näherung ist, verschwindet immer Q_{mn}^* aus Symmetriegründen, wenn nicht

$$i = i', \quad (47)$$

wo i und i' die Impulsmomente (in Einheiten $h/2\pi$) des Neutronenzustands u_n und des Protonenzustands v_m darstellen. Der Auswahlregel (47) entspricht, falls die individuellen Zustände keine gute Näherung sind, die allgemeinere

$$I = I', \quad (48)$$

wo I und I' die Impulsmomente des Kerns vor und nach dem β-Zerfall bedeuten.

Die Auswahlregeln (47) und (48) sind bei weitem nicht so scharf wie die Auswahlregeln der Optik. Es gibt hauptsächlich zwei Prozesse, wodurch ein Durchbrechen dieser Auswahlregeln möglich ist:

a) Formel (26) ist durch Vernachlässigung der Variationen von ψ_s und φ_σ innerhalb der Kernausdehnung erhalten worden. Falls man aber ψ_s und φ_σ im Bereich des Kerns nicht als Konstante betrachtet, so erhält man die Möglichkeit von β-Übergängen auch in Fällen, wo Q^*_{mn} verschwindet.

Es ist leicht einzusehen, daß die Intensität solcher Übergänge zur Intensität der erlaubten Prozesse größenordnungsmäßig im Verhältnis $(\varrho/\lambda)^2$ steht, wo λ die de Broglie-Wellenlänge der leichten Teilchen dargestellt. Man bemerke hierzu, daß, bei gleicher Energie, die kinetische Energie der Elektronen am Orte des Kerns wegen der elektrostatischen Anziehung erheblich größer ist als die der Neutrinos; die größte Wirkung rührt also von der Variation von ψ_s her. Eine Abschätzung der Intensität dieser verbotenen Prozesse zeigt, daß sie rund 100 mal schwächer sein müssen als die nach (48) erlaubten Übergänge, für welche β-Teilchen der gleichen Energie emittiert werden.

Ein Merkmal für verbotene Übergänge dieses Typs könnte man nicht nur in der verhältnismäßig längeren Lebensdauer, sondern auch in der verschiedenen Form der Energieverteilungskurve der β-Strahlen erblicken; man findet nämlich, daß für diese Übergänge die Verteilungskurve für kleine Energien tiefer liegen muß als im normalen Falle.

b) Eine zweite Möglichkeit von nach (48) verbotenen Übergängen folgt aus der am Ende von Ziffer 3 bemerkten Tatsache, daß, falls man die Geschwindigkeit der schweren Kernbestandteile nicht gegen die Lichtgeschwindigkeit vernachlässigt, zum Wechselwirkungsglied (12) noch weitere von der Größenordnung v/c hinzutreten. Falls man etwa auch für die schweren Teilchen eine relativistische Wellengleichung vom Diracschen Typus annimmt, könnte man z. B. zu (12) Terme wie

$$gQ\,(\alpha_x A_1 + \alpha_y A_2 + \alpha_z A_3) + \text{komplex konjugiert} \qquad (49)$$

addieren, wo $\alpha_x \alpha_y \alpha_z$ die Diracschen Matrizen für das schwere Teilchen bedeuten und A_1, A_2, A_3 die Raumkomponenten des von (11) definierten Vierervektors sind. Das Glied (49) würde zu (12) in demselben Verhältnis stehen wie die Terme eV bzw. $e\,(\alpha, U)$ ($V =$ skalares Potential; $U =$ Vektorpotential) zu der Diracschen Hamilton-Funktion.

Ein Wechselwirkungsglied wie (49) würde natürlich auch verbotene Übergänge ermöglichen, mit einer relativen Intensität von der Größenordnung $(v/c)^2$ in bezug auf die der erlaubten Übergänge. Dies gibt also eine zweite Möglichkeit für das Vorhandensein von Übergängen, die etwa 100 mal schwächer sind als die normalen.

10. Vergleich mit der Erfahrung.

Formel (45) gibt eine Beziehung zwischen dem maximalen Impuls der emittierten β-Strahlen und der Lebensdauer der β-strahlenden Substanz. In dieser Beziehung tritt zwar noch ein unbekanntes Element auf, nämlich das Integral

$$\int v_m^* \, u_n \, d\,\tau, \tag{50}$$

für dessen Auswertung eine Kenntnis der Eigenfunktionen des Protons und des Neutrons im Kern notwendig wäre. Im Falle der erlaubten Übergänge ist jedoch (50) von der Größenordnung 1. Man kann also erwarten, daß das Produkt

$$\tau F\,(\eta_0) \tag{51}$$

für alle erlaubten Übergänge dieselbe Größenordnung hat. Falls aber der betreffende Übergang verboten ist, ist die Lebensdauer rund 100 mal größer als im normalen Falle und auch das Produkt (51) wird entsprechend größer.

In der Tabelle 2 sind die Produkte (51) für die radioaktiven Elemente zusammengestellt, für welche man genügende Daten über das kontinuierliche β-Spektrum hat.

Tabelle 2.

Element	τ (Stunden)	η_0	$F\,(\eta_0)$	$\tau\,F\,(\eta_0)$
U X_2	0,026	5,4	115	3,0
Ra B	0,64	2,04	1,34	0,9
Th B	15,3	1,37	0,176	2,7
Th C''	0,076	4,4	44	3,3
Ac C''	0,115	3,6	17,6	2,0
Ra C	0,47	7,07	398	190
Ra E	173	3,23	10,5	1800
Th C	2,4	5,2	95	230
Ms Th_2	8,8	6,13	73	640

Aus der Tabelle sind die zwei erwarteten Gruppen ohne weiteres erkennbar; eine solche Einteilung ist übrigens bereits von Sargent[1]) auf empirischem Wege festgestellt worden. Die Werte von η_0 sind aus der genannten Arbeit von Sargent genommen (zum Vergleich bemerke man, daß: $\eta_0 = (H\varrho)_{max}/1700$). Die von Sargent als nicht zuverlässig angegebenen Werte von η_0 passen nicht besonders gut in die Einteilung; für $U X_1$ hat man $\tau = 830$; $\eta_0 = 0,76$; $F\,(\eta_0) = 0,0065$; $\tau F\,(\eta_0) = 5,4$; dies Element scheint also zur ersten Gruppe zu passen. Für Ac B hat man

---

[1]) B. W. Sargent, Proc. Roy. Soc. London (A) **139**, 659, 1933.

die folgenden Daten: $\tau = 0{,}87$; $\eta_0 = 1{,}24$; $F(\eta_0) = 0{,}102$; $\tau F(\eta_0) = 0{,}09$, also ein τF-Wert etwa zehnmal kleiner als die der ersten Gruppe. Für RaD hat man $\tau = 320000$; $\eta_0 = 0{,}38$ (sehr unsicher); $F(\eta_0) = 0{,}00011$; $\tau F(\eta_0) = 35$. RaD liegt also ungefähr in der Mitte zwischen den beiden Gruppen. Ich habe keine Daten über die anderen β-strahlenden Elemente MsTh$_1$, UY, Ac, AcC, UZ, RaC'' gefunden.

Aus den Daten der Tabelle 2 kann man eine, wenn auch sehr grobe, Abschätzung der Konstante g gewinnen. Nimmt man etwa an, daß in den Fällen wo (50) gleich Eins wird, man $\tau F(\eta_0) = 1$ hat (d. h., in Sekunden, $= 3600$), so bekommt man aus (45):

$$g = 4 \cdot 10^{-50}\ \text{cm}^3 \cdot \text{erg}.$$

Dieser Wert gibt natürlich nur die Größenordnung von g.

Zusammenfassend kann man sagen, daß dieser Vergleich von Theorie und Erfahrung eine so gute Übereinstimmung gibt, wie man nur erwarten

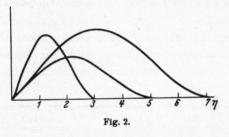

Fig. 2.

konnte. Die bei den als experimentell unsicheren Elementen RaD und AcB festgestellten Abweichungen können wohl teilweise durch Ungenauigkeit der Messungen erklärt werden, teilweise auch durch etwas abnorm große aber gar nicht unplausible Schwankungen des Matrixelements (50). Man hat weiter zu bemerken, daß man aus der den β-Zerfall begleitenden γ-Strahlung schließen kann, daß die meisten β-Zerfälle zu verschiedenen Endzuständen des Protons führen können, wodurch wieder Schwankungen in dem $\tau F(\eta_0)$-Wert erklärt werden können.

Wir wenden uns jetzt zur Frage nach der Form der Geschwindigkeitsverteilungskurve der emittierten β-Strahlen. Für den Fall der erlaubten Übergänge ist die Verteilungskurve als Funktion von η (d. h. bis auf den Faktor 1700, von $H\varrho$) durch (44) gegeben. Verteilungskurven für verschiedene Werte von η_0 sind in der Fig. 2 zusammengestellt, wobei für die Bequemlichkeit der Zeichnung die Ordinateneinheit in den verschiedenen Fällen passend gewählt worden ist. Diese Kurven zeigen eine befriedigende Ähnlichkeit etwa zu den von Sargent[1]) zusammengestellten Verteilungskurven. Nur in dem Teil der Kurve kleiner Energie liegen die Kurven von Sargent etwas tiefer als die theoretischen. Dies ist deutlicher in der

---

[1]) B. W. Sargent, Proc. Cambridge Phil. Soc. **28**, 538, 1932.

Fig. 3 zu sehen, wo als Abszisse die Energie an Stelle des Impulses genommen worden ist. Hierzu muß man jedoch bemerken, daß die experimentelle Kenntnis des Verteilungsgesetzes für kleine Energien besonders unsicher ist[1]). Übrigens hat man für die verbotenen Übergänge auch theoretisch Kurven zu erwarten, die im Gebiet kleiner Energie tiefer liegen als die der Fig. 2 und 3. Dieser letzte Punkt ist

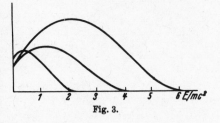

Fig. 3.

besonders für den Fall der experimentell verhältnismäßig gut bekannten Kurve des RaE zu beachten. Aus der Tabelle 2 sieht man nämlich, daß RaE einen sehr großen $\tau F(\eta_0)$-Wert hat; der β-Zerfall des RaE ist also gewiß verboten und wird sogar möglicherweise nur erst in zweiter Näherung erlaubt. Ich hoffe in einer nächsten Mitteilung etwas Genaueres über den Verlauf der Energieverteilungskurven für die verbotenen Übergänge sagen zu können.

Zusammenfassend darf man wohl sagen, daß die Theorie in der hier angegebenen Form in Übereinstimmung mit den allerdings nicht immer besonders genauen experimentellen Daten ist. Sollte man übrigens auch bei einem näheren Vergleich von Theorie und Erfahrung zu Widersprüchen kommen, so wäre es noch möglich, die Theorie abzuändern, ohne ihre begrifflichen Fundamente zu berühren. Man könnte nämlich Gleichung (9) behalten und eine verschiedene Wahl der $c_{s\sigma}$ treffen. Dies könnte insbesondere zu einer Abänderung der Auswahlregel (48) führen und eine andere Form der Energieverteilungskurve sowie der Abhängigkeit der Lebensdauer von der maximalen Energie ergeben. Ob eine solche Änderung notwendig sein wird, kann jedoch erst durch eine weitere Entwicklung der Theorie und möglicherweise auch durch eine Verschärfung der experimentellen Daten gezeigt werden.

---

[1]) Vgl z. B. E. Rutherford, B. Ellis u. J. Chadwick, Radiations from Radioactive Substances, Cambridge 1932. Siehe insbesondere S. 407.

4

(Untersuchungen zur Molekularstrahlmethode aus dem Institut für physikalische Chemie der Hamburgischen Universität. Nr. 24.)

Über die magnetische Ablenkung von Wasserstoffmolekülen und das magnetische Moment des Protons. I.

Von **R. Frisch** und **O. Stern** in Hamburg.

Mit 12 Abbildungen. (Eingegangen am 27. Mai 1933.)

Strahlen aus Wasserstoffmolekülen wurden nach der Methode von Gerlach und Stern magnetisch abgelenkt und so ihr magnetisches Moment bestimmt. Die Messungen an Parawasserstoff ergaben das von der Rotation des Moleküls herrührende magnetische Moment zu etwa 1 Kernmagneton ($^1/_{1840}$ Bohrmaneton) pro Rotationsquant. Die Messungen an Orthowasserstoff ergaben das magnetische Moment des Protons zu 2 bis 3 Kernmagnetonen (nicht 1 Kernmagneton, wie bisher vermutet wurde).

In den bisherigen Arbeiten des hiesigen Instituts ist seit jeher[1]) betont worden, daß die Molekularstrahlmethode die Möglichkeit gibt, sehr kleine Momente zu messen, die anderen Methoden nicht zugänglich sind. Den ersten Versuch in dieser Richtung stellt die Arbeit von Knauer und Stern[2]) dar, in der das magnetische Moment des H_2O-Moleküls in der erwarteten Größenordnung ($\sim ^1/_{1000}$ Bohrmagneton) nachgewiesen wurde. Während aber damals selbst die Messung der Größenordnung nur durch einen besonderen Kunstgriff (Intensitätsmultiplikator) möglich war, gibt die inzwischen erfolgte Entwicklung der Molekularstrahlmethode, insbesondere der Methoden zur Intensitätsmessung[3]), die Möglichkeit einer quantitativen Messung solch kleiner Momente.

Gerade die Untersuchung des H_2 war schon lange beabsichtigt und zwar aus folgenden Gründen. Erstens sollte die Messung bei H_2 *experimentell* besonders gut durchführbar sein: Denn man kann bei Wasserstoff Strahlen von sehr tiefer Temperatur verwenden und damit besonders große Ablenkung erreichen, da die Ablenkung ceteris paribus der absoluten Temperatur umgekehrt proportional ist; außerdem war gerade bei Wasserstoff eine empfindliche und quantitative Meßmethode für die Intensität des Strahles im hiesigen Institut gut durchgearbeitet und vielfach erprobt. Zweitens sind aber die Versuche bei Wasserstoff auch vom *theoretischen*

---

[1]) Bereits U. z. M. Nr. 1; O. Stern, ZS. f. Phys. **39**, 751, 1926.

[2]) U. z. M. Nr. 3; F. Knauer u. O. Stern, ZS. f. Phys. **39**, 780, 1926.

[3]) U. z. M. Nr. 10; F. Knauer u. O. Stern, ZS. f. Phys. **53**, 766, 1929; U. z. M. Nr. 14; J. B. Taylor, ebenda **57**, 242, 1929.

64

Über die magnetische Ablenkung von Wasserstoffmolekülen usw. I.   **5**

Standpunkt besonders interessant, namentlich seit der Entdeckung des Ortho- und Parawasserstoffs. Vor allem bietet sich die Möglichkeit einer Messung des magnetischen Moments des Protons, einer Größe, die experimentell bisher nicht zugänglich war, dabei aber ihrer Art nach, als eine Eigenschaft der positiven Elementarladung, besonderes Interesse beansprucht.

Das mechanische Moment des Protons ist mit großer Sicherheit bekannt; es ist gleich dem des Elektrons $= \dfrac{1}{2}\dfrac{h}{2\pi}$. Das magnetische Moment des Elektrons ist $2\dfrac{e}{2mc}\cdot\dfrac{1}{2}\cdot\dfrac{h}{2\pi}$ (ein Bohrmagneton $= 0{,}9\cdot 10^{-20}$ CGS für ein Elektron bzw. 5600 CGS pro Mol); nimmt man an, daß für das magnetische Moment des Protons dieselbe Formel gilt (eine Annahme, die durch die Diracsche Theorie des Elektrons nahegelegt wird), so würde dieses im Verhältnis der Massen, also 1840 mal kleiner sein ($0{,}5\cdot 10^{-23}$ CGS für ein Proton bzw. 3 CGS pro Mol). Wir wollen diese Größe im folgenden wie bisher (U. z. M. Nr. 1, l. c.) als *ein Kernmagneton* bezeichnen.

Der unmittelbare Zweck der vorliegenden Arbeit war also die Untersuchung des Wasserstoffs mit dem Ziele einer Bestimmung des Protonenmoments. Darüber hinaus aber sollte ganz allgemein eine Apparatur zur Messung von magnetischen Momenten von der Größenordnung Kernmagneton entwickelt werden. In erster Linie sind Messungen von magnetischen Kernmomenten für Fragen der Kernstruktur von Wichtigkeit und könnten die Bestimmungen aus der Hyperfeinstruktur der Spektrallinien kontrollieren und ergänzen. Außerdem gibt es noch andere Fälle, wo Momente dieser Größenordnung auftreten, z. B. bei der Rotation von Molekülen, diamagnetische Momente usw.

Experimentelle Anordnung. Die experimentelle Anordnung war die übliche bei der magnetischen Ablenkung von Molekularstrahlen, nur mußte infolge des kleinen magnetischen Moments der Strahl sehr lang und schmal gemacht werden, und die Inhomogenität recht groß, also auch die Höhe des Strahles sehr klein, um gut meßbare Ablenkungen zu erhalten. Fig. 1 gibt einen schematischen Überblick über die Anordnung. Die Gesamtlänge des Strahles betrug etwa 30 cm, und zwar die Entfernung vom Ofenspalt zum Abbildespalt knapp 15 cm, die Länge des Feldes 10 cm und der Abstand des Auffängerspaltes vom Feldende 5 cm. Die Polschuhe zur Erzeugung des inhomogenen Feldes hatten die übliche Schneide-Furcheform; die Breite der Furche war 1 mm, der Abstand der Schneide von der Furchenebene 0,5 mm. Sie erzeugten eine Inhomogenität $\partial H/\partial s$ von etwa $2{,}2\cdot 10^5$ Gauß pro Zentimeter.

6 R. Frisch und O. Stern,

Aus diesen Daten berechnet sich die Ablenkung nach der Formel:

$$s_\alpha = \frac{1}{2} g t^2 = \frac{1}{2} \frac{M}{M} \frac{\partial H}{\partial s} \frac{l^2}{\alpha^2} = \frac{M}{4RT} \frac{\partial H}{\partial s} l^2 \begin{pmatrix} M = 3\,CGS \\ R = 8,3 \cdot 10^{-7} \\ l^2 = 200 \end{pmatrix}.$$

Bei einer Strahltemperatur von 90° abs. beträgt für Moleküle mit der wahrscheinlichsten Geschwindigkeit α die Ablenkung 0,044 mm für ein Kernmagneton. Es wurden Strahlen verschiedener Breite verwendet, bis herab zu etwa 0,03 mm. Da die Strahlen nicht monochromatisiert waren, sondern Maxwellverteilung hatten, und das Aufspaltungsbild ziemlich

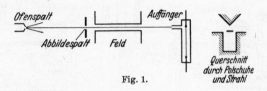

Fig. 1.

kompliziert ist, bekommt man keine wirkliche Aufspaltung in die einzelnen Komponenten, sondern das gesuchte magnetische Moment muß aus der Intensitätsverteilung der abgelenkten Moleküle erschlossen werden.

Infolge der großen Länge und geringen Höhe des Strahles war seine Intensität außerordentlich klein; in einem Auffänger mit idealem Spalt würde der durch den Strahl erzeugte Druck nur etwa $2 \cdot 10^{-8}$ mm betragen, entsprechend einem Galvanometerausschlag von 1,6 cm. Wir mußten daher den üblichen Kunstgriff anwenden, den Spalt kanalförmig zu gestalten, und mußten dabei das Verhältnis Kanallänge zu Kanalbreite besonders groß machen; der Faktor $\varkappa$, um den der Druck durch diese Maßnahme vergrößert wird, betrug in unserem Falle etwa 50, der Ausschlag also etwa 80 cm, was genügende Intensität auch für die abgelenkten Moleküle ergibt. Dieser hohe $\varkappa$-Faktor in Verbindung mit der kleinen Spaltöffnung hat aber zur Folge, daß es sehr lange dauert, bis der Enddruck im Manometer praktisch erreicht wird; diese Zeit hätte bei den üblichen Manometern mit etwa 20 cm³ Volumen etwa $\frac{1}{2}$ Stunde betragen. Wir mußten deshalb Manometer mit sehr viel kleinerem Volumen konstruieren. Die von uns angewandten Manometer hatten nur etwa $\frac{1}{2}$ cm³ Volumen, wodurch die Einstellzeit auf den 40. Teil, also auf $\frac{3}{4}$ Minute heruntergedrückt wurde.

Experimentelle Einzelheiten (siehe Fig. 2 und 3). Der „Ofen" bestand aus einem Kupferrohr, an dessen vorderem Ende der Ofenspalt saß, aus dem die H_2-Moleküle in das Vakuum eintraten, der also die Strahlenquelle

darstellt. Die H_2-Zufuhr erfolgte vom anderen Ende aus durch ein dünn-
wandiges Neusilberrohr (zur thermischen Isolierung des Ofens), das an

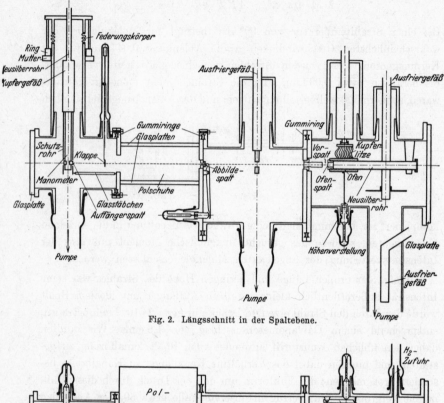

Fig. 2. Längsschnitt in der Spaltebene.

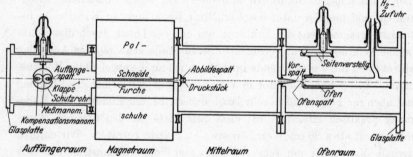

Fig. 3. Längsschnitt senkrecht zur Spaltebene.

einem Ende am Apparatgehäuse festgeklemmt war. Eine seitliche Ab-
zweigung aus biegsamem Bleirohr diente zur Verbindung mit dem H_2-
Vorratsgefäß; auf das rückwärtige Ende des Neusilberrohres war ein Glas-
fenster aufgekittet, um den Ofenspalt von hinten beleuchten zu können.

8   R. Frisch und O. Stern,

Die Elastizität des Neusilberrohres gestattete es, den Ofen mittels zweier Mikrometerschrauben sowohl in der Höhe als auch seitwärts um kleine Beträge zu verschieben. Zur Kühlung des Ofens war an ihm ein Band aus Kupferlitze angelötet (Gesamtquerschnitt $\sim 20\,mm^2$, Länge etwa 2 cm), an dessen anderem Ende ein mit Woodschem Metall gefüllter Kupfernapf angelötet war; in diesen Napf tauchte ein Kupferzapfen am Boden eines Neusilbergefäßes, das mit flüssiger Luft gefüllt werden konnte. Durch das Einschmelzen mittels Woodschem Metall wurde der erforderliche gute Wärmekontakt erreicht.

Der *Vorspalt* stand etwa 6 mm vor dem Ofenspalt, so daß die Moleküle vom Ofenspalt aus nur diese kurze Strecke in dem relativ hohen Druck

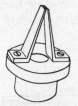

Fig. 4. Vorspalt.

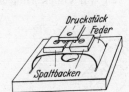

Fig. 5. Abbildespalt.

(einige 10^{-4} mm) im Ofenraum laufen mußten; hinter dem Vorspalt, im Mittelraum, wurde durch eine zweite Pumpe wesentlich besseres Vakuum (1 bis $2 \cdot 10^{-5}$ mm) aufrechterhalten. Der Vorspalt war ebenso wie der Ofenspalt nur 0,02 mm breit, so daß eine eventuelle Verbreiterung des Strahles durch Streuung im Ofenraum ohne Einfluß blieb. Man kann gewissermaßen den Vorspalt als die eigentliche Strahlenquelle ansehen; Streuung im Mittelraum spielte bei dem geringen Druck darin keine Rolle mehr. Durch diesen Kunstgriff gelang es Strahlen zu erhalten, deren Intensitätsverteilung („Form") recht genau der Geometrie der Spaltanordnung entsprach und die insbesondere praktisch keine „Schwänze" hatten. Um möglichst wenig durch reflektierte Moleküle gestört zu werden, wurde der Vorspalt schnabelförmig ausgebildet (Fig. 4); die Spaltbacken waren dünne geschliffene Stahlstreifen, die federnd gegeneinander drückten und durch kleine Stückchen Platinfolie im richtigen Abstand (0,02 mm) gehalten wurden; seitlich wurden sie mit Aluminiumfolie abgeschlossen (mittels Picein). Der Vorspalt wurde an der richtigen Stelle festgeschraubt (siehe unter Justierung) und konnte während des Versuchs nicht verschoben werden.

Der *Abbildespalt* war so eingerichtet, daß man seine Breite während des Versuchs verändern konnte; das erleichterte einmal das Auffinden

68

Über die magnetische Ablenkung von Wasserstoffmolekülen usw. I.   9

des Strahles, zweitens erwies es sich als sehr vorteilhaft, daß man Aufspaltungsversuche mit verschieden breiten Strahlen ohne großen Zeitverlust vornehmen konnte. Die beiden Spaltbacken waren auf einem federnden Blechstreifen montiert (Fig. 5) und konnten durch Druck mittels eines geeignet geformten Druckstückes einander bis zur Berührung genähert werden; Feder und Druckstück waren natürlich durchbrochen, um den Strahl durchzulassen. Das Druckstück saß am kürzeren Ende eines zweiarmigen Hebels, auf dessen längeres Ende eine Mikrometerschraube wirkte. So konnte der Spalt sehr feinfühlig von 0,2 mm bis zu beliebig geringer Breite verstellt werden.

Der *Auffangespalt* war ebenso wie der Ofenspalt 0,02 mm breit und 0,5 mm hoch. Er war kanalförmig ausgebildet mit einer Kanallänge von

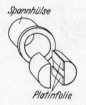

Fig. 6. Auffangespalt,   Fig. 7. Meßmanometer.
auseinandergenommen.

4 mm; Einzelheiten seiner Konstruktion siehe Fig. 6. Aus diesen Daten berechnet sich der $\varkappa$-Faktor zu[1])

$$\varkappa = \frac{l}{b} \cdot \frac{1}{0,5 + 2,3 \log \dfrac{2\,a}{b}} \doteq \sim 45$$

(b = Spaltbreite, a = Spalthöhe, l = Kanallänge). Dieser ungewöhnlich große $\varkappa$-Faktor (bisher wurde er selten größer als 10 gewählt) in Verbindung mit der kleinen Spaltöffnung bedingte, wie schon oben dargelegt, die Konstruktion von Manometern mit besonders kleinem Volumen.

Das *Meßmanometer* bestand aus einem Stück Messingrohr von 2 mm lichter Weite und etwa 11 cm Länge, in dem axial der Hitzdraht ausgespannt war (Nickelband $3 \times 50\,\mu$, 10 cm lang). Seine Zuleitungen aus Platindraht (eine davon als Spiralfeder ausgebildet, um ihn zu spannen) waren durch konische Messingtopfen durchgelötet, die in die konisch erweiterten Enden des Messingrohres eingesetzt waren, isoliert durch dünne Galalithringe. Der Auffängerspalt war in eine seitliche Bohrung eingesetzt. Das übliche, ganz gleich gebaute Kompensationsmanometer war mit dem

---

[1]) Berechnet nach M. v. Smoluchowski, Ann. d. Phys. **33**, 1559, 1910.

10   R. Frisch und O. Stern,

Meßmanometer der Länge nach verlötet, um besten Temperaturausgleich
zu erzielen. Die Manometer waren an dem Boden eines kupfernen Kühlgefäßes angeschraubt, das oben einen langen Hals aus Neusilberrohr hatte
und für die Messung mit flüssiger Luft gefüllt wurde. Ein kupfernes Schutz-
rohr, das ebenfalls am Boden des Kühlgefäßes angeschraubt war, schützte
die Manometer vor Wärmezustrahlung; ein kleines Loch ließ den Molekularstrahl zum Auffangespalt durchtreten.

Um die Intensitätsverteilung im Strahl ausmessen zu können, mußte
der Auffänger quer durch den Strahl durchbewegt werden; doch betrug

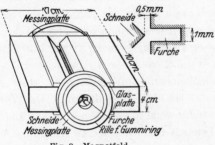

Fig. 8.   Magnetfeld.

die ganze erforderliche Verschiebung nur einige Zehntel
Millimeter, so daß eine geringe
elastische Verbiegung des Neusilberrohres völlig ausreichte.
Die Verschiebung erfolgte durch
eine Mikrometerschraube, unter
Zwischenschaltung eines beiderseits zugespitzten Druckstiftes,
zur Vermeidung von Reibung

und Wärmezufuhr.   Um ein Ausweichen der Manometer in der zur
Verschiebung senkrechten Richtung (in Richtung des Strahles) zu ver-
hindern, waren sie in dieser Richtung durch ein beiderseits zugespitztes
Glasstäbchen abgestützt.

Um die bei Kühlung erfolgende Längenänderung des Kühlgefäßes
und der Manometer selbst (etwa 1 mm) zu kompensieren, konnten die
Manometer der Höhe nach verschoben werden; zu diesem Zweck war in
dem das Kühlgefäß tragenden äußeren Messingrohr ein Federungskörper
aus gewelltem Tombakrohr eingeschaltet; der obere Teil des Rohres war
durch vier Stäbe mit einem Ring verbunden, der auf dem unteren Teil
verschiebbar war; der Ring saß auf einer großen Mutter auf und konnte
durch Drehen derselben in der Höhe verschoben werden.

Das *inhomogene Magnetfeld* wurde von zwei Polschuhen gebildet,
deren Form und Dimensionen aus Fig. 8 hervorgehen. Sie waren an den
Stirnseiten mit Messingplatten verschraubt, an die wiederum die an-
schließenden Gehäuseteile angeschraubt wurden; oben und unten waren sie
mit Glasplatten abgedeckt, die mit Picein angekittet waren; ebenso waren
alle anderen Fugen mit Picein überschmolzen. Mit dieser einfachen Konstruktion wurde eine völlig ausreichende Abdichtung erreicht, mit dem

70

Über die magnetische Ablenkung von Wasserstoffmolekülen usw. I.   **11**

weiteren Vorteil, daß man durch die Glasplatten die Polschuhe der ganzen Länge nach überblicken konnte.

Die übrigen *Gehäuseteile* (vgl. Fig. 2 und 3), nämlich Ofenraum, Mittelraum und Auffängerraum, waren durch Flansche mit Gummiringdichtung miteinander bzw. mit dem Magnetraum verbunden. Der *Ofenraum* enthielt den oben beschriebenen Ofen mit Kühl- und Verschiebeeinrichtung sowie ein Kühlgefäß zum Ausfrieren von Dämpfen; er wurde durch eine große (dreistufige) Stahlpumpe ausgepumpt (Sauggeschwindigkeit mit Verbindungsrohr 10 Liter/sec für Luft, also etwa 36 Liter/sec für H_2). Der *Mittelraum* war so eingerichtet, daß man für Versuche mit „monochromatischen" Strahlen ein System von rasch rotierenden Zahnscheiben einbauen konnte, das nur Moleküle eines engen Geschwindigkeitsbereiches passieren läßt. Zu diesem Zweck war er ziemlich geräumig (9 cm Durchmesser) und der Strahl lief nahe an der Wand (1 cm Abstand). Bei den vorliegenden Versuchen wurde noch ohne „Monochromator" gearbeitet; statt seiner wurde nur eine Platte eingesetzt, die den Vorspalt trug. Der Vorspalt ragte durch ein Loch in der Zwischenwand in den Ofenraum; der Luftspalt zwischen der Platte, die den Vorspalt trug, und der Zwischenwand wurde mit Ramsayfett abgedichtet, so daß der Vorspalt selbst die einzige Verbindung zwischen Mittelraum und Ofenraum bildete. Auch der Mittelraum enthielt ein Kühlgefäß. Er wurde durch eine zweite Pumpe evakuiert (Sauggeschwindigkeit mit Saugleitung und Kühlfalle etwa 1 Liter/sec für Luft). *Magnetraum* und *Auffängerraum* wurden gemeinsam durch eine dritte Pumpe ähnlicher Sauggeschwindigkeit evakuiert; da sie mit dem Mittelraum nur durch einen kleinen Kanal (etwa 1 mm Durchmesser, 2 mm lang) kommunizierten, konnte in ihnen während des Versuchs Hängevakuum aufrechterhalten werden. Der Abbildespalt saß dicht am Anfang des Magnetfeldes. Im Auffängerraum befand sich außer den schon besprochenen Manometern die elektromagnetisch betätigte Klappe zum Absperren des Strahles, die am Ende eines langen Hebels saß, an dessen anderem Ende, in hinreichendem Abstand vom Magnetfeld, ein Eisenanker befestigt war.

Die *Justierung* geschah teils optisch, teils mit den Molekularstrahlen selbst. Zunächst wurde der Abbildespalt am Magnetfeld befestigt, und zwar so, daß seine Mittellinie zwischen der Furchenebene und der Schneide lag, dicht (etwa 0,1 mm) über der Furchenebene; das war mittels eines Mikroskops leicht zu kontrollieren. Dann wurde der Mittelraum mit dem Vorspalt montiert; ein Lichtstrahl (Wolframpunktlampe) wurde durch Vorspalt und Abbildespalt geschickt und der Vorspalt so lange verschoben, bis der Lichtstrahl das Feld genau parallel zu den Polschuhkanten durchlief. Nun wurde die Punktlampe an ihrem Ort belassen, der Ofenraum montiert und der Ofen so lange verschoben, bis der Strahl seine größte Helligkeit hatte. Damit waren diese drei Spalte in eine Linie gestellt und richtig zum Feld justiert. Ihre gegenseitige Parallellage wurde dadurch gesichert, daß jeder einzeln ins Lot gestellt wurde; darauf kam es nicht so sehr genau an, da der Strahl ja nur $^1/_2$ mm hoch, das Verhältnis Länge zu Breite also nur etwa 25 war; ein Fehler von 1^0 in der Vertikalstellung, der sicher nicht vorkam, hätte noch nicht viel geschadet.

Die Justierung des Auffängers und die Feinjustierung des Ofens erfolgte mit dem Molekularstrahl selbst. Insbesondere mußte der Auffangespalt so gestellt werden, daß die Moleküle wirklich durch den Kanal durchlaufen konnten, ohne die Wand zu treffen; da die Kanallänge das 200fache der Kanalbreite war, setzte schon eine Verdrehung um $^1/_{200} = 0,3^0$ den Druck im Auffänger auf die Hälfte herab; man mußte also auf wenige Bogenminuten genau zielen; zu diesem Zweck war ein langer kräftiger Arm mit Mikrometerverschiebung am Auffängerschliff befestigt.

Ergebnisse. Es wurden zuerst Strahlen aus gewöhnlichem H_2 bei tiefer Temperatur (flüssiger Luft) untersucht. Zur Deutung dieser Versuche ist folgendes zu sagen: Gewöhnlicher Wasserstoff besteht aus 25% Parawasserstoff und 75% Orthowasserstoff. Beim Para-H_2 stehen die beiden Protonen antiparallel, er sollte also kein vom Kernspin herrührendes magnetisches Moment haben. Dagegen ist zu erwarten, daß die Rotation der Moleküle ein magnetisches Moment erzeugt. Bei der Temperatur der flüssigen Luft haben aber die Para-H_2-Moleküle fast alle (99%) die Rotationsquantenzahl 0; der Para-H_2 sollte also bei dieser Temperatur kein magnetisches Moment haben. Wir haben das durch Versuche an reinem Para-H_2 bestätigt.

Beim Ortho-H_2 stehen die beiden Protonen parallel, er hat also aus diesem Grunde ein magnetisches Moment vom Betrag 2 Protonenmomente. Außerdem gibt wiederum die Rotation einen Beitrag zum magnetischen Moment; und in diesem Falle ist dieser Beitrag nicht durch Erniedrigung der Temperatur wegzuschaffen, da der niedrigste Rotationszustand vom Ortho-H_2 die Quantenzahl 1 hat. Da die Kopplung zwischen den beiden Momenten (Rotation und Kernspin) sehr klein und in den zur Aufspaltung benutzten Feldern von etwa 20000 Gauß sicher völlig aufgehoben ist, ist für einen Ortho-H_2-Strahl einheitlicher Geschwindigkeit bei tiefer Temperatur das Aufspaltungsbild Fig. 9 zu erwarten. Jedes der beiden

72

Über die magnetische Ablenkung von Wasserstoffmolekülen usw. I.   **13**

Momente hat drei Einstellungen im Feld (entsprechend der Quantenzahl 1); in der Figur ist angenommen, daß das Rotationsmoment viel kleiner ist als das Kernmoment. Bei den wirklich verwendeten Strahlen mit Maxwell-Ferteilung der Geschwindigkeiten entspricht jedem Strich der obigen vigur (außer dem Mittelstrich) eine Maxwellkurve; die gemessene Intensitätsverteilung ist die Überlagerung dieser Kurven. Unter S_R bzw. S_P ist im Folgenden immer die Ablenkung für Moleküle der wahrscheinlichsten Geschwindigkeit verstanden.

Prinzipiell könnte man aus der gemessenen Intensitätsverteilung die beiden Unbekannten S_R und S_P (s. Fig. 9) errechnen; doch würde das eine sehr hohe Genauigkeit der Messungen voraus-

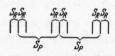

setzen. Wir haben daher die eine Unbekannte, das Rotationsmoment, d. h. S_R, auf folgendem Wege be-stimmt. Wir haben reinen Para-H_2[1]) außer bei der Temperatur der flüssigen Luft auch bei höheren

Fig. 9. Aufspaltungsbild von Orthowasserstoff

Temperaturen (festes CO_2, d. h. 195⁰ abs. und Zimmer-temperatur, d. h. 292⁰ abs.) untersucht. Bei der Temperatur der flüssigen Luft war er, wie erwähnt, unmagnetisch[2]); bei höheren Temperaturen zeigte er ein Moment, das von den dann auftretenden höheren Rotationsquantenzuständen herrührt. Wir haben

Fig. 10. Aufspaltungsbild für Parawasserstoff; oben $n = 2$, unten $n = 4$

die Häufigkeit dieser Quantenzustände nach der Boltzmannformel berechnet; bezeichnet man die Rotationsquantenzahl mit n, so ergibt für $T = 195⁰$ abs. die Rechnung 73% Moleküle mit $n = 0$, und 27% mit $n = 2$; bei Zimmertemperatur (292⁰ abs.) findet man 52,5% mit $n = 0$, 46,1% mit $n = 2$, und 1,4% mit $n = 4$. Unter der Voraussetzung, daß die auftretenden Komponenten des magnetischen Moments (vgl. Fig. 10) alle ganzzahlige Vielfache eines Grundmoments sind, das $n = 1$ entspricht, können wir dieses Grund-moment μ_R aus den Messungen entnehmen; es ergibt sich zu etwa ein Kernmagneton, eher etwas kleiner.

Auf die Art der Ausrechnung des Moments soll noch etwas näher ein-gegangen werden. Es wurde zunächst die Form des unaufgespaltenen Strahles ausgemessen. Unter Form des Strahles verstehen wir immer die Form der Kurve, die man erhält, wenn man den Auffänger quer durch den

---

[1]) Hergestellt von Herrn I. Estermann, wofür wir ihm besten Dank schuldig sind.

[2]) Ein kleiner Betrag von magnetischen Molekülen kam offenbar von einer geringen Verunreinigung (3 bis 4%) mit Ortho-H_2.

14 R. Frisch und O. Stern,

Strahl durchbewegt und die gemessene Intensität als Funktion der Auf-
fängerverschiebung aufträgt; Beispiel siehe Fig. 11. Die Form des Strahles
stimmte recht gut mit der geometrisch zu erwartenden Form überein,
falls das remanente Feld des Magneten durch einen schwachen Gegenstrom
(0,2 Amp.) beseitigt wurde. Sodann wurde das Feld eingeschaltet und die
Strahlform neuerdings ausgemessen (Fig. 11 und 12). Wie daraus das

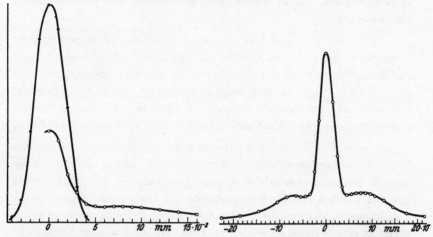

Fig. 11. Beispiel einer Messung
an gewöhnlichem Wasserstoff bei 95⁰ abs.;
Stahlform ohne Feld (●) und mit Feld (○).

Fig. 12. Vollständiges Aufspaltungsbild
von gewöhnlichem Wasserstoff bei 95⁰ abs.;
die Unsymmetrie ist apparativ bedingt.

Moment berechnet wurde, sei an dem Beispiel „Para-H_2 bei 195⁰ abs."
besprochen.

Wir greifen irgendeinen Punkt der gemessenen Kurve heraus, dessen
Abstand von der Mitte hinreichend groß gegen die Strahlbreite ist (praktisch
genügt ein Abstand gleich der Strahlbreite). Wir wissen nun, daß 73%
der Moleküle die Rotationsquantenzahl $n = 0$ haben, also unmagnetisch
sind und überhaupt nichts zur Intensität an dieser Stelle beitragen. Von
den restlichen 27% wird ein Fünftel ebenfalls nicht abgelenkt (s. Fig. 10,
oben), während zwei Fünftel in zwei Strahlen aufgespalten werden, genau
wie ein Strahl von Silberatomen, nur mit dem sehr viel kleineren Moment μ_R,
wie es durch die Rotation eines H_2-Moleküls mit einem Rotationsquant
erzeugt wird, und die restlichen zwei Fünftel genau so, nur mit dem doppelten
Moment.

Wir können nun dem Rotationsmoment versuchsweise irgendeinen
Wert erteilen, z. B. ein Kernmagneton; daraus können wir s_α, die Ablenkung
für Moleküle mit der wahrscheinlichsten Geschwindigkeit, berechnen

74

Über die magnetische Ablenkung von Wasserstoffmolekülen usw. I.   15

und daraus unter Berücksichtigung der Maxwellverteilung die Intensität in dem betreffenden Abstand von der Strahlmitte[1]). Für diese Berechnung haben wir angenommen, daß der ursprüngliche Strahl Rechtecksform besitzt mit einer Breite, die gleich der Halbwertsbreite des gemessenen Strahls ist; diese Vereinfachung ist in unserem Falle völlig unbedenklich, wie wir durch Kontrollrechnung sichergestellt haben. Es wurde nun μ_R so lange variiert, bis die berechneten Intensitäten mit den gemessenen übereinstimmten.

Ein zweiter Weg, auf dem wir diese Werte verifizierten, war der, die Intensität in der Mitte des Strahles mit und ohne Feld zu messen. Auch hier benutzten wir bei der Berechnung wieder die Vereinfachung, den unabgelenkten Strahl als rechteckig zu behandeln, was hier ebenfalls unbedenklich war, da wir diese Methode vor allem bei sehr breiten Strahlen verwendeten, die wirklich mit großer Annäherung Rechtecksform aufwiesen.

In prinzipiell der gleichen Weise wurden die Messungen an gewöhnlichem Wasserstoff ausgewertet. Wir nehmen an, daß μ_R auch das magnetische Rotationsmoment des einquantigen Ortho-H_2 ist. Rechnen wir mit diesem Wert des Rotationsmoments aus unseren Messungen an gewöhnlichem H_2 das von den Protonen herrührende Moment aus, so ergibt es sich zu etwa 5 Kernmagnetonen pro Ortho-H_2-Molekül. Das magnetische Moment eines Protons wäre also danach nicht ein Kernmagneton, sondern etwa 2 bis 3 Kernmagnetonen. Dieser Zahlenwert ist nicht sehr genau; doch scheint ein Wert von der Größe 1 mit den Messungen nicht vereinbar zu sein.

Zum Rotationsmoment sei noch folgendes gesagt: Wir machten anfangs nur Messungen mit gewöhnlichem H_2 und versuchten, für das Rotationsmoment einen theoretisch berechneten Wert zu verwenden. Auf Veranlassung von Herrn Fermi hatte Herr Bethe die Freundlichkeit, das elektrische Trägheitsmoment des H_2-Moleküls abzuschätzen. Aus seinen Rechnungen ergab sich unter der Annahme, daß das H_2-Molekül wie ein starrer Körper rotiert, für das Rotationsmoment ein Wert von etwa 3 Kernmagnetonen (für die Rotationsquantenzahl eins). Erst später kamen wir darauf, daß man in der oben beschriebenen Weise durch Messungen an reinem Parawasserstoff das Rotationsmoment direkt experimentell bestimmen kann. Es ergab sich, wie erwähnt, ein Wert von höchstens einem Kernmagneton. Da diese Diskrepanz weit außerhalb der Fehlergrenzen

---

[1]) Siehe z. B. U. z. M. Nr. 5; O. Stern, ZS. f. Phys. **41**, 563, 1927.

16   R. Frisch und O. Stern.

sowohl der theoretischen Abschätzung als auch unserer Messungen lag,
wandten wir uns neuerdings an Herrn Fermi, der dann folgendes heraus-
brachte: Die Annahme, daß das Wasserstoffmolekül wie ein starrer Körper
rotiert, ist unzutreffend; man muß sich vielmehr vorstellen, daß die Elek-
tronenhülle bei der Rotation zurückbleibt („rutscht"). Eine Abschätzung
dieses Effektes, die Herr Wick[1]) auf Anregung von Herrn Fermi vornahm,
ergab, daß das Rotationsmoment zwischen 0,35 und 0,92 Kernmagnetonen
liegen sollte. Das ist mit unseren Messungen vereinbar; wir möchten ver-
muten, daß der wirkliche Wert näher an der oberen Grenze liegt.

---

[1]) ZS. f. Phys. **85**, 25, 1933.

204

Zur Quantentheorie des Atomkernes.

Von **G. Gamow**, z. Zt. in Göttingen.

Mit 5 Abbildungen. (Eingegangen am 2. August 1928.)

Es wird der Versuch gemacht, die Prozesse der α-Ausstrahlung auf Grund der Wellenmechanik näher zu untersuchen und den experimentell festgestellten Zusammenhang zwischen Zerfallskonstante und Energie der α-Partikel theoretisch zu erhalten.

§ 1. Es ist schon öfters * die Vermutung ausgesprochen worden, daß im Atomkern die nichtcoulombschen Anziehungskräfte eine sehr wichtige Rolle spielen. Über die Natur dieser Kräfte können wir viele Hypothesen machen.

Es können die Anziehungen zwischen den magnetischen Momenten der einzelnen Kernbauelemente oder die von elektrischer und magnetischer Polarisation herrührenden Kräfte sein.

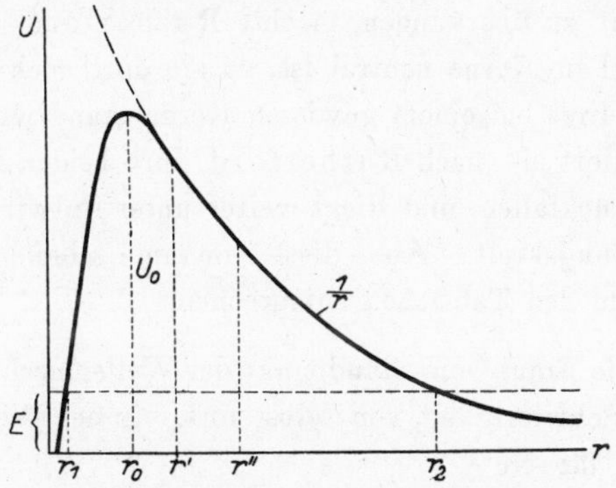

Fig. 1.

Jedenfalls nehmen diese Kräfte mit wachsender Entfernung vom Kern sehr schnell ab, und nur in unmittelbarer Nähe des Kernes überwiegen sie den Einfluß der Coulombschen Kraft.

Aus Experimenten über Zerstreuung der α-Strahlen können wir schließen, daß, für schwere Elemente, die Anziehungskräfte bis zu einer Entfernung $\sim 10^{-12}$ cm noch nicht merklich sind. So können wir das auf Fig. 1 gezeichnete Bild für den Verlauf der potentiellen Energie annehmen.

Hier bedeutet r'' die Entfernung, bis zu welcher experimentell nachgewiesen ist, daß Coulombsche Anziehung allein existiert. Von r' beginnen die Abweichungen (r' ist unbekannt und vielleicht viel kleiner als r'') und bei r_0 hat die U-Kurve ein Maximum. Für $r < r_0$ herrschen schon die Anziehungskräfte vor, in diesem Gebiet würde das Teilchen den Kernrest wie ein Satellit umkreisen.

---

* J. Frenkel, ZS. f. Phys. **37**, 243, 1926; E. Rutherford, Phil. Mag. **4**, 580, 1927; D. Enskog, ZS. f. Phys. **45**, 852, 1927.

Diese Bewegung ist aber nicht stabil, da seine Energie positiv ist, und nach einiger Zeit wird das α-Teilchen wegfliegen (α-Ausstrahlung). Hier aber begegnen wir einer prinzipiellen Schwierigkeit.

Um wegzufliegen, muß das α-Teilchen eine Potentialschwelle von der Höhe U_0 (Fig. 1) überwinden, seine Energie darf nicht kleiner als U_0 sein. Aber die Energie der α-Partikel ist, wie experimentell nachgewiesen ist, viel kleiner. Z. B. findet man * bei der Untersuchung der Streuung von Ra C'-α-Strahlen, als sehr schnelle Partikel, an Uran, daß für den Urankern das Coulombsche Gesetz bis zu einer Entfernung von $3{,}2 \cdot 10^{-12}$ cm gilt. Andererseits haben die von Uran selbst emittierten α-Partikeln eine Energie, die auf der Abstoßungskurve einem Kernabstand von $6{,}3 \cdot 10^{-12}$ cm (r_2 in Fig. 1) entspricht. Soll eine α-Partikel, die aus dem Inneren des Kernes kommt, wegfliegen, so müßte sie das Gebiet zwischen r_1 und r_2 durchlaufen, wo ihre kinetische Energie negativ wäre, was nach klassischen Vorstellungen natürlich unmöglich ist.

Um diese Schwierigkeit zu überwinden, machte Rutherford ** die Annahme, daß die α-Partikel im Kerne neutral ist, da sie dort noch zwei Elektronen enthalten soll. Erst bei einem gewissen Kernabstand jenseits des Potentialmaximums verliert sie, nach Rutherford, ihre beiden Elektronen, die in den Kern zurückfallen, und fliegt weiter unter Entwirkung der Coulombschen Abstoßungskraft. Aber diese Annahme scheint sehr unnatürlich und dürfte kaum den Tatsachen entsprechen.

§ 2. Betrachten wir die Frage vom Standpunkt der Wellenmechanik, so fällt die oben erwähnte Schwierigkeit von selbst fort. In der Wellenmechanik nämlich gibt es für ein Teilchen immer eine von Null verschiedene Übergangswahrscheinlichkeit, von einem Gebiet in ein anderes Gebiet gleicher Energie, das durch eine beliebig, aber endlich hohe Potentialschwelle von dem ersten getrennt ist ***.

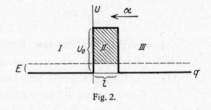

Fig. 2.

Wie wir weiter sehen werden, ist die Wahrscheinlichkeit eines solchen Überganges allerdings sehr klein, und zwar um so kleiner, je

* Rutherford, l. c., S. 581.
** Derselbe, l. c., S. 584.
*** Siehe z. B. Oppenheimer, Phys. Rev. **31**, 66, 1928; Nordheim, ZS. f. Phys. **46**, 833, 1927.

206   G. Gamow,

höher die zu überwindende Potentialschwelle ist. Um diese Tatsache
zu erläutern, wollen wir ein einfaches Beispiel untersuchen.

Wir haben eine rechteckige Potentialschwelle und wir wollen die
Lösung der Schrödingerschen Gleichung finden, welche den Durchgang
der Partikel von rechts nach links darstellt. Für die Energie E schreiben
wir die Wellenfunktion ψ in der folgenden Form:

$$\psi = \boldsymbol{\Psi}(q) \cdot e^{+\frac{2\pi i E}{h}t},$$

wo $\boldsymbol{\Psi}(q)$ der Amplitudengleichung:

$$\frac{\partial^2 \psi}{\partial q^2} + \frac{8\pi^2 m}{h^2}(E - U)\boldsymbol{\Psi} = 0 \tag{1}$$

genügt.

Für das Gebiet I haben wir die Lösung

$$\boldsymbol{\Psi}_{\mathrm{I}} = A\cos(kq + \alpha),$$

wo A und α zwei beliebige Konstanten sind und

$$k = \frac{2\pi\sqrt{2m}}{h} \cdot \sqrt{E} \tag{2a}$$

bedeutet. In dem Gebiet II lautet die Lösung

$$\boldsymbol{\Psi}_{\mathrm{II}} = B_1 e^{-k'q} + B_2 e^{+k'q},$$

wo

$$k' = \frac{2\pi\sqrt{2m}}{h}\sqrt{U_0 - E} \tag{2b}$$

ist.

An der Grenze $q = 0$ gelten die Bedingungen:

$$\psi_{\mathrm{I}}(0) = \psi_{\mathrm{II}}(0) \quad \text{und} \quad \left[\frac{\partial \boldsymbol{\Psi}_{\mathrm{I}}}{\partial q}\right]_{q=0} = \left[\frac{\partial \boldsymbol{\Psi}_{\mathrm{II}}}{\partial q}\right]_{q=0},$$

woraus wir leicht

$$B_1 = \frac{A}{2\sin\vartheta}\sin(\alpha + \vartheta); \quad B_2 = -\frac{A}{2\sin\vartheta}\sin(\alpha - \vartheta)$$

erhalten, wo

$$\sin\vartheta = \frac{1}{\sqrt{1 + \left(\frac{k}{k'}\right)^2}}$$

ist.

Die Lösung im Gebiet II lautet daher:

$$\boldsymbol{\Psi}_{\mathrm{II}} = \frac{A}{2\sin\vartheta}\left[\sin(\alpha + \vartheta) \cdot e^{-k'q} - \sin(\alpha - \vartheta)e^{+k'q}\right].$$

In III haben wir wieder:

$$\boldsymbol{\Psi}_{\mathrm{III}} = C\cos(kq + \beta).$$

An der Grenze $q = l$ haben wir aus den Grenzbedingungen:

$$\frac{A}{2\sin\vartheta}\left[\sin(\alpha + \vartheta)\,e^{-\,lk'} - \sin(\alpha - \vartheta)\,e^{+\,lk'}\right] = C\cos(kl + \beta)$$

und

$$\frac{A}{2\sin\vartheta}\,k'\left[-\sin(\alpha + \vartheta)\,e^{-\,lk'} - \sin(\alpha - \vartheta)\,e^{+\,lk'}\right] = -\,kC\cos(kl + \beta).$$

So ist:

$$C^2 = \frac{A^2}{4\sin^2\vartheta}\left\{\left[1 + \left(\frac{k'}{k}\right)^2\right]\sin^2(\alpha - \vartheta)\cdot e^{2\,lk'}\right.$$

$$-\left[1 - \left(\frac{k'}{k}\right)^2\right]2\sin(\alpha - \vartheta)\sin(\alpha + \vartheta)$$

$$\left.+ \left[1 + \left(\frac{k'}{k}\right)^2\right]\sin^2(\alpha + \vartheta)\,e^{-2\,lk'}\right\}. \tag{3}$$

Die Ausrechnung des β ist für uns nicht von Interesse. Uns interessiert nur der Fall, wo lk' sehr groß ist, so daß wir nur das erste Glied in (3) zu berücksichtigen brauchen.

So haben wir die folgende Lösung:

Links: Rechts:

$$A\cos(kq + \alpha) \;\ldots\; A\,\frac{\sin(\alpha - \vartheta)}{2\sin\vartheta}\left[1 + \left(\frac{k'}{k}\right)^2\right]^{\frac{1}{2}}\cdot e^{+\,lk'}\cos(kq + \beta).$$

Wenn wir jetzt $\alpha - \dfrac{\pi}{2}$ statt α schreiben, die erhaltene Lösung mit i multiplizieren und beide Lösungen addieren, so erhalten wir links:

$$\boldsymbol{\Psi} = A\,e^{i\,(kq + \alpha)}, \tag{4a}$$

rechts aber:

$$\boldsymbol{\Psi} = \frac{A}{2\sin\vartheta}\left[1 + \left(\frac{k'}{k}\right)^2\right]^{\frac{1}{2}}\cdot e^{+\,lk'}\{\sin(\alpha - \vartheta)\cos(kq + \beta)$$

$$-\,i\cos(\alpha + \vartheta)\cos(kq + \beta')\}, \tag{4b}$$

wo β' die neue Phase ist.

Multiplizieren wir diese Lösung mit $e^{2\pi i\frac{E}{h}t}$, so erhalten wir für ψ links die (von rechts nach links) laufende Welle, rechts aber den komplizierten, von der stehenden Welle wenig abweichenden Schwingungsprozeß mit einer sehr großen ($e^{lk'}$) Amplitude. Das bedeutet nichts anderes, als daß die von rechts kommende Welle teils reflektiert und teils durchgegangen ist.

208   G. Gamow,

So sehen wir, daß die Amplitude der durchgegangenen Welle um so kleiner ist, je kleiner die Gesamtenergie E ist, und zwar spielt der Faktor:

$$e^{-lk'} = e^{-\frac{2\pi \cdot \sqrt{2m}}{h}\sqrt{U_0 - E} \cdot l}$$

in dieser Abhängigkeit die wichtigste Rolle.

§ 3. Jetzt können wir das Problem für zwei symmetrische Potentialschwellen (Fig. 3) lösen. Wir werden zwei Lösungen suchen.

Eine Lösung soll für positive q gelten und für $q > q_0 + l$ die Welle:

$$A\, e^{i\left(\frac{2\pi E}{h}t - kq + \alpha\right)}$$

geben. Die andere Lösung gilt für negative q und gibt für $q < -(q_0 + l)$ die Welle

$$A \cdot e^{i\left(\frac{2\pi}{h}Et + q\varkappa' - \alpha\right)}.$$

Dann können wir die beiden Lösungen an der Grenze $q = 0$ nicht stetig aneinanderfügen, denn wir haben hier zwei Grenzbedingungen zu

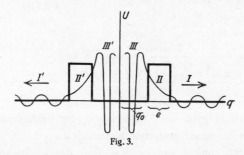

Fig. 3.

erfüllen und nur eine Konstante α zur Verfügung. Die physikalische Ursache dieser Unmöglichkeit ist, daß die aus diesen zwei Lösungen konstruierte ψ-Funktion dem Erhaltungssatz

$$\frac{\partial}{\partial t}\int_{-(q_0 + l)}^{+(q_0 + l)} \psi \overline{\psi}\, dq = 2 \cdot \frac{-h}{4\pi i m}[\psi\, \text{grad}\, \overline{\psi} - \overline{\psi}\, \text{grad}\, \psi]_I$$

nicht genügt.

Um diese Schwierigkeit zu überwinden, müssen wir annehmen, daß die Schwingungen gedämpft sind, und E komplex setzen:

$$E = E_0 + i\frac{h\lambda}{4\pi},$$

wo E_0 die gewöhnliche Energie ist und λ das Dämpfungsdekrement (Zerfallskonstante). Dann sehen wir aber aus den Relationen (2a) und (2b), daß auch k und k' komplex sein sollen, d. h., daß die Amplitude unserer Wellen auch von der Koordinate q expotentiell abhängt. Z. B. für die laufende Welle wird die Amplitude in Richtung der Wellenausbreitung wachsen. Das bedeutet aber nichts weiter als daß, wenn die Schwingung am Ausgangspunkt der Welle gedämpft ist, die Amplitude des früher ausgegangenen Wellenstückes größer sein muß. Wir können jetzt α so wählen, daß die Grenzbedingungen erfüllt werden. Aber die strenge Lösung interessiert uns nicht. Wenn λ im Vergleich mit $\dfrac{E}{h}$ klein ist

$\left(\text{für Ra C' ist } \dfrac{E}{h} \eqsim \dfrac{10^{-5}}{10^{-27}\sec} = 10^{+22}\sec^{-1} \text{ und } \lambda_! = 10^{+5}\sec^{-1}\right),$

so ist die Änderung der $\Psi(q)$ sehr klein, und wir können einfach die alte Lösung mit $e^{-\frac{\alpha}{2}t}$ multiplizieren.

Dann lautet der Erhaltungssatz:

$$\frac{\partial}{\partial t} e^{-\lambda t} \int_{-(q_0+l)}^{+(q_0+l)} \Psi_{\mathrm{II,\,III}}^{(q)} \cdot \Psi_{\mathrm{II,\,III}}^{(q)} \, dq = -2 \cdot \frac{A^2 h}{4\pi i m} \cdot 2 i k \cdot e^{-\lambda t},$$

woraus

$$\lambda = \frac{4hk\sin^2\vartheta}{\pi m\left[1+\left(\dfrac{k'}{k^0}\right)^2\right]2(l+q_0)\varkappa} \cdot e^{-\frac{4\pi l\sqrt{2m}}{h}\sqrt{U_0-E}}, \tag{5}$$

folgt, wo $\varkappa$ eine Zahl von der Größenordnung 1 ist.

Diese Formel gibt die Abhängigkeit der Zerfallkonstante von der Zerfallsenergie für unser einfaches Kernmodell.

§ 4. Jetzt können wir zu dem Falle des wirklichen Kernes übergehen.

Wir können die entsprechende Wellengleichung nicht lösen, da wir den genauen Potentialverlauf in der Nähe des Kernes nicht kennen. Aber einige, für unser einfaches Modell erhaltene, Ergebnisse können wir auch auf den wirklichen Kern ohne genaue Kenntnis des Potentialverlaufs übertragen.

Wie gewöhnlich im Falle der Zentralkraft, werden wir die Lösung in Polarkoordinaten suchen, und zwar in der Form

$$\Psi = u(\theta, \varphi)\,\chi(r).$$

Für u erhalten wir die Kugelfunktionen, und χ muß der Differentialgleichung:

$$\frac{\partial^2 \chi}{\partial r^2} + \frac{2}{r}\frac{d\chi}{dr} + \frac{8\pi^2 m}{h^2}\left[E - U - \frac{h^2}{8\pi^2 m} \cdot \frac{n(n+1)}{r^2}\right]\chi = 0$$

210 G. Gamow,

genügen, wo n die Ordnung der Kugelfunktion ist. Wir können $n = 0$ annehmen, denn wenn $n > 0$, würde das wirklich sein, als ob die potentielle Energie vergrößert wäre, und infolgedessen wird für diese Schwingungen

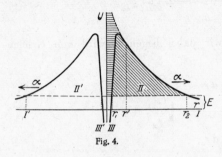

Fig. 4.

die Dämpfung viel kleiner. Die Partikel muß zuerst in den Zustand $n = 0$ übergehen und kann erst dann wegfliegen.

Es ist sehr gut möglich, daß derartige Übergänge die γ-Strahlen verursachen, welche stets α-Emission begleiten. Der wahrscheinliche Verlauf von U ist in Fig. 4 wiedergegeben.

Für große Werte von r werden wir für χ die Lösung

$$\chi_{\mathrm{I}} = \frac{A}{r}\, e^{\,i\left(\frac{2\pi E}{h}t - kr\right)}$$

annehmen.

Obgleich man die genaue Lösung des Problems in diesem Falle nicht erhalten kann, können wir doch sagen, daß in den Gebieten I und III χ im Mittel nicht rasch $\left(\text{in dreidimensionalem Falle etwa wie } \frac{1}{r}\right)$ abnehmen wird.

Im Gebiet III wird aber χ expotentiell abnehmen, und zwar können wir in Analogie mit unserem einfachen Falle erwarten, daß der Zusammenhang zwischen Amplitudenabnahme und E durch den Faktor:

$$e^{-\frac{2\pi\sqrt{2m}}{h}\int\limits_{r_1}^{r_2}\sqrt{U-E}\,dr}$$

angenähert gegeben ist.

Bei Anwendung des Erhaltungssatzes können wir wieder die Formel:

$$\lambda = D \cdot e^{-\frac{2\pi\sqrt{2m}}{h}\int\limits_{r_1}^{r_2}\sqrt{U-E}\,dr} \qquad (6)$$

schreiben, wo D von den besonderen Eigenschaften des Kernmodells abhängt. Die Abhängigkeit des D von E können wir neben der der expotentiellen Abhängigkeit des zweiten Faktors vernachlässigen.

Wir können auch statt des Integrals

$$\int\limits_{r_1}^{r_2}\sqrt{U-E}\,dr$$

angenähert das Integral:

$$\int_0^{\frac{2\,Z\,e^2}{E}} \sqrt{\frac{2\,Z\,e^2}{r} - E}\;.\,dr$$

setzen.

Der relative Fehler, den wir dabei begehen, wird von Größenordnung

$$\frac{\displaystyle\int_0^{r_1} \sqrt{\frac{1}{r}}\,dr}{\displaystyle\int_0^{r_2} \sqrt{\frac{1}{r}}\,dr} = \sqrt{\frac{r'}{r_2}}\,.$$

Da $\dfrac{r'}{r_2}$ klein ist, so wird dieser Fehler nicht sehr groß. Da bei den verschiedenen radioaktiven Elementen E nicht sehr verschiedene Werte hat, schreiben wir angenähert:

$$\lg \lambda = \lg D - \frac{4\,\pi\,\sqrt{2\,m}}{h} \left\{ \int_0^{\frac{2\,Z\,e^2}{E_0}} \sqrt{\frac{2\,Z\,e^2}{r} - E_0}\,dr + \frac{\partial}{\partial E} \int_0^{\frac{2\,Z\,e^2}{E}} \sqrt{\frac{2\,Z\,e^2}{r} - E}\,dr \,.\, \varDelta E \right\}$$

oder

$$\lg \lambda = \mathrm{Const}_E + B_E \,.\, \varDelta E,$$

wo

$$B = -\frac{4\,\pi\,\sqrt{2\,m}}{h}\,\frac{\partial}{\partial E} \int_0^{\frac{2\,Z\,e^2}{E}} \sqrt{\frac{2\,Z\,e^2}{r} - E}\,dr = \frac{4\,\pi\,\sqrt{2\,m}}{2\,h} \int_0^{\frac{2\,Z\,e^2}{E}} \frac{dr}{\sqrt{\frac{2\,Z\,e^2}{r} - E}}\,dr\,.$$

Setzen wir:

$$\varrho = \frac{E}{2\,Z\,e^2}\,r,$$

so ist:

$$\left. b = \frac{4\,\pi\,\sqrt{2\,m}\,.\,2\,Z\,e^2}{2\,h\,E^{3/2}} \int_0^1 \frac{d\varrho}{\sqrt{\frac{1}{\varrho} - 1}} = \frac{\pi^2\,\sqrt{2\,m}\,.\,2\,Z\,e^2}{h\,E^{3/2}}\,. \right\} \quad (7)$$

Nun wollen wir diese Formel mit den experimentellen Tatsachen vergleichen. Es ist bekannt*, daß, wenn wir als Abszisse die Energie

---

* Geiger und Nuttall, Phil. Mag. **23**, 439, 1912; Swinne, Phys. ZS. **13**, 14, 1912.

212   G. Gamow, Zur Quantentheorie des Atomkernes.

der α-Partikel, als Ordinate den Logarithmus der Zerfallkonstante auf-
tragen, alle Punkte für eine bestimmte radioaktive Familie auf einer
Geraden liegen. Für verschiedene Familien erhält man verschiedene
parallele Gerade. Die empirische Formel lautet:

$$\lg \lambda = \text{Const} + b\,E,$$

wo b eine allen radioaktiven Familien gemeinsame Konstante ist.

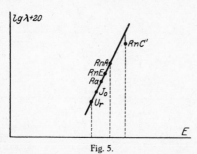

Fig. 5.

Der experimentelle Wert von b
(aus **Ra A** und **Ra** berechnet) ist

$$b_{\text{exper.}} = 1{,}02 \cdot 10^{+7}.$$

Wenn wir aber in unsere Formel
den Energiewert für Ra A einsetzen,
so gibt die Rechnung

$$b_{\text{theor.}} = 0{,}7 \cdot 10^{+7}\,{}^{*}.$$

Die Übereinstimmung der Größenordnung zeigt, daß die Grund-
annahme der Theorie richtig sein dürfte. Nach unserer Theorie müssen
gewisse Abweichungen von dem linearen Gesetz bestehen: mit wachsender
Energie muß b abnehmen, d. h., daß $\log \lambda$ etwas langsamer als E abnehmen muß. Hiermit stimmten die Messungen von Jacobsen[**],
welcher für RaC', dessen α-Strahlung sehr energiereich ist, als Zerfallskonstante den Wert $8{,}4 \cdot 10^5$ erhält, während aus dem linearen Gesetz
der Wert $5 \cdot 10^7$ folgt.

Zum Schluß möchte ich noch meinem Freund N. Kotschin meinen
besten Dank aussprechen für die freundliche Besprechung der mathe-
matischen Fragen. Auch Herrn Prof. Born möchte ich für die Er-
laubnis, in seinem Institut zu arbeiten, herzlich danken.

Göttingen, Institut für theoretische Physik, 29. Juli 1928.

---

[*] Für andere Elemente erhalten wir angenähert denselben Wert, da Z für
verschiedene radioaktive Elemente nur wenig verschieden ist.
[**] Jacobsen, Phil. Mag. **47**, 23, 1924.

Über den Nachweis und das Verhalten der bei der Bestrahlung des Urans mittels Neutronen entstehenden Erdalkalimetalle[1].

Von O. HAHN und F. STRASSMANN, Berlin-Dahlem.

In einer vor kurzem an dieser Stelle erschienenen vorläufigen Mitteilung[2] wurde angegeben, daß bei der Bestrahlung des Urans mittels Neutronen außer den von MEITNER, HAHN und STRASSMANN im einzelnen beschriebenen Trans-Uranen — den Elementen 93 bis 96 — noch eine ganze Anzahl anderer Umwandlungsprodukte entstehen, die ihre Bildung offensichtlich einem sukzessiven zweimaligen α-Strahlenzerfall des vorübergehend entstandenen Urans 239 verdanken. Durch einen solchen Zerfall muß aus dem Element mit der Kernladung 92 ein solches mit der Kernladung 88 entstehen, also ein Radium. In der genannten Mitteilung wurden in einem noch als vorläufig bezeichneten Zerfallsschema 3 derartiger isomerer Radiumisotope mit ungefähr geschätzten Halbwertszeiten und ihren Umwandlungsprodukten, nämlich drei isomeren Actiniumisotopen, angegeben, die ihrerseits offensichtlich in Thorisotope übergehen.

Zugleich wurde auf die zunächst unerwartete Beobachtung hingewiesen, daß diese unter α-Strahlenabspaltung über ein Thorium sich bildenden Radiumisotope nicht nur mit schnellen, sondern auch mit verlangsamten Neutronen entstehen.

Der Schluß, daß es sich bei den Anfangsgliedern dieser drei neuen isomeren Reihen um Radiumisotope handelt, wurde darauf begründet, daß diese Substanzen sich mit Bariumsalzen abscheiden lassen und alle Reaktionen zeigen, die dem Element Barium eigen sind. Alle anderen bekannten Elemente, angefangen von den Trans-Uranen über das Uran, Protactinium, Thorium bis zum Actinium haben andere chemische Eigenschaften als das Barium und lassen sich leicht von ihm trennen. Dasselbe trifft zu für die Elemente unterhalb Radium, also etwa Wismut, Blei, Polonium, Ekacäsium.

Es bleibt also, wenn man das Barium selbst außer Betracht läßt, nur das Radium übrig.

Im folgenden soll kurz die Abscheidung des Isotopengemisches und die Gewinnung der einzelnen

Glieder beschrieben werden. Aus dem Aktivitätsverlauf der einzelnen Isotope ergibt sich ihre Halbwertszeit und lassen sich die daraus entstehenden Folgeprodukte ermitteln. Die letzteren werden in dieser Mitteilung aber im einzelnen noch nicht beschrieben, weil wegen der sehr komplexen Vorgänge — es handelt sich um mindestens 3, wahrscheinlich 4 Reihen mit je 3 Substanzen — die Halbwertszeiten aller Folgeprodukte bisher noch nicht erschöpfend festgestellt werden konnten.

Als Trägersubstanz für die „Radiumisotope" diente naturgemäß immer das Barium. Am nächstliegenden war die Fällung des Bariums als Bariumsulfat, das neben dem Chromat schwerstlösliche Bariumsalz. Nach früheren Erfahrungen und einigen Vorversuchen wurde aber von der Abscheidung der „Radiumisotope" mit Bariumsulfat abgesehen; denn diese Niederschläge reißen neben geringen Mengen Uran nicht unbeträchtliche Mengen von Actinium- und Thoriumisotopen mit, also auch die mutmaßlichen Umwandlungsprodukte der Radiumisotope, und erlauben daher keine Reindarstellung der Ausgangsglieder. Statt der quantitativen, sehr oberflächenreichen Sulfatfällung wurde daher das in starker Salzsäure sehr schwer lösliche BaChlorid als Fällungsmittel gewählt; eine Methode, die sich bestens bewährt hat.

Bei der energetisch nicht leicht zu verstehenden Bildung von Radiumisotopen aus Uran beim Beschießen mit langsamen Neutronen war eine besonders gründliche Bestimmung des chemischen Charakters der neu entstehenden künstlichen Radioelemente unerläßlich. Durch die Abtrennung einzelner analytischer Gruppen von Elementen aus der Lösung des bestrahlten Urans wurde außer der großen Gruppe der Transurane eine Aktivität stets bei den Erdalkalien (Trägersubstanz Ba), den seltenen Erden (Trägersubstanz La) und bei Elementen der vierten Gruppe des Periodischen Systems (Trägersubstanz Zr) gefunden. Eingehender untersucht wurden zunächst die Bariumfällungen, die offensichtlich die Anfangsglieder der beobachteten isomeren Reihen enthielten. Es soll gezeigt werden, daß Transurane, Uran, Protactinium, Thorium und Actinium

[1] Aus dem Kaiser Wilhelm-Institut für Chemie in Berlin-Dahlem. Eingegangen 22. Dezember 1938.
[2] O. HAHN u. F. STRASSMANN, Naturwiss. 26, 756 (1938).

sich stets leicht und vollständig von der mit Barium ausfallenden Aktivität trennen lassen.

1. Zu diesem Zweck wurden aus einem bestrahlten Uran mittels Schwefelwasserstoff die Transurane mit Platinsulfid zusammen abgeschieden und in Königswasser gelöst. Aus dieser Lösung wurde Bariumchlorid mit Salzsäure gefällt. Aus dem Filtrat des Bariumniederschlages wurde das Platin nochmals mit Schwefelwasserstoff gefällt. Das Bariumchlorid war inaktiv, das Platinsulfid hatte noch eine Aktivität von ∞500 Teilchen/Minute. Entsprechende Versuche mit den langlebigeren Transuranen hatten das gleiche Ergebnis.

2. Eine Fällung von Bariumchlorid aus 10 g nichtbestrahltem Uranylnitrat, das im Gleichgewicht mit $UX_1 + UX_2$ (Thor- und Protactiniumisotope) war und eine Aktivität von ∞ 400000 Teilchen/Minute hat,

herrühren, wenn man das Barium selbst als allzu unwahrscheinlich vorerst außer Betracht läßt.

Wir gehen jetzt kurz auf die mit Bariumchlorid erhaltenen Aktivitätskurven ein, die einerseits zu Aussagen über die Anzahl der „Radiumisotope" führen und außerdem deren Halbwertszeiten zu bestimmen erlauben.

Die Fig. 1 zeigt den Aktivitätsverlauf des aktiven Bariumchlorids nach viertägiger Bestrahlung des Urans. Die Kurve a gibt die Messungen über die ersten 70 Stunden; die Kurve b die Meßwerte für das gleiche Präparat über 800 Stunden fortgeführt. Der Maßstab der unteren Kurve ist zehnmal kleiner als der für die obere. Die anfänglich schnelle Abnahme wird allmählich langsamer und geht nach etwa 12 Stunden in eine langsame Zunahme über. Nach ungefähr 120 Stunden beginnt dann wieder eine sehr allmähliche Aktivitätsabnahme; sie erfolgt exponentiell mit einer Halbwertszeit von rund 13 Tagen.

Der Verlauf der Kurven zeigt deutlich, daß hier mehrere Substanzen vorliegen müssen. Man kann aber nicht ohne weiteres sagen, welches die Körper sind: ob mehrere „Radiumisotope" oder ein „Radiumisotop" mit einer Reihe von Folgeprodukten den Aktivitätsverlauf bestimmen.

Es sei hier gleich vorweggenommen, daß die schon in der ersten Mitteilung angegebenen drei isomeren „Radiumisotope" bestätigt wurden. Sie seien vorerst als RaII und RaIII und RaIV bezeichnet (wegen eines mutmaßlichen RaI s. weiter unten).

Ihr Nachweis und die Ermittlung ihrer Halbwertszeiten wird an Hand der folgenden Figuren kurz dargestellt.

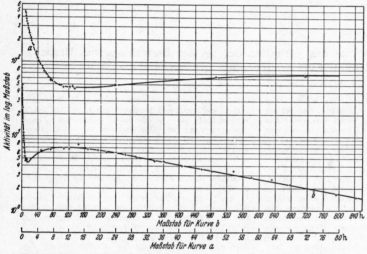

Fig. 1. Die drei Ra-Isotope nach langer Bestrahlung. a = Ra [4 Tage bestrahlt] über 70 Std. gemessen. b = obere Kurve im Maßstabe 1 : 10 über 800 Std. gemessen.

zeigte eine Aktivität von ∞ 14 Teilchen/Minute, war also ebenfalls praktisch inaktiv; d. h. weder Uran, noch Protactinium, noch Thorium fallen mit dem kristallisierenden Bariumchlorid aus.

3. Schließlich wurde noch aus der Lösung eines Actiniumpräparates (MsTh₂) von ∞ 2500 Teilchen pro Minute ein Bariumchloridniederschlag gefällt, der ∞ 3 Teilchen/Minute aufwies, also praktisch ebenfalls inaktiv war.

In ähnlicher Weise wurden die aus bestrahltem Uran gefällten stark aktiven Bariumchloridniederschläge sorgfältig geprüft, doch waren Sulfidniederschläge aus neutraler, schwach essigsaurer oder schwach mineralsaurer Lösung des aktiven Bariums praktisch inaktiv, während die Lanthan- und Zirkonfällungen nur Aktivitäten hatten, deren Entstehung aus der Aktivität der Bariumfällungen leicht nachgewiesen werden konnte.

Die einfache Fällung mit $BaCl_2$ aus stark salzsaurer Lösung gestattet natürlich keine Unterscheidung zwischen Barium und Radium. Nach diesen hier nur sehr summarisch aufgezählten Reaktionen kann die mit den Bariumsalzen abgeschiedene Aktivität nur von Radium

Die Fig. 2 bringt die Auswertung einer „Radium"-Abfallskurve nach 6 Minuten langer Bestrahlung des Urans.

Die Kurve a gibt die direkt gemessene Aktivität, 215 Minuten lang gemessen. Sie setzt sich zusammen aus der Aktivität von zwei „Radium"-Isotopen RaII und RaIII (vgl. Fig. 3) und einer geringen Actiniumaktivität, die sich aus RaII bildet. Diese als AcII bezeichnete Substanz hat, wie andere Versuche, auf die hier nicht eingegangen wird, gezeigt haben, eine Halbwertszeit von etwa 2½ Stunden. Die theoretische Zunahmekurve für ein solches aus RaII entstehendes Actiniumisotop ist in der Figur als Kurve b wiedergegeben. Als Halbwertszeit für das RaII ist dabei schon der Wert von 14 Minuten vorweggenommen. Zieht man die Werte der Kurve b von denen der Kurve a ab, dann ergibt sich die Kurve c der Figur. Diese Aktivität rührt nun praktisch nur noch von Radiumisotopen her, und zwar in der Hauptsache von dem kurzlebigen RaII und in untergeordnetem Maße dem längerlebigen RaIII. Letzteres hat, wie sich weiter unten aus Fig. 3 ergibt, eine Halbwertszeit von ungefähr 86 Minuten. Den Aktivitätsverlauf von RaIII zeigt die Kurve d der

Fig. 2. Zieht man d von c ab, so erhält man schließlich in Kurve e die Aktivität des reinen Ra II. Die Abnahme erfolgt exponentiell mit einer Halbwertszeit von 14 Minuten. Dieser Wert dürfte sicher innerhalb $\pm$ 2 Minuten richtig sein.

Wir kommen nun zum Nachweis der Existenz und zur Bestimmung der Halbwertszeit von Ra III. Wird ein Uranpräparat eine Stunde oder ein paar Stunden lang bestrahlt, dann findet man außer der anfänglichen schnellen Aktivitätsabnahme eine sehr erhebliche Aktivität, die in etwa 100—110 Minuten zur Hälfte abklingt und schließlich noch langsamer wird. Um zu beweisen, daß diese Aktivität im wesentlichen ebenfalls einem Radiumisotop zuzuschreiben ist, wurde folgendermaßen vorgegangen. Aus dem bestrahlten Uranpräparat wurde das „Radium" mittels Bariumchlorid abgetrennt; $2^{1}/_{2}$ Stunden später wurde das Bariumchlorid wieder aufgelöst und erneut gefällt. Das kurzlebige Ra II ist in dieser Zeit vollständig zerfallen, und das aus dem Ra II in dem Bariumchlorid entstandene Ac II ($2^{1}/_{2}$ Stunden Halbwertszeit) wird bei der Umkristallisation des Bariumchlorids entfernt. Das Bariumchlorid ist noch erheblich aktiv; es liegt also noch ein „Radiumisotop" vor. Es wird hier also so vorgegangen, wie dies von MEITNER, STRASSMANN und HAHN schon bei der Aufklärung der aus dem Thorium entstehenden künstlichen Umwandlungsprodukte durchgeführt worden war[1]. Der nunmehr erhaltene Aktivitätsverlauf ist in Fig. 3, Kurve a wiedergegeben.

Die Abnahme erfolgt während der ersten Stunden fast rein exponentiell mit $\infty 86$ Minuten Halbwertszeit; eine kleine Restaktivität bleibt übrig. Sie besteht wohl zweifellos aus einem aus dem Ra III sich bildenden langlebigen „Actiniumisotop", dessen mutmaßlicher Aktivitätsverlauf sich aus der Abweichung der Kurve a von einem rein exponentiellen Zerfall ungefähr erschließen läßt. Die dabei erhaltene Aktivitätskurve ist als Kurve b in der Fig. 3 wiedergegeben. (Daß bei dem Zerfall des Ra III ein „Actiniumisotop" von verhältnismäßig langer Lebensdauer entsteht, wurde auch chemisch nachgewiesen.) Zieht man b von a ab, so erhält man die Kurve c für das nunmehr reine Ra III. Sie zeigt einen sehr schönen exponentiellen Abfall mit einer Halbwertszeit von 86 Minuten. Dieser Wert dürfte wohl innerhalb $\pm$ 6 Minuten richtig sein.

Wir kommen jetzt noch zu dem dritten, hier als Ra IV bezeichneten „Radiumisotop". Der spätere Verlauf der Kurve b in Fig. 1 ergab eine mit etwa 12 bis 13 Tagen Halbwertszeit zerfallende Substanz. Daß diese langsamere Aktivitätsabnahme im wesentlichen von einem „Radiumisotop" herrührt, wurde auf ganz ähnliche Weise bewiesen wie beim Ra III. Läßt man

ein lange bestrahltes Uran nach der Entfernung von der Neutronenquelle etwa einen Tag stehen, dann zerfallen die Isotope Ra II und Ra III vollständig. Macht man jetzt eine Bariumfällung, kristallisiert vorsichtshalber noch einmal um, dann kann eine beim Bariumchlorid gefundene Aktivität nur noch von einem weiteren „Radiumisotop" herrühren. Derartige Aktivitäten wurden, auch nach tagelangem Stehen, immer gefunden. Ihr Aktivitätsverlauf ist sehr charakteristisch. Die Aktivität nimmt während mehrerer Tage allmählich zu, erreicht ein Maximum und verschwindet dann mit einer Halbwertszeit von rund 300 Stunden (12,5 Tage).

In der Fig. 4 sind einige solcher Kurven wiedergegeben. Das Präparat der Kurve c war aus einem unverstärkt bestrahlten Uranpräparat abgetrennt, die

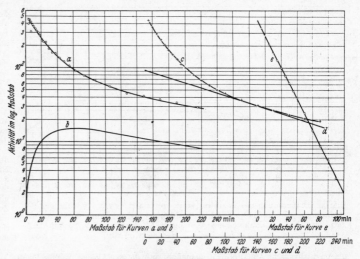

Fig. 2. Bestimmung der Halbwertszeit von Ra II (kurze Bestrahlung). $a =$ Ra nach 6 Min. Bestrahlung: Direkte Abfallskurve. $b =$ theoret. Zunahmekurve d. 2,5-Std.-Ac aus Ra II, H.Z. $= 14$ Min. $c = a$ [Ra] $- b$ [Zunahme 2,5 Std.] $d =$ Ra III, H.Z. $= 86$ Min. $e = c - d =$ Ra II; ergibt 14 Min. H.Z.

anderen Kurven beziehen sich auf entsprechende Bariumfällungen aus verstärkt bestrahltem Uran. (Über den Verstärkungsfaktor läßt sich aus den Kurven nichts aussagen, weil hier nicht unter gleichen geometrischen Bedingungen gearbeitet wurde. Bei gleichen Bedingungen, gleicher bestrahlter Uranmenge usw., fanden wir einen Verstärkungsfaktor von rund 7.) Der Verlauf der 3 Kurven ist sehr ähnlich. Die Zunahme der Aktivität erfolgt mit einer Halbwertszeit von weniger als 40 Stunden, die Abnahme mit rund 300 Stunden. Zweifellos ist die Halbwertszeit dieses langlebigen „Ra IV" aber etwas kürzer als 300 Stunden, denn außer der für die anfängliche Zunahme im wesentlichen verantwortlichen Entstehung von Ac IV aus Ra IV entsteht aus dem Ac IV wohl noch ein langlebiges „Thorisotop", so daß die Halbwertszeit des Ra IV nicht ganz genau festgelegt ist. Ein Wert von 250—300 Stunden wird wohl der Wahrheit nahe kommen. Aus den Kurven a, b und c sieht man auch deutlich, daß die β-Strahlung des Ra IV wesentlich absorbierbarer ist als die seiner Folgeprodukte; sonst könnte kein so starker Anstieg erfolgen.

---

[1] L. MEITNER, F. STRASSMANN u. O. HAHN, Z. Physik *109*, 538 (1938).

14　HAHN u. STRASSMANN: Über den Nachweis und das Verhalten der Erdalkalimetalle.　[Die Natur-wissenschaften

Die im vorhergehenden gebrachten Ergebnisse zusammenfassend haben wir also drei als Ra II, Ra III und Ra IV bezeichnete isomere Erdalkalimetalle festgestellt. Ihre Halbwertszeiten sind 14 ± 2 Minuten, 86 ± 6 Minuten, 250—300 Stunden. Es wird aufgefallen sein, daß der 14-Minuten-Körper nicht als Ra I, die weiteren Isomeren nicht als Ra II und Ra III bezeichnet worden sind. Der Grund liegt darin, daß wir an ein noch instabileres „Ra" glauben, obgleich es bisher nicht nachgewiesen wurde. In unserer ersten Mitteilung über die neuen Umwandlungsprodukte haben wir ein Actinium von etwa 40 Minuten Halbwertszeit angegeben und

Die große Gruppe der „Transurane" steht bisher in keinem erkennbaren Zusammenhang mit diesen Reihen.

Die in dem vorliegenden Schema mitgeteilten Umwandlungsreihen sind in ihren *genetischen* Beziehungen wohl zweifellos als richtig anzusehen. Von den am Ende der isomeren Reihen als „Thorium" angegebenen Endgliedern haben wir auch schon einige nachweisen können. Aber da über ihre einzelnen Halbwertszeiten noch keine genauen Angaben gemacht werden können, haben wir bei ihnen vorerst überhaupt auf eine Angabe verzichtet.

Nun müssen wir aber noch auf einige neuere Untersuchungen zu sprechen kommen, die wir der seltsamen Ergebnisse wegen nur zögernd veröffentlichen. Um den Beweis für die chemische Natur der mit dem Barium abgeschiedenen und als „Radiumisotope" bezeichneten Anfangsglieder der Reihen über jeden Zweifel hinaus zu erbringen, haben wir mit den aktiven Bariumsalzen fraktionierte Kristallisationen und fraktionierte Fällungen vorgenommen, in der

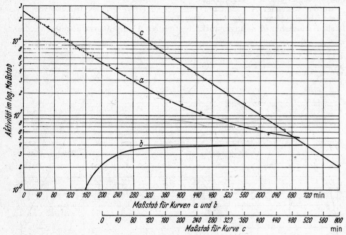

Fig. 3. Bestimmung der Halbwertszeit von Ra III nach 2,5-stündiger Bestrahlung. a = Ra III [2,5 Std. bestrahlt]. 3 Std. n. Bestr. wurde Ac abgetrennt. b = ∞Zunahmekurve v. langem Ac aus Ra III v. 86 Min. H.Z. c = a − b = Ra III. H.Z. = ∞86 Min.

als nächstliegende Annahme die gemacht, daß dieses instabile Actiniumisotop aus dem instabilsten Radiumisotop entsteht. Nun haben wir in der Zwischenzeit festgestellt, daß das aus dem 14-Minuten-Radium (früher 25 Minuten) entstehende „Actinium" eine ungefähre Halbwertszeit von 2,5 Stunden hat (früher mit 4 Stunden angegeben). Das obenerwähnte instabilere Actiniumisotop ist aber ebenfalls vorhanden. Seine Halbwertszeit ist etwas kleiner als früher angegeben, — wohl unter 30 Minuten. Da dieses „Actiniumisotop" weder aus dem 14-Minuten-, noch aus dem 86-Minuten-Körper, noch aus dem langlebigen „Ra" entstehen kann, — da außerdem dieses „Actiniumisotop" schon nach 5 Minuten langer Bestrahlung des Urans nachweisbar ist, ist die einfachste Annahme für seine Entstehung ein „Radiumisotop", dessen Halbswertszeit kürzer als 1 Minute sein muß. Mit einer größeren Halbwertszeit als eine Minute hätten wir es nämlich nachweisen müssen; wir haben sehr danach gesucht. Wir bezeichnen deshalb diese bisher unbekannte, mit einer stärkeren Strahlenquelle wohl zweifellos nachweisbare Muttersubstanz des instabilsten „Actiniumisotops" als „Ra I".

Das in unserer ersten Mitteilung gebrachte Schema muß dadurch eine gewisse Korrektur erfahren. Das folgende Schema trägt dieser Änderung Rechnung und gibt für die Anfangsglieder der Reihen die nunmehr genauer bestimmten Halbwertszeiten:

Weise, wie sie für die Anreicherung (oder auch Abreicherung) des Radiums in Bariumsalzen bekannt sind.

Bariumbromid reichert das Radium bei fraktionierter Kristallisation stark an, Bariumchromat bei nicht zu schnellem Herauskommen der Kriställchen noch mehr. Bariumchlorid reichert weniger stark an als das Bromid, Bariumkarbonat reichert etwas ab. Entsprechende Versuche, die wir mit unseren von Folgeprodukten gereinigten aktiven Bariumpräparaten gemacht haben, *verliefen ausnahmslos negativ: Die Aktivität blieb gleichmäßig auf alle Bariumfraktionen verteilt*, wenigstens soweit wir dies innerhalb der nicht ganz geringen Versuchsfehlermöglichkeiten angeben können. Es wurden dann ein paar Fraktionierungsversuche mit dem Radiumisotop ThX und mit dem Radiumisotop MsTh₁ gemacht. Sie verliefen genau so, wie man aus früheren Erfahrungen mit dem Radium erwarten sollte. Es wurde dann die „Indikatorenmethode" auf ein Gemisch des gereinigten langlebigen „Ra" mit reinem, radiumfreien MsTh₁ angewandt: das Gemisch mit Bariumbromid als Trägersubstanz wurde fraktioniert kristallisiert. *Das MsTh₁ wurde angereichert, das „Ra IV" nicht*, sondern seine Aktivität blieb bei gleichem Bariumgehalt der Fraktionen wieder gleich. Wir kommen zu dem Schluß: Unsere „Radiumisotope" haben die Eigenschaften des Bariums; als Chemiker müßten wir eigentlich sagen, bei den neuen Körpern handelt es sich nicht um Radium, sondern um Barium;

denn andere Elemente als Radium oder Barium kommen nicht in Frage.

Schließlich haben wir auch einen Indikatorversuch mit unserem rein abgeschiedenen „Ac II" (H.Z. rund 2,5 Stunden) und dem reinen Actiniumisotop MsTh₂ gemacht. Wenn unsere „Ra-Isotope" kein Radium sind, dann sind die „Ac-Isotope" auch kein Actinium, sondern sollten Lanthan sein. Nach dem Vorgehen von Mme Curie[1] haben wir eine Fraktionierung von Lanthanoxalat, das die beiden aktiven Substanzen enthielt, aus salpetersaurer Lösung vorgenommen. Das MsTh₂ fand sich, wie von Mme Curie angegeben, in den Endfraktionen stark angereichert. Bei unserem „Ac II" war von einer Anreicherung am Ende nichts zu merken. In Übereinstimmung mit Curie und Savitch[2] über ihren allerdings nicht einheitlichen 3,5-Stunden-Körper finden wir also, daß das aus unserem aktiven Erdalkalimetall durch β-Strahlenemission entstehende Erdmetall kein Actinium ist. Den von Curie und Savitch angegebenen Befund, daß sie die Aktivität im Lanthan anreicherten, der also gegen eine Gleichheit mit Lanthan spricht, wollen wir noch genauer experimentell prüfen, da bei dem dort vorliegenden Gemisch eine Anreicherung vorgetäuscht sein könnte.

Ob die aus den „Ac-La-Präparaten" entstehenden, als „Thor" bezeichneten Endglieder unserer Reihen sich als Cer herausstellen, wurde noch nicht geprüft.

Was die „Trans-Urane" anbelangt, so sind diese Elemente ihren niedrigeren Homologen Rhenium, Osmium, Iridium, Platin zwar chemisch verwandt, mit ihnen aber nicht gleich. Ob sie etwa mit den noch niedrigeren Homologen Masurium, Ruthenium, Rhodium, Palladium chemisch gleich sind, wurde noch

nicht geprüft. Daran konnte man früher ja nicht denken. Die Summe der Massenzahlen Ba + Ma, also z. B. 138 + 101, ergibt 239!

Als Chemiker müßten wir aus den kurz dargelegten Versuchen das oben gebrachte Schema eigentlich umbenennen und statt Ra, Ac, Th die Symbole Ba, La, Ce einsetzen. Als der Physik in gewisser Weise nahestehende „Kernchemiker" können wir uns zu diesem, allen bisherigen Erfahrungen der Kernphysik widersprechenden, Sprung noch nicht entschließen. Es könnten doch noch vielleicht eine Reihe seltsamer Zufälle unsere Ergebnisse vorgetäuscht haben.

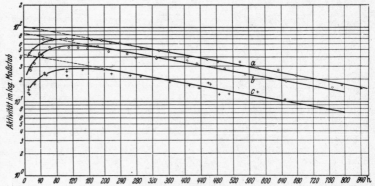

Fig. 4. Bestimmung der Halbwertszeit von Ra IV bei verschiedener Bestrahlungszeit und -art.

a = Ra IV [∾4 Tage bestrahlt] ⎫ verstärkt,
b = Ra IV [∾2,6 „   „   ] ⎭ ⟵ Ra-Abtrennung 15 Std. nach Ende
c = Ra IV [∾2,6 „   „   ] unverstärkt,   der Bestrahlung,

a H.Z. = ∾311 Stunden,   b H.Z. = ∾310 Stunden,   c H.Z. = ∾300 Stunden.

Es ist beabsichtigt, weitere Indikatorenversuche mit den neuen Umwandlungsprodukten durchzuführen. Insbesondere soll auch eine gemeinsame Fraktionierung der aus Thor durch Bestrahlen mit schnellen Neutronen entstehenden, von Meitner, Strassmann und Hahn[1] untersuchten Radiumisotope mit unseren aus dem Uran entstandenen Erdalkalimetallen versucht werden. An Stellen, denen starke künstliche Strahlenquellen zur Verfügung stehen, könnte dies allerdings wesentlich leichter geschehen.

Zum Schlusse danken wir Frl. Cl. Lieber und Frl. I. Bohne für ihre wirksame Hilfe bei den sehr zahlreichen Fällungen und Messungen.

[1] Mme Pierre Curie, J. Chim. physique etc. **27**, 1 (1930).
[2] I. Curie u. P. Savitch, C. r. Acad. Sci. Paris **206**, 1643 (1938).

[1] L. Meitner, F. Strassmann u. O. Hahn, l. c.

Besprechungen.

STUBBE, H., **Spontane und strahleninduzierte Mutabilität.** (Probleme der theoretischen und angewandten Genetik und deren Grenzgebiete, redigiert von W. F. Reinig.) Leipzig: Georg Thieme 1937. 190 S. und 12 Abbild. 13 cm × 21 cm. Preis kart. RM 6.80.

Nach 10 Jahren emsiger Arbeit hat die experimentelle Mutationsforschung heute ein Stadium erreicht, das dazu berechtigt, die Ergebnisse dieser Forschungen zusammenfassend darzustellen und einem größeren Leserkreis näherzubringen. Nachdem erst kürzlich von Timoféeff-Ressovsky eine Darstellung der experimentellen Mutationsforschung gegeben wurde,

behandelt Stubbe dieses Gebiet in den „Problemen der Genetik". Er versucht, sich auf die spontane und die strahleninduzierte Mutabilität zu beschränken, doch ergeben sich verschiedene Schwierigkeiten, wenn man die spontane Mutabilität behandeln will, ohne auf die Frage der Abhängigkeit der Mutationsrate von verschiedenen physiologischen Bedingungen einzugehen. Wenn auch erst wenig brauchbare Ergebnisse auf diesem Gebiet vorliegen, so kann doch an den hier aufgeworfenen Problemen nicht vorbeigegangen werden. Sie werden denn auch vom Verf. angeschnitten bei der Behandlung der Abhängigkeit der „Mutationsrate"

APRIL 1, 1932 *PHYSICAL REVIEW* VOLUME 40

THE PRODUCTION OF HIGH SPEED LIGHT IONS WITHOUT THE USE OF HIGH VOLTAGES

By Ernest O. Lawrence and M. Stanley Livingston

University of California

(Received February 20, 1932)

Abstract

The study of the nucleus would be greatly facilitated by the development of sources of high speed ions, particularly protons and helium ions, having kinetic energies in excess of 1,000,000 volt-electrons; for it appears that such swiftly moving particles are best suited to the task of nuclear excitation. The straightforward method of accelerating ions through the requisite differences of potential presents great experimental difficulties associated with the high electric fields necessarily involved. The present paper reports the development of a method that avoids these difficulties by means of the multiple acceleration of ions to high speeds without the use of high voltages. The method is as follows: Semi-circular hollow plates, not unlike duants of an electrometer, are mounted with their diametral edges adjacent, in a vacuum and in a uniform magnetic field that is normal to the plane of the plates. High frequency oscillations are applied to the plate electrodes producing an oscillating electric field over the diametral region between them. As a result during one half cycle the electric field accelerates ions, formed in the diametral region, into the interior of one of the electrodes, where they are bent around on circular paths by the magnetic field and eventually emerge again into the region between the electrodes. The magnetic field is adjusted so that the time required for traversal of a semi-circular path within the electrodes equals a half period of the oscillations. In consequence, when the ions return to the region between the electrodes, the electric field will have reversed direction, and the ions thus receive second increments of velocity on passing into the other electrode. Because the path radii within the electrodes are proportional to the velocities of the ions, the time required for a traversal of a semi-circular path is independent of their velocities. Hence if the ions take exactly one half cycle on their first semi-circles, they do likewise on all succeeding ones and therefore spiral around in resonance with the oscillating field until they reach the periphery of the apparatus. Their final kinetic energies are as many times greater than that corresponding to the voltage applied to the electrodes as the number of times they have crossed from one electrode to the other. This method is primarily designed for the acceleration of light ions and in the present experiments particular attention has been given to the production of high speed protons because of their presumably unique utility for experimental investigations of the atomic nucleus. Using a magnet with pole faces 11 inches in diameter, a current of 10^{-9} ampere of 1,220,000 volt-protons has been produced in a tube to which the maximum applied voltage was only 4000 volts. There are two features of the developed experimental method which have contributed largely to its success. First there is the focussing action of the electric and magnetic fields which prevents serious loss of ions as they are accelerated. In consequence of this, the magnitudes of the high speed ion currents obtainable in this indirect manner are comparable with those conceivably obtainable by direct high voltage methods. Moreover, the focussing action results in the generation of very narrow beams of ions—less than 1 mm cross-sectional diameter—which are ideal for experimental studies of collision processes. Of hardly less importance is the second feature of the method which is the simple and highly effective means for the correction of the magnetic field along the paths of the ions. This makes it possible, indeed easy, to operate the tube effectively

20   E. O. LAWRENCE AND M. S. LIVINGSTON

with a very high amplification factor (i.e., ratio of final equivalent voltage of accelerated ions to applied voltage). In consequence, this method in its present stage of development constitutes a highly reliable and experimentally convenient source of high speed ions requiring relatively modest laboratory equipment. Moreover, the present experiments indicate that this indirect method of multiple acceleration now makes practicable the production in the laboratory of protons having kinetic energies in excess of 10,000,000 volt-electrons. With this in mind, a magnet having pole faces 114 cm in diameter is being installed in our laboratory.

INTRODUCTION

THE classical experiments of Rutherford and his associates[1] and Pose[2] on artificial disintegration, and of Bothe and Becker[3] on excitation of nuclear radiation, substantiate the view that the nucleus is susceptible to the same general methods of investigation that have been so successful in revealing the extra-nuclear properties of the atom. Especially do the results of their work point to the great fruitfulness of studies of nuclear transitions excited artificially in the laboratory. The development of methods of nuclear excitation on an extensive scale is thus a problem of great interest; its solution is probably the key to a new world of phenomena, the world of the nucleus.

But it is as difficult as it is interesting, for the nucleus resists such experimental attacks with a formidable wall of high binding energies. Nuclear energy levels are widely separated and, in consequence, processes of nuclear excitation involve enormous amounts of energy—millions of volt-electrons.

It is therefore of interest to inquire as to the most promising modes of nuclear excitation. Two general methods present themselves; excitation by absorption of radiation (gamma radiation), and excitation by intimate nuclear collisions of high speed particles.

Of the first it may be said that recent experimental studies [4,5] of the absorption of gamma radiation in matter show, for the heavier elements, variations with atomic number that indicate a quite appreciable nuclear effect. This suggests that nuclear excitation by absorption of radiation is perhaps a not infrequent process, and therefore that the development of an intense artificial source of gamma radiation of various wave-lengths would be of considerable value for nuclear studies. In our laboratory, as elsewhere, this being attempted.

But the collision method appears to be even more promising, in consequence of the researches of Rutherford and others cited above. Their pioneer investigations must always be regarded as really great experimental achievements, for they established definite and important information about nuclear processes of great rarity excited by exceedingly weak beams of bombarding particles—alpha-particles from radioactive sources. Moreover, and this is the point to be emphasized here, their work has shown strikingly the

[1] See Chapter 10 of Radiations from Radioactive Substances by Rutherford, Chadwick and Ellis.

[2] H. Pose, Zeits. f. Physik 64, 1 (1930).

[3] W. Bothe and H. Becker, Zeits. f. Physik 66, 1289 (1930).

[4] G. Beck, Naturwiss. 18, 896 (1930).

[5] C. Y. Chao, Phys. Rev. 36, 1519 (1930).

great fruitfulness of the kinetic collision method and the importance of the development of intense artificial sources of alpha-particles. Of course it cannot be inferred from their experiments that alpha-particles are the most effective nuclear projectiles: the question naturally arises whether lighter or heavier particles of given kinetic energy would be more effective in bringing about nuclear transitions.

A beginning has been made on the theoretical study of the nucleus and a partial answer to this question has been obtained. Gurney and Condon[6] and Gamow[7] have independently applied the ideas of the wave mechanics to radioactivity with considerable success. Gamow[8] has further considered along the same lines the penetration into the nucleus of swiftly moving charged particles (with excitation of nuclear transitions in mind) and has concluded that, for a given kinetic energy, the lighter the particle the greater is the probability that it will penetrate the nuclear potential wall. This result is not unconnected with the smaller momentum and consequent longer wave-length of the ligher particles; for it is well-known that transmission of matter waves through potential barriers becomes greater with increasing wave-lengths.

If the probability of nuclear excitation by a charged particle were mainly dependent on its ability to penetrate the nuclear potential wall, electrons would be the most effective. However, there is considerable evidence that nuclear excitation by electrons is negligible. It suffices to mention here the current view that the average density of the extra-nuclear electrons is quite great in the region of the nucleus, i.e., that the nucleus is quite transparent to electrons; in other words, there are no available stable energy levels for them.

On the other hand, there is evidence that there are definite nuclear levels for protons as well as alpha-particles;[9] indeed, there is some justification for the view that the general principles of the quantum mechanics are applicable in the nucleus to protons and alpha particles. It is not possible at the present time to estimate the relative excitation probabilities of the protons and alpha particles that succeed in penetrating the nucleus. However, it does seem likely that the greater penetrability of the proton* is an advantage outweighing any differences in their excitation characteristics. Protons thus appear to be most suited to the task of nuclear excitation.

Though at present the relative efficacy of protons and alpha-particles cannot be established with much certainty, it does seem safe to conclude at least that the most efficacious nuclear projectiles will prove to be swiftly moving ions, probably of low atomic number. In consequence it is important to develop methods of accelerating ions to speeds much greater than have heretofore been produced in the laboratory.

[6] Gurney and Condon, Phys. Rev. **33**, 127 (1929).

[7] Gamow, Zeits. f. Physik **51**, 204 (1928).

[8] Gamow, Zeits. f. Physik **52**, 514 (1929).

[9] J. Chadwick, J. E. R. Constable, E. C. Pollard, Proc. Roy. Soc. **A130**, 463 (1930).

* According to Gamow's theory a one million volt-proton has as great a penetrating power as a sixteen million volt alpha-particle.

The importance of this is generally recognized and several laboratories are developing techniques of the production and the application to vacuum tubes of high voltages for the generation of high speed electrons and ions. Highly significant progress in this direction has been made by Coolidge,[10] Lauritsen,[11] Tuve, Breit, Hafstad, Dahl,[12] Brasch and Lange,[13] Cockroft and Walton,[14] Van de Graaff[15] and others, who have developed several distinct techniques which have been applied to voltages of the order of magnitude of one million.

These methods involving the direct utilization of high voltages are subject to certain practical limitations. The experimental difficulties go up rapidly with increasing voltage; there are the difficulties of corona and insulation and also there is the problem of design of suitable high voltage vacuum tubes.

Because of these difficulties we have thought it desirable to develop methods for the acceleration of charged particles that do not require the use of high voltages. Our objective is two fold: first, to make the production of particles having kinetic energies of the order of magnitude of one million volt-electrons a matter that can be carried through with quite modest laboratory equipment and with an experimental convenience that, it is hoped, will lead to a widespread attack on this highly important domain of physical phenomena; and second, to make practicable the production of particles having kinetic energies in excess of those producible by direct high voltage methods—perhaps in the range of 10,000,000 volt-electrons and above.

A method for the multiple acceleration of ions to high speeds, primarily designed for heavy ions, has recently been described in this journal.[16] The present paper is a report of the development of a method for the multiple acceleration of light ions.[17] Particular attention has been given to the acceleration of protons because of their apparent unique utility in nuclear studies. In the present work relatively large currents of 1,220,000 volt-protons have been generated and there is foreshadowed in the not distant future the production of 10,000,000 volt-protons.

The Experimental Method

In the method for the multiple acceleration of ions to high speeds, recently described,[16] the ions travel through a series of metal tubes in synchronism with an applied oscillating electric potential. It is so arranged that as an

[10] W. D. Collidge, Am. Inst. E. Eng. **47,** 212 (1928).

[11] C. C. Lauritsen and R. D. Bennett, Phys. Rev. **32,** 850 (1928).

[12] M. A. Tuve, G. Breit, L. R. Hafstad and O. Dahl, Phys. Rev. **35,** 66 (1930); M. A. Tuve, L. R. Hafstad, O. Dahl, Phys. Rev. **39,** 384, (1932).

[13] A. Brasch and J. Lange, Zeits. f. Physik **70,** 10 (1931).

[14] J. J. Cockroft and E. T. S. Walton, Proc. Roy. Soc. **A129,** 477 (1930).

[15] R. S. Van de Graaff, Schenectady Meeting American Physical Society, 1931.

[16] D. H. Sloan and E. O. Lawrence, Phys. Rev. **38,** 2021 (1931).

·[17] This method was first described before the September, 1930, meeting of the National Academy of Sciences (Lawrence and Edlefsen, Science **72,** 376–377 (1930)). Later before the American Physical Society (Lawrence and Livingston, Phys. Rev. **37,** 1707, (1931)) results of a preliminary study of the practicability of the method were given. Further work was reported in a Letter to the Editor of the Physical Review (Lawrence and Livingston, Phys. Rev. **38,** 834 (1931).

ion travels from the interior of one tube to the interior of the next there is always an accelerating field, and the final velocity of the ion on emergence from the system corresponds approximately to a voltage as many times greater than the applied voltage between adjacent tubes as there are tubes. The method is most conveniently used for the acceleration of heavy ions; for light ions travel faster and hence require longer systems of tubes for any given frequency of applied oscillations.

The present experimental method makes use of the same principle of repeated acceleration of the ions by a similar sort of resonance with an oscillating electric field, but has overcome the difficulty of the cumbersomely long accelerating system by causing, with the aid of a magnetic field, the ions to circulate back and forth from the interior of one electrode to the interior of another.

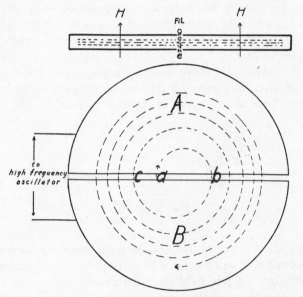

Fig. 1. Diagram of experimental method for multiple acceleration of ions.

This may be seen most readily by an outline of the experimental arrangement (Fig. 1). Two electrodes A, B in the form of semi-circular hollow plates are mounted in a vacuum tube in coplanar fashion with their diametral edges adjacent. By placing the system between the poles of a magnet, a magnetic field is introduced that is normal to the plane of the plates. High frequency electric oscillations are applied to the plates so that there results an oscillating electric field in the diametral region between them.

With this arrangement it is evident that, if at one moment there is an ion in the region between the electrodes, and electrode A is negative with respect to electrode B, then the ion will be accelerated to the interior of the former. Within the electrode the ion traverses a circular path because of the magnetic field, and ultimately emerges again between the electrodes; this is indicated in the diagram by the arc a .. b. If the time consumed by the ion in making the

24   E. O. LAWRENCE AND M. S. LIVINGSTON

semi-circular path is equal to the half period of the electric oscillations, the electric field will have reversed and the ion will receive a second acceleration, passing into the interior of electrode B with a higher velocity. Again it travels on a semi-circular path ($b .. c$), but this time the radius of curvature is greater because of the greater velocity. For all velocities (neglecting variation of mass with velocity) the radius of the path is proportional to the velocity, so that the time required for traversal of a semi-circular path is independent of the ion's velocity. Therefore, if the ion travels its first half circle in a half cycle of the oscillations, it will do likewise on all succeeding paths. Hence it will circulate around on ever widening semi-circles from the interior of one electrode to the interior of the other, gaining an increment of energy on each crossing of the diametral region that corresponds to the momentary potential difference between the electrodes. Thus, if, as was done in the present experiments, high frequency oscillations having peak values of 4000 volts are applied to the electrodes, and protons are caused to spiral around in this way 150 times, they will receive 300 increments of energy, acquiring thereby a speed corresponding to 1,200,000 volts.

It is well to recapitulate these remarks in quantitative fashion. Along the circular paths within the electrodes the centrifugal force of an ion is balanced by the magnetic force on it, i.e., in customary notation,

$$\frac{mv^2}{r} = \frac{Hev}{c} \tag{1}$$

It follows that the time for traversal of a semi-circular path is

$$t = \frac{\pi r}{v} = \frac{\pi mc}{He} \tag{2}$$

which is independent of the radius r of the path and the velocity v of the ion. The particle of mass m and charge e thus may be caused to travel in phase with the oscillating electric field by suitable adjustment of the magnetic field H: the relation between the wave-length λ of the oscillations and the corresponding synchronizing magnetic field H is in consequence

$$\lambda = \frac{2\pi mc^2}{He} \tag{3}$$

Thus for protons and a magnetic field of 10,000 gauss the corresponding wavelength is 19.4 meters; for heavier particles the proper wave-length is proportionately longer.*

It is easily shown also that the energy V in volt-electrons of the charged particles arriving at the periphery of the apparatus on a circle of radius r is

---

* It should be mentioned that, for a given wave-length, the ions resonate with the oscillations when magnetic fields of 1/3, 1/5, etc., of that given by Eq. (3) are used. Such types of resonance were observed in the earlier experimental studies. In the present experiments, however, the high speed ions resulting from the primary type of resonance only were able to pass through the slit system to the collector, because of the high deflecting voltages used.

$$V = 150 \frac{H^2 r^2}{c^2} \frac{e}{m} \tag{4}$$

Thus, the theoretical maximum producible energy varies as the square of the radius and the square of the magnetic field.

EXPERIMENTAL ARRANGEMENT

The experimental arrangement is shown diagrammatically in some detail in Fig. 2. Fig. 3 is a photograph of the brass vacuum tube with cover removed showing the filament, the accelerating electrode, the deflecting plates and slit system, the probe in front of the first slit mounted on a ground joint and the Faraday collector behind the last slit. An external view of the apparatus is shown in Fig. 4. Here the tube is shown between the magnet pole faces, connected with the oscillator, the vacuum system and hydrogen generator. This gives a good general idea of the modest extent of the equipment involved for the generation of protons having energies somewhat in excess of 1,000,000 volt-electrons. The control panel and electrometer, being on the other side, are not shown in the picture. The description of the apparatus follows.

The accelerating system. Though there are obvious advantages in applying the high frequency potentials with respect to ground to both accelerating electrodes, in the present experiments it was found convenient to apply the high frequency voltage to only one of the electrodes, as indicated in Fig. 2. This electrode was a semi-circular hollow brass plate 24 cm in diameter and 1 cm thick. The sides of the hollow plate were of thin brass so that the interior of the plate had approximately these dimensions. It was mounted on a water-cooled copper re-entrant tube which in turn passed through a copper to glass seal. The electrode insulated in this way was mounted in an evacuated brass box having internal dimensions 2.6 cm by 28.6 cm by 28.6 cm, there being thus a lateral clearance between the electrode and walls of the brass chamber of 8 mm.

The brass box itself constituted the other electrode of the accelerating system. Across the mid-section of the brass chamber parallel to the diametral edge of the electrode A was placed a brass dividing wall S with slits of the same dimensions as the opening of the nearby electrode. This arrangement gave rise to the same type of oscillating electric fields as would have been produced had there been used two insulated semi-circular electrodes with their diametral edges adjacent and parallel.

The source of ions. An ideal source of ions is one that delivers to the diametral region between the electrodes large quantities of ions with low components of velocity normal to the plane of the accelerators. This requirement has most conveniently been met in the present experiments merely by having a filament placed above the diametral region from which a stream of electrons pass down along the magnetic lines of force, generating ions of gases in the tube. The ions so formed are pulled out sideways by the oscillating electric field. The electrons are not drawn out because of their very small radii of curvature in the magnetic field. Thus, the beam of electrons is col-

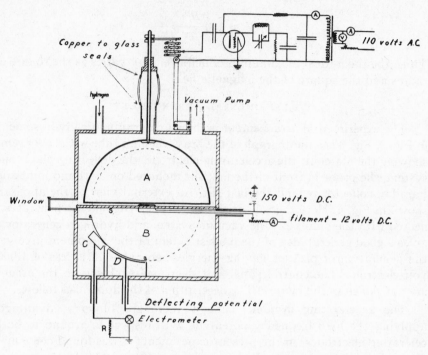

Fig. 2. Diagram of apparatus for the multiple acceleration of ions.

Fig. 3. Tube for the multiple acceleration of light ions—with cover removed.

limated and the ions are formed with negligible initial velocities right in the region where they are wanted. The oscillating electric field immediately draws them out and takes them on their spiral paths to the periphery. This arrangement is diagrammatically shown in the upper part of Fig. 1.

Fig. 4. External view of apparatus for generation of 1,220,000 volt protons.

The magnetic field. This experimental method requires a highly uniform magnetic field normal to the plane of the accelerating system. For example, if the ions are to circulate around 100 times, thereby gaining energy corresponding to 200 times the applied voltage, it is necessary that the magnetic field be uniform to a fraction of one percent. A general consideration of the matter leads one to the conclusion that, if possible, the magnetic field should be constant to about 0.1 percent from the center outward. Though this presumably difficult requirement has been met easily by an empirical method of field correction, the magnet used in the present experiments has pole faces machined as accurately as could be done conveniently. Its design was quite similar to that of Curtis.[18] The pole faces were 11 inches in diameter and the gap separation was $1\frac{1}{2}$ inches. Armco iron was used throughout the magnetic circuit. The magnetomotive force was provided by two coils of number 14 double cotton covered wire of 2,000 turns each. No water cooling was incorporated, for the magnet was not intended for high fields. In practice the magnet would give a field of 14,000 gauss for considerable periods without overheating. The pole faces were made parallel to about 0.2 percent and so it was to be expected that the magnetic field produced would be highly uniform. Exploration with a bismuth spiral confirmed this expectation, since it failed to show an appreciable variation of the magnetic field in the region between the poles, excepting within an inch of the periphery.

The collector system. In planning a suitable arrangement for collecting the high speed ions at the periphery of the apparatus, it was clearly desirable to devise something that would collect the high speed ions only and which would also measure their speeds. One might regard it as legitimate to suppose that the magnetic field itself and the distance of the collector from the center of the system would determine the speeds of the ions collected. This would be true provided there were no scattering and reflection of ions. To eliminate these extraneous effects a set of 1 mm slits was arranged on a circle $a .. a$, as shown in Fig. 2, of radius about 12 percent greater than the circle, indicated by the dotted line in the figure, having its center at the center of the tube and a radius of 11.5 cm. The two circles were tangent at the first slit as shown. The ions on arrival at the first slit would be traveling presumably on circles approximately like the dotted line, and hence would not be able to pass through the second and third slits to the Faraday collector C. Electrostatic deflecting plates D, separated by 2 mm, were placed between the first two slits, making possible the application of electrostatic fields to increase the radius of curvature of the paths of the high speed ions sufficiently to allow them to enter the collector. By applying suitable high potentials to the deflecting system in this way, only correspondingly high speed ions were registered.

The collector currents were measured by an electrometer shunted with a suitable high resistance leak.

The oscillator. The high frequency oscillations applied to the electrode were supplied by a 20 kilowatt Federal Telegraph water-cooled power tube in a "tuned plate tuned grid" circuit, for which the diagram of Fig. 2 is self-explanatory.

THE FOCUSSING ACTIONS

When one considers the circulation of the ions around many times as they are accelerated to high speeds in this way, one wonders whether in practice an appreciable fraction of those starting out can ever be made to

[18] L. F. Curtis, Jour. Op. Soc. Am. **13**, 73 (1926).

arrive at the periphery and to pass through a set of slits perhaps 1 mm wide and 1 cm long. The paths of the ions in the course of their acceleration would be several meters, and, because of the unavoidable spreading effects of space charge, thermal velocities and contact electromotive forces, as well as inhomogeneities of the applied fields, it would appear that the effective solid angle of the peripheral slit for the ions starting out would be exceedingly small.

Fortunately, however, this does not turn out to be the case. The electric and magnetic fields have been so arranged that they provide extremely strong focussing actions on the spiraling ions, which keep them circulating close to the median plane of the accelerating system.

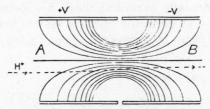

Fig. 5. Diagram indicating the focussing action of the electric field between the accelerating electrodes.

Fig. 5 shows the focussing action of the electric fields. There is depicted a cross-section of the diametral region between the accelerating electrodes with the nature of the field indicated by lines of force. There is shown also a dotted line which represents qualitatively the path of an ion as it passes from the interior of one electrode to the interior of the other. It is seen that, since it is off the median plane in electrode A, on crossing to B it receives an inward displacement towards the median plane. This is because of the existence of the curvature of the field, which over certain regions has an appreciable component normal to the plane, as indicated. If the velocity of the ion is very

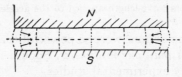

Fig. 6. Diagram indicating focussing action of magnetic field.

high in comparison to the increment of velocity gained in going from plate A to plate B, its displacement towards the center will be relatively small and, to the first approximation, it may be described as due to the ion having been accelerated inward on the first half of its path across and accelerated outward by an equal amount during the remainder of its journey, the net result being a displacement of the ion towards the center without acquiring a net transverse component of velocity. In general, however, the outward acceleration during the second half will not quite compensate the inward acceleration of the first, resulting in a gain of an inward component of velocity as well as an inward displacement. In any event, as the ion spirals around it will migrate back and forth across the median plane and will not be lost to the walls of the tube.

The magnetic field also has a focussing action. Fig. 6 shows diagrammatically the form of the field produced by the magnet. In the central region of the
pole faces the magnetic field is quite uniform and normal to the plane of the
faces; but out near the periphery the field has a curvature. Ions traveling on
circles near the periphery experience thereby magnetic forces, indicated by
the arrows. If the circular path is on the median plane then the magnetic
force is towards the center in that plane. If the ion is traveling in a circle off
the median plane, then there is a component of magnetic force that accelerates it towards the median plane, thereby giving effectively a focussing action.

We have experimentally examined these two focussing actions, using a
probe in front of the first slit of the collector system that could be moved up
and down across the beam by means of a ground joint (see Fig. 3). It was

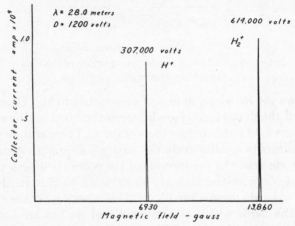

Fig. 7. Ion current to Faraday collector as a function of the magnetic field with oscillations of 28 meters wave-length applied to the accelerating electrodes.

found that the focussing actions were so powerful that *the beam of high speed
ions had a width of less than one millimeter.* Such a narrow beam of ions of
course is ideal for many experimental studies.

As a further test of the focussing action of the two fields, the median
plane of the accelerating system was lowered 3 mm with respect to the plane
of symmetry of the magnetic field. It was found that the high speed ion beam
at the periphery traveled in a plane that was between the planes of symmetry
of the two fields showing that both focussing actions were operative and at the
periphery were of the same order of magnitude.

EXPERIMENTAL RESULTS

As a typical example there is shown in Fig. 7 a plot of the ion current to
the Faraday collector as a function of the magnetic field for applied oscillations of wave-length 28 meters and with hydrogen in the tube. It is seen that
there are only two narrow ranges of magnetic field strength over which ion
currents are observed; both correspond exactly to expectations, the one at

6930 gauss involving the resonance of protons, the other, hydrogen molecule ions.

For each wave-length used, *the magnetic field giving the greatest current to the collector agreed precisely with the theoretically expected value*. This is illustrated in Fig. 8 where the curves represent the theoretical hyperbolic relations between wave-length and magnetic field (Eq. 3) for protons and hydrogen molecule ions, and the circles represent the experimental observations. The magnetic fields were measured with a bismuth spiral and the oscillation wave-lengths were determined with a General Radio wavemeter. No effort was made to obtain considerable precision in these measurements, and in consequence their accuracy was hardly greater than 1 percent.

The variation with applied high frequency voltage of the widths of the resonance peaks agreed also with theoretical expectations. It was found that as the voltage was reduced the peaks became sharper, and indeed, with voltages such that the ions were required to spiral around fifty and more times to reach

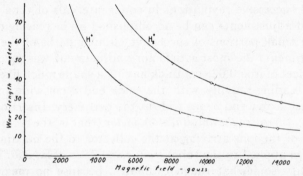

Fig. 8. Magnetic fields producing resonance of ions with oscillations of various wavelengths: the curves are the theoretical relations (Eq. (3)) for H^+ and H_2^+ ions and the circles are the experimental observations.

the periphery, the ion currents diminished practically to zero when the magnetic field was changed a few tenths of one percent from the optimum value. This sharpness of resonance is understandable when it is remembered that the time required for an ion to execute one of its semi-circular paths is inversely proportional to the magnetic field. If, for example, the magnetic field were one percent greater or less than the resonance value, the ions would find themselves completely out of phase with the oscillations after having made fifty revolutions in the tube. In Fig. 7 the peaks exhibit an appreciable width, and indeed they extend over a one percent range of magnetic field. In most of the experiments, however, the ions circulated around many more times resulting in peaks of such restricted breadth as scarcely to be discernible in a diagram of this sort.

It is of course evident that the upper limit to the number of times the ions will circulate is determined by the degree of uniformity of the average value of the magnetic field along the spiral paths. Indeed, it would seem difficult to construct a magnet with pole faces giving fields of sufficient uniformity to allow more than 100 accelerations of the ions. But happily there is a very simple empirical way of correcting for the lack of uniformity of the field, that makes possible a surprisingly large voltage amplification. This is accomplished by insertion of thin sheets of iron between the tube and the magnet; either in the central region or out towards the periphery, as may be needed. If the magnetic field is, on the average, slightly less out towards the periphery so that the ions lag in phase more and more with respect to the oscillations as they spiral around, they may be brought back into step again by the insertion near the periphery of a strip of iron of suitable width, thickness and extension. If, on the other hand, the ions tend to get ahead in phase in this region, an effective correction can be made by inserting a suitable iron sheet in the central region.

It should be emphasized in this connection that the requirement is not that the magnetic field has to be uniform everywhere to the extent indicated above; small deviations from uniformity are allowable provided that the average value of the magnetic field over the paths of the ions is such that they traverse successive revolutions in equal intervals of time. Thus, small magnetic field adjustments can be accomplished by increasing or decreasing the field over small portions of successive circular paths of the ions. In the present experiments the most satisfactory adjustment was made by the insertion of a sheet of iron 0.025 cm thick having a shape much like an exclamation point extending radially with the thick end 8 cm wide in the central region and the narrow end 3 cms wide at the periphery. Insertion of this correcting "shim" *increased the amplification factor* (that is, the ratio of the equivalent voltage of the ions arriving at the collector to the maximum high frequency voltage applied to the tube) *from about 75 to about 300*. These figures are of necessity somewhat rough estimates, because no means were conveniently at hand to measure the high frequency voltages applied to the tube. Our estimates are based solely on sparking distances in air, and hence it is not unlikely that the voltage amplifications were even greater.

The greatest voltage amplification was obtained when generating the highest speed ions, 1,220,000 volt-protons. In all our work we have found the experimental method to be increasingly effective in this regard, as in others, as we go to higher voltages.

For example, the optimum pressure of hydrogen in the tube has been found to increase from less than 10^{-4} mm of Hg when generating 200,000 volt-protons to more than 10^{-3} mm when producing 1,000,000 volt-protons. By the optimum pressure is meant the pressure that gives the largest current to the collector for a given electron emission from the filament. The reason for this is, of course, connected with the fact that the effective mean free path of the spiralling particles increases with voltage.

Examples of the observed variation with voltage on the deflecting plates of the ion currents to the collector are shown in Fig. 9. Each curve is for a particular resonance condition; curve A, for example, was obtained when protons resonated with 37.5 meter oscillations in a magnetic field of 5180 gauss, thereby theoretically resulting in the arrival of 172,000 volt-protons at the first slit of the collector system. The wave-lengths used and the theoretically expected equivalent voltages of the ions generated in each instance is indicated in the figure. It is seen that, the higher the equivalent voltage of the ions, the higher was the required deflecting voltage to obtain the maximum ion currents to the collector. Indeed, within the experimental error, the optimum deflecting voltage was proportional to the theoretical kinetic energies of the ions (calculated from Eq. (4)) and was quite independent of the magnitude of the high frequency voltage applied to the accelerating electrode. *These observations constitute incontrovertible evidence that the ions arriving at the collector actually had the high speeds theoretically expected. The observed absolute magnitudes of the deflecting voltages also agreed with theoretical calculations* within the experimental uncertainty of the paths of the ions before entering the deflecting system. Because of the considerable width of the ion source (the filament was 2.5 cm long) the effective center of

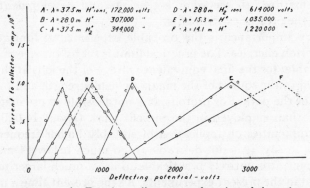

Fig. 9. Ion currents to the Faraday collector as a function of the voltage applied to the deflecting plates. The optimum deflecting voltages are seen to be proportional to the theoretically calculated kinetic energies of the ions (indicated in the figure in volts), thus proving that the ions arriving at the collector actually have the theoretically expected high speeds.

the circular paths of the ions at the periphery was quite broad. This fact together with the slit widths accounted for the absolute range of deflecting voltages over which ion currents reached the collector.

DISCUSSION

The present experiments have accomplished one of the objectives set forth in the introduction, namely, the development of a convenient method for the production of protons having kinetic energies of the order of magnitude of 1,000,000 volt-electrons. It is well to emphasize two particular features that have contributed more than anything else to the effectiveness of the method: the *focussing actions of the electric and magnetic fields*, and the *simple means of empirically correcting the magnetic field* by the introduction of suitable iron strips. The former has solved the practical problem of generation of intense high speed ion beams of restricted cross-section so much desired in studies of collision processes. The latter has eliminated the problem of uniformity of magnetic field, making possible voltage amplifications of more than 300. This in turn has practically eliminated any difficulties associated

E. O. LAWRENCE AND M. S. LIVINGSTON

with generation and application to the accelerating electrodes of required high frequency voltages. In consequence, we have here a source of high speed light ions that is readily constructed and assembled in a relatively small laboratory space out of quite modest laboratory equipment. The beam of ions so produced has valuable characteristics of convenience and flexibility for many experimental investigations; there are obvious advantages of a steady beam of high speed ions of but one millimeter diameter generated in an apparatus on an ordinary laboratory table. Moreover, the apparatus evolved in the present work is in no respects capricious, but functions always in a satisfactorily predictable fashion. This is illustrated by the fact that the accelerating tube can be taken apart and reassembled, and then within a few hours after re-evacuation steady beams of 1,200,000 protons can always be obtained.

But it is perhaps of even more interest to inquire as to the practical limitations of the method; to see what extensions and developments are foreshadowed by the present experiments.

Of primary importance is the probable experimental limitation on the producible proton energies. The practical limit is set by the size of the electromagnet available; for the final equivalent voltage of the ions at the periphery is proportional to the square of the magnetic field strength and to the square of the radius of the path. For protons, it is not feasible to use magnetic fields much greater than employed in the present work (about 14,000 gauss) because of the difficulties of application of suitably higher frequency oscillations—that is to say, it is not desirable to go much below 14 meters wavelength. However, it is entirely practicable to use a much larger magnet than that employed in the present experiments. At the present time a magnet having pole faces 114 cm in diameter is being installed in our laboratory. As will be seen from Eq. (4), a magnetic field of 14,000 gauss over such a large region *makes possible the production of* 25,000,000 *volt-protons.*

Of course, it may be argued that there are other difficulties which preclude ever reaching such a range of energies. For example, there is the question of whether it is possible to obtain such a great amplification factor that the high frequency voltages necessarily applied to the accelerating electrodes are low enough to be realizable in practice. In the present experiments an amplification of 300 was obtained with no great effort, and it would seem that with more careful correction of the field this amplification could be considerably increased at higher voltages. In the higher range of speeds the variation of mass with velocity begins to be appreciable, but presents no difficulty as it can be allowed for by suitable alteration of the magnetic field in the same empirical manner as is done to correct its otherwise lack of uniformity.

Assuming then a voltage amplification of 500, the production of 25,000,-000 volt-protons would require 50,000 volts at a wave-length of 14 meters applied across the accelerators; thus, 25,000 volts on each accelerator with respect to ground. *It does appear entirely feasible to do this*, although to be sure a considerable amount of power would have to be supplied because of the capacity of the system.

Of similar interest is the matter of maximum obtainable beam intensities. In the present experiments no efforts have been made to obtain high intensities and the collector currents have usually been of the order of magnitude of 10^{-9} amp. Using the present method of generation of the ions, there are two factors that can be drawn upon to increase the yield of high speed ions—the electron emission and the pressure of hydrogen in the tube. The electron emission can easily be increased from 10 to 100 times over that used in the present experiments. The effective free paths of the protons increase with voltage so that, as was found to be the case, the maximum usable pressure of hydrogen is governed by the setting in of a high frequency discharge in the tubes due to the voltage on the accelerators. This appears to occur at a pressure greater than 10^{-3} mm of Hg; the reason the critical pressure is so high is probably to be associated with the quenching action of the magnetic field. These considerations make it seem reasonable to expect that, using the present ion source, *high speed ion currents of as much as 0.1 microampere can readily be obtained.*

At all events, it seems that the focussing of the spiralling ions is so effective that a quite considerable portion of those starting out arrive at the collector and that the beam intensity is determined largely by the source. *This method of multiple acceleration is capable of yields of the same order of magnitude as would conceivably result from the direct application of high voltages.*

For a given experimental arrangement the energy of the ions arriving at the collector varies inversely as their masses and directly as their charges. Thus, the large magnet mentioned above makes possible the production of 12,500,000 volt hydrogen molecule ions and doubly charged helium ions (alpha-particles) as well as 25,000,000 volt-protons. Moreover, generating the theoretically maximum value of ion energies becomes much easier with increasing atomic weight because the wave-length of the applied high frequency oscillations increases in a like ratio. For example, using a magnetic field of 14,000 gauss over a region 114 cm in diameter, 2,800,000 volt nitrogen ions could be generated by applying 123 meter oscillations. Broadly speaking, then, the apparatus is well adapted to the production of ions of all the elements up to atomic weight 25 having kinetic energies in excess of 1,000,000 volt-electrons.

We wish to express our gratitude and thanks to the Committee-on-Grants-in-Aid of the National Research Council, the Federal Telegraph Company through the courtesy of Dr. Leonard F. Fuller, Vice-President, the Research Corporation, and the Chemical Foundation for their generous assistance which has made these experiments possible.

[669]

LXXIX. *The Scattering of α and β Particles by Matter and the Structure of the Atom.* By Professor E. RUTHERFORD, *F.R.S., University of Manchester* *.

§ 1.   IT is well known that the α and β particles suffer deflexions from their rectilinear paths by encounters with atoms of matter.   This scattering is far more marked for the β than for the α particle on account of the much smaller momentum and energy of the former particle. There seems to be no doubt that such swiftly moving particles pass through the atoms in their path, and that the deflexions observed are due to the strong electric field traversed within the atomic system.   It has generally been supposed that the scattering of a pencil of α or β rays in passing through a thin plate of matter is the result of a multitude of small scatterings by the atoms of matter traversed.   The observations, however, of Geiger and Marsden † on the scattering of α rays indicate that some of the α particles must suffer a deflexion of more than a right angle at a single encounter.   They found, for example, that a small fraction of the incident α particles, about 1 in 20,000, were turned through an average angle of 90° in passing through a layer of gold-foil about ·00004 cm. thick, which was equivalent in stopping-power of the α particle to 1·6 millimetres of air.   Geiger ‡ showed later that the most probable angle of deflexion for a pencil of α particles traversing a gold-foil of this thickness was about 0°·87.   A simple calculation based on the theory of probability shows that the chance of an α particle being deflected through 90° is vanishingly small.   In addition, it will be seen later that the distribution of the α particles for various angles of large deflexion does not follow the probability law to be expected if such large deflexions are made up of a large number of small deviations. It seems reasonable to suppose that the deflexion through a large angle is due to a single atomic encounter, for the chance of a second encounter of a kind to produce a large deflexion must in most cases be exceedingly small.   A simple calculation shows that the atom must be a seat of an intense electric field in order to produce such a large deflexion at a single encounter.

Recently Sir J. J. Thomson § has put forward a theory to

---

* Communicated by the Author.   A brief account of this paper was communicated to the Manchester Literary and Philosophical Society in February, 1911.
† Proc. Roy. Soc. lxxxii. p. 495 (1909).
‡ Proc. Roy. Soc. lxxxiii. p. 492 (1910).
§ Camb. Lit. & Phil. Soc. xv. pt. 5 (1910).

670 Prof. E. Rutherford *on the*

explain the scattering of electrified particles in passing through small thicknesses of matter. The atom is supposed to consist of a number N of negatively charged corpuscles, accompanied by an equal quantity of positive electricity uniformly distributed throughout a sphere. The deflexion of a negatively electrified particle in passing through the atom is ascribed to two causes—(1) the repulsion of the corpuscles distributed through the atom, and (2) the attraction of the positive electricity in the atom. The deflexion of the particle in passing through the atom is supposed to be small, while the average deflexion after a large number *m* of encounters was taken as $\sqrt{m} \cdot \theta$, where θ is the average deflexion due to a single atom. It was shown that the number N of the electrons within the atom could be deduced from observations of the scattering of electrified particles. The accuracy of this theory of compound scattering was examined experimentally by Crowther * in a later paper. His results apparently confirmed the main conclusions of the theory, and he deduced, on the assumption that the positive electricity was continuous, that the number of electrons in an atom was about three times its atomic weight.

The theory of Sir J. J. Thomson is based on the assumption that the scattering due to a single atomic encounter is small, and the particular structure assumed for the atom does not admit of a very large deflexion of an α particle in traversing a single atom, unless it be supposed that the diameter of the sphere of positive electricity is minute compared with the diameter of the sphere of influence of the atom.

Since the α and β particles traverse the atom, it should be possible from a close study of the nature of the deflexion to form some idea of the constitution of the atom to produce the effects observed. In fact, the scattering of high-speed charged particles by the atoms of matter is one of the most promising methods of attack of this problem. The development of the scintillation method of counting single α particles affords unusual advantages of investigation, and the researches of H. Geiger by this method have already added much to our knowledge of the scattering of α rays by matter.

§ 2. We shall first examine theoretically the single encounters † with an atom of simple structure, which is able to

---

* Crowther, Proc. Roy. Soc. lxxxiv. p. 226 (1910).

† The deviation of a particle throughout a considerable angle from an encounter with a single atom will in this paper be called "single" scattering. The deviation of a particle resulting from a multitude of small deviations will be termed "compound" scattering.

produce large deflexions of an α particle, and then compare the deductions from the theory with the experimental data available.

Consider an atom which contains a charge $\pm Ne$ at its centre surrounded by a sphere of electrification containing a charge $\mp Ne$ supposed uniformly distributed throughout a sphere of radius R. e is the fundamental unit of charge, which in this paper is taken as $4\cdot65 \times 10^{-10}$ E.S. unit. We shall suppose that for distances less than 10^{-12} cm. the central charge and also the charge on the α particle may be supposed to be concentrated at a point. It will be shown that the main deductions from the theory are independent of whether the central charge is supposed to be positive or negative. For convenience, the sign will be assumed to be positive. The question of the stability of the atom proposed need not be considered at this stage, for this will obviously depend upon the minute structure of the atom, and on the motion of the constituent charged parts.

In order to form some idea of the forces required to deflect an α particle through a large angle, consider an atom containing a positive charge Ne at its centre, and surrounded by a distribution of negative electricity Ne uniformly distributed within a sphere of radius R. The electric force X and the potential V at a distance r from the centre of an atom for a point inside the atom, are given by

$$X = Ne\left(\frac{1}{r^2} - \frac{r}{R^3}\right)$$

$$V = Ne\left(\frac{1}{r} - \frac{3}{2R} + \frac{r^2}{2R^3}\right).$$

Suppose an α particle of mass m and velocity u and charge E shot directly towards the centre of the atom. It will be brought to rest at a distance b from the centre given by

$$\tfrac{1}{2}mu^2 = NeE\left(\frac{1}{b} - \frac{3}{2R} + \frac{b^2}{2R^3}\right).$$

It will be seen that b is an important quantity in later calculations. Assuming that the central charge is $100\,e$, it can be calculated that the value of b for an α particle of velocity $2\cdot09 \times 10^9$ cms. per second is about $3\cdot4 \times 10^{-12}$ cm. In this calculation b is supposed to be very small compared with R. Since R is supposed to be of the order of the radius of the atom, viz. 10^{-8} cm., it is obvious that the α particle before being turned back penetrates so close to the central charge, that the field due to the uniform distribution of negative electricity may be neglected. In general, a simple calculation shows that for all deflexions greater than a degree, we may without sensible error suppose the deflexion due to the field of the central charge alone. Possible single deviations due to the negative electricity, if distributed in the form of corpuscles, are not taken into account at this stage of the theory. It will be shown later that its effect is in general small compared with that due to the central field.

Consider the passage of a positive electrified particle close to the centre of an atom. Supposing that the velocity of the particle is not appreciably changed by its passage through the atom, the path of the particle under the influence of a repulsive force varying inversely as the square of the distance will be an hyperbola with the centre of the atom S as the external focus. Suppose the particle to enter the atom in the direction PO (fig. 1), and that the direction of motion

Fig. 1.

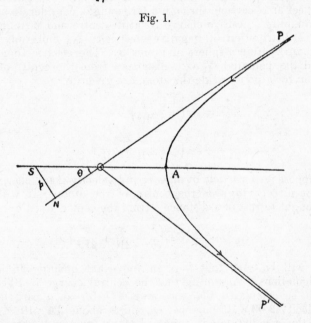

on escaping the atom is OP′. OP and OP′ make equal angles with the line SA, where A is the apse of the hyperbola. $p = SN$ = perpendicular distance from centre on direction of initial motion of particle.

Let angle $POA = \theta$.

Let $V =$ velocity of particle on entering the atom, v its velocity at A, then from consideration of angular momentum

$$pV = SA \cdot v.$$

From conservation of energy

$$\tfrac{1}{2}m V^2 = \tfrac{1}{2}mv^2 + \frac{NeE}{SA},$$

$$v^2 = V^2\left(1 - \frac{b}{SA}\right).$$

Since the eccentricity is $\sec \theta$,

$$SA = SO + OA = p \operatorname{cosec} \theta(1 + \cos \theta)$$

$$= p \cot \theta/2,$$

$$p^2 = SA(SA - b) = p \cot \theta/2(p \cot \theta/2 - b),$$

$$\therefore \quad b = 2p \cot \theta.$$

The angle of deviation ϕ of the particle is $\pi - 2\theta$ and

$$\cot \phi/2 = \frac{2p}{b} \text{*} \quad \ldots \ldots \ldots \quad (1)$$

This gives the angle of deviation of the particle in terms of b, and the perpendicular distance of the direction of projection from the centre of the atom.

For illustration, the angle of deviation ϕ for different values of p/b are shown in the following table:—

p/b	10	5	2	1	·5	·25	·125
ϕ	5°·7	11°·4	28°	53°	90°	127°	152°

§ 3. *Probability of single deflexion through any angle.*

Suppose a pencil of electrified particles to fall normally on a thin screen of matter of thickness t. With the exception of the few particles which are scattered through a large angle, the particles are supposed to pass nearly normally through the plate with only a small change of velocity. Let $n =$ number of atoms in unit volume of material. Then the number of collisions of the particle with the atom of radius R is $\pi R^2 nt$ in the thickness t.

---

* A simple consideration shows that the deflexion is unaltered if the forces are attractive instead of repulsive.

The probabilty m of entering an atom within a distance p of its centre is given by

$$m = \pi p^2 nt.$$

Chance dm of striking within radii p and $p + dp$ is given by

$$dm = 2\pi pnt \cdot dp = \frac{\pi}{4} ntb^2 \cot \phi/2 \, \mathrm{cosec}^2 \, \phi/2 \, d\phi, \quad (2)$$

since
$$\cot \phi/2 = 2p/b.$$

The value of dm gives the *fraction* of the total number of particles which are deviated between the angles ϕ and $\phi + d\phi$.

The fraction ρ of the total number of particles which are deflected through an angle greater than ϕ is given by

$$\rho = \frac{\pi}{4} ntb^2 \cot^2 \phi/2. \quad \cdots \cdots \quad (3)$$

The fraction ρ which is deflected between the angles ϕ_1 and ϕ_2 is given by

$$\rho = \frac{\pi}{4} ntb^2 \left(\cot^2 \frac{\phi_1}{2} - \cot^2 \frac{\phi_2}{2} \right). \quad \cdots \quad (4)$$

It is convenient to express the equation (2) in another form for comparison with experiment. In the case of the α rays, the number of scintillations appearing on a *constant* area of a zinc sulphide screen are counted for different angles with the direction of incidence of the particles. Let $r =$ distance from point of incidence of α rays on scattering material, then if Q be the total number of particles falling on the scattering material, the number y of α particles falling on unit area which are deflected through an angle ϕ is given by

$$y = \frac{Qdm}{2\pi r^2 \sin \phi \cdot d\phi} = \frac{ntb^2 \cdot Q \cdot \mathrm{cosec}^4 \, \phi/2}{16r^2}. \quad \cdots \quad (5)$$

Since $b = \dfrac{2NeE}{mu^2}$, we see from this equation that the number of α particles (scintillations) per unit area of zinc sulphide screen at a given distance r from the point of incidence of the rays is proportional to

(1) $\cosec^4 \phi/2$ or $1/\phi^4$ if ϕ be small ;
(2) thickness of scattering material t provided this is small ;
(3) magnitude of central charge Ne ;
(4) and is inversely proportional to $(mu^2)^2$, or to the fourth power of the velocity if m be constant.

In these calculations, it is assumed that the α particles scattered through a large angle suffer only one large deflexion. For this to hold, it is essential that the thickness of the scattering material should be so small that the chance of a second encounter involving another large deflexion is very small. If, for example, the probability of a single deflexion ϕ in passing through a thickness t is 1/1000, the probability of two successive deflexions each of value ϕ is $1/10^6$, and is negligibly small.

The angular distribution of the α particles scattered from a thin metal sheet affords one of the simplest methods of testing the general correctness of this theory of single scattering. This has been done recently for α rays by Dr. Geiger *, who found that the distribution for particles deflected between 30° and 150° from a thin gold-foil was in substantial agreement with the theory. A more detailed account of these and other experiments to test the validity of the theory will be published later.

§ 4. *Alteration of velocity in an atomic encounter.*

It has so far been assumed that an α or β particle does not suffer an appreciable change of velocity as the result of a single atomic encounter resulting in a large deflexion of the particle. The effect of such an encounter in altering the velocity of the particle can be calculated on certain assumptions. It is supposed that only two systems are involved, viz., the swiftly moving particle and the atom which it traverses supposed initially at rest. It is supposed that the principle of conservation of momentum and of energy applies, and that there is no appreciable loss of energy or momentum by radiation.

* Manch. Lit. & Phil. Soc. 1910.

Let m be mass of the particle,

v_1 = velocity of approach,

v_2 = velocity of recession,

M = mass of atom,

V = velocity communicated to atom as result of encounter.

Let OA (fig. 2) represent in magnitude and direction the momentum mv_1 of the entering particle, and OB the momentum of the receding particle which has been turned through an angle AOB=ϕ. Then BA represents in magnitude and direction the momentum MV of the recoiling atom.

Fig. 2.

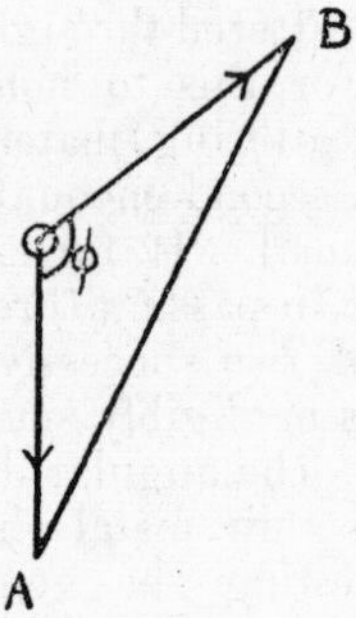

$$(MV)^2 = (mv_1)^2 + (mv_2)^2 - 2m^2 v_1 v_2 \cos\phi. \quad (1)$$

By the conservation of energy

$$MV^2 = mv_1{}^2 - mv_2{}^2. \quad . \quad . \quad (2)$$

Suppose $M/m = K$ and $v_2 = \rho v_1$, where ρ is < 1.

From (1) and (2),

$$(K+1)\rho^2 - 2\rho\cos\phi = K - 1,$$

or

$$\rho = \frac{\cos\phi}{K+1} + \frac{1}{K+1}\sqrt{K^2 - \sin^2\phi}.$$

Consider the case of an α particle of atomic weight 4, deflected through an angle of 90° by an encounter with an atom of gold of atomic weight 197.

Since K=49 nearly,

$$\rho = \sqrt{\frac{K-1}{K+1}} = \cdot979,$$

or the velocity of the particle is reduced only about 2 per cent. by the encounter.

In the case of aluminium $K = 27/4$ and for $\phi = 90°$ $\rho = \cdot86$.

It is seen that the reduction of velocity of the α particle becomes marked on this theory for encounters with the lighter atoms. Since the range of an α particle in air or other matter is approximately proportional to the cube of the velocity, it follows that an α particle of range 7 cms. has its range reduced to 4·5 cms. after incurring a single deviation of 90° in traversing an aluminium atom. This is of a magnitude to be easily detected experimentally. Since the value of K is very large for an encounter of a β particle with an atom, the reduction of velocity on this formula is very small.

Some very interesting cases of the theory arise in considering the changes of velocity and the distribution of scattered particles when the α particle encounters a light atom, for example a hydrogen or helium atom. A discussion of these and similar cases is reserved until the question has been examined experimentally.

§ 5. *Comparison of single and compound scattering.*

Before comparing the results of theory with experiment, it is desirable to consider the relative importance of single and compound scattering in determining the distribution of the scattered particles. Since the atom is supposed to consist of a central charge surrounded by a uniform distribution of the opposite sign through a sphere of radius R, the chance of encounters with the atom involving small deflexions is very great compared with the chance of a single large deflexion.

This question of compound scattering has been examined by Sir J. J. Thomson in the paper previously discussed (§ 1). In the notation of this paper, the average deflexion ϕ_1 due to the field of the sphere of positive electricity of radius R and quantity Ne was found by him to be

$$\phi_1 = \frac{\pi}{4} \cdot \frac{NeE}{mu^2} \cdot \frac{1}{R}.$$

The average deflexion ϕ_2 due to the N negative corpuscles supposed distributed uniformly throughout the sphere was found to be

$$\phi_2 = \frac{16}{5} \frac{eE}{mu^2} \cdot \frac{1}{R} \sqrt{\frac{3N}{2}}.$$

The mean deflexion due to both positive and negative electricity was taken as

$$(\phi_1{}^2 + \phi_2{}^2)^{1/2}.$$

In a similar way, it is not difficult to calculate the average deflexion due to the atom with a central charge discussed in this paper.

Since the radial electric field X at any distance r from the centre is given by

$$X = Ne\left(\frac{1}{r^2} - \frac{r}{R^3}\right),$$

it is not difficult to show that the deflexion (supposed small) of an electrified particle due to this field is given by

$$\theta = \frac{b}{p}\left(1 - \frac{p^2}{R^2}\right)^{3/2},$$

where p is the perpendicular from the centre on the path of the particle and b has the same value as before. It is seen that the value of θ increases with diminution of p and becomes great for small values of ϕ.

Since we have already seen that the deflexions become very large for a particle passing near the centre of the atom, it is obviously not correct to find the average value by assuming θ is small.

Taking R of the order 10^{-8} cm., the value of p for a large deflexion is for α and β particles of the order 10^{-11} cm. Since the chance of an encounter involving a large deflexion is small compared with the chance of small deflexions, a simple consideration shows that the average small deflexion is practically unaltered if the large deflexions are omitted. This is equivalent to integrating over that part of the cross section of the atom where the deflexions are small and neglecting the small central area. It can in this way be simply shown that the average small deflexion is given by

$$\phi_1 = \frac{3\pi}{8}\frac{b}{R}.$$

This value of ϕ_1 for the atom with a concentrated central charge is three times the magnitude of the average deflexion for the same value of Ne in the type of atom examined by Sir J. J. Thomson. Combining the deflexions due to the electric field and to the corpuscles, the average deflexion is

$$(\phi_1{}^2 + \phi_2{}^2)^2 \quad \text{or} \quad \frac{b}{2R}\left(5{\cdot}54 + \frac{15{\cdot}4}{N}\right)^{1/2}.$$

It will be seen later that the value of N is nearly proportional to the atomic weight, and is about 100 for gold. The effect due to scattering of the individual corpuscles expressed by the second term of the equation is consequently small for heavy atoms compared with that due to the distributed electric field.

Neglecting the second term, the average deflexion per atom is $\dfrac{3\pi b}{8R}$. We are now in a position to consider the relative effects on the distribution of particles due to single and to compound scattering. Following J. J. Thomson's argument, the average deflexion θ_t after passing through a thickness t of matter is proportional to the square root of the number of encounters and is given by

$$\theta_t = \frac{3\pi b}{8R} \sqrt{\pi R^2 . n . t} = \frac{3\pi b}{8} \sqrt{\pi n t},$$

where n as before is equal to the number of atoms per unit volume.

The probability p_1 for compound scattering that the deflexion of the particle is greater than ϕ is equal to $e^{-\phi^2/\theta_t^2}$.

Consequently
$$\phi^2 = -\frac{9\pi^3}{64} b^2 \, n t \log p_1.$$

Next suppose that single scattering alone is operative. We have seen (§ 3) that the probability p_2 of a deflexion greater than ϕ is given by

$$p_2 = \frac{\pi}{4} b^2 . n . t \cot^2 \phi/2.$$

By comparing these two equations
$$p_2 \log p_1 = -\cdot 181 \phi^2 \cot^2 \phi/2,$$
ϕ is sufficiently small that
$$\tan \phi/2 = \phi/2,$$
$$p_2 \log p_1 = -\cdot 72.$$
If we suppose $p_2 = \cdot 5$, then $p_1 = \cdot 24$.

If $\qquad p_2 = \cdot 1, \qquad p_1 = \cdot 0004.$

It is evident from this comparison, that the probability for any given deflexion is always greater for single than for compound scattering. The difference is especially marked when only a small fraction of the particles are scattered through any given angle. It follows from this result that the distribution of particles due to encounters with the atoms is for small thicknesses mainly governed by single scattering. No doubt compound scattering produces some effect in equalizing the distribution of the scattered particles; but its effect becomes relatively smaller, the smaller the fraction of the particles scattered through a given angle.

§ 6. *Comparison of Theory with Experiments.*

On the present theory, the value of the central charge Ne is an important constant, and it is desirable to determine its value for different atoms. This can be most simply done by determining the small fraction of α or β particles of known velocity falling on a thin metal screen, which are scattered between ϕ and $\phi + d\phi$ where ϕ is the angle of deflexion. The influence of compound scattering should be small when this fraction is small.

Experiments in these directions are in progress, but it is desirable at this stage to discuss in the light of the present theory the data already published on scattering of α and β particles.

The following points will be discussed :—

(a) The "diffuse reflexion" of α particles, *i. e.* the scattering of α particles through large angles (Geiger and Marsden).

(b) The variation of diffuse reflexion with atomic weight of the radiator (Geiger and Marsden).

(c) The average scattering of a pencil of α rays transmitted through a thin metal plate (Geiger).

(d) The experiments of Crowther on the scattering of β rays of different velocities by various metals.

(a) In the paper of Geiger and Marsden (*loc. cit.*) on the diffuse reflexion of α particles falling on various substances it was shown that about 1/8000 of the α particles from radium C falling on a thick plate of platinum are scattered back in the direction of the incidence. This fraction is deduced on the assumption that the α particles are uniformly scattered in all directions, the observations being made for a deflexion of about 90°. The form of experiment is not very suited for accurate calculation, but from the data available it can be shown that the scattering observed is about that to be expected on the theory if the atom of platinum has a central charge of about 100 e.

(b) In their experiments on this subject, Geiger and Marsden gave the relative number of α particles diffusely reflected from thick layers of different metals, under similar conditions. The numbers obtained by them are given in the table below, where z represents the relative number of scattered particles, measured by the number of scintillations per minute on a zinc sulphide screen.

Metal.	Atomic weight.	z.	$z/A^{3/2}$.
Lead	207	62	208
Gold	197	67	242
Platinum	195	63	232
Tin	119	34	226
Silver	108	27	241
Copper	64	14·5	225
Iron	56	10·2	250
Aluminium ...	27	3·4	243

Average 233

On the theory of single scattering, the fraction of the total number of α particles scattered through any given angle in passing through a thickness t is proportional to $n.A^2t$, assuming that the central charge is proportional to the atomic weight A. In the present case, the thickness of matter from which the scattered α particles are able to emerge and affect the zinc sulphide screen depends on the metal. Since Bragg has shown that the stopping power of an atom for an α particle is proportional to the square root of its atomic weight, the value of nt for different elements is proportional to $1/\sqrt{A}$. In this case t represents the greatest depth from which the scattered α particles emerge. The number z of α particles scattered back from a thick layer is consequently proportional to $A^{3/2}$ or $z/A^{3/2}$ should be a constant.

To compare this deduction with experiment, the relative values of the latter quotient are given in the last column. Considering the difficulty of the experiments, the agreement between theory and experiment is reasonably good [*].

The single large scattering of α particles will obviously affect to some extent the shape of the Bragg ionization curve for a pencil of α rays. This effect of large scattering should be marked when the α rays have traversed screens of metals of high atomic weight, but should be small for atoms of light atomic weight.

(c) Geiger made a careful determination of the scattering of α particles passing through thin metal foils, by the scintillation method, and deduced the most probable angle

[*] The effect of change of velocity in an atomic encounter is neglected in this calculation.

through which the α particles are deflected in passing through known thicknesses of different kinds of matter.

A narrow pencil of homogeneous α rays was used as a source. After passing through the scattering foil, the total number of α particles deflected through different angles was directly measured. The angle for which the number of scattered particles was a maximum was taken as the most probable angle. The variation of the most probable angle with thickness of matter was determined, but calculation from these data is somewhat complicated by the variation of velocity of the α particles in their passage through the scattering material. A consideration of the curve of distribution of the α particles given in the paper (*loc. cit.* p. 496) shows that the angle through which half the particles are scattered is about 20 per cent greater than the most probable angle.

We have already seen that compound scattering may become important when about half the particles are scattered through a given angle, and it is difficult to disentangle in such cases the relative effects due to the two kinds of scattering. An approximate estimate can be made in the following way : — From (§ 5) the relation between the probabilities p_1 and p_2 for compound and single scattering respectively is given by

$$p_2 \log p_1 = -\cdot 721.$$

The probability q of the combined effects may as a first approximation be taken as

$$q = (p_1{}^2 + p_2{}^2)^{1/2}.$$

If $q = \cdot 5$, it follows that

$$p_1 = \cdot 2 \quad \text{and} \quad p_2 = \cdot 46.$$

We have seen that the probability p_2 of a single deflexion greater than ϕ is given by

$$p_2 = \frac{\pi}{4} n . t . b^2 \cot^2 \phi/2.$$

Since in the experiments considered ϕ is comparatively small

$$\frac{\phi \sqrt{p_2}}{\sqrt{\pi n t}} = b = \frac{2 N e E}{m u^2}.$$

Geiger found that the most probable angle of scattering of the α rays in passing through a thickness of gold equivalent in stopping power to about ·76 cm. of air was 1° 40′. The angle ϕ through which half the α particles are turned thus corresponds to 2° nearly.

$$t = \cdot 00017 \text{ cm.} ; \quad n = 6 \cdot 07 \times 10^{22} ;$$

$$u \text{ (average value)} = 1 \cdot 8 \times 10^9.$$

$$E/m = 1 \cdot 5 \times 10^{14} . \text{ e.s. units} ; \quad e = 4 \cdot 65 \times 10^{-10}.$$

Taking the probability of single scattering $=\cdot 46$ and substituting the above values in the formula, the value of N for gold comes out to be 97.

For a thickness of gold equivalent in stopping power to $2 \cdot 12$ cms. of air, Geiger found the most probable angle to be $3° 40'$. In this case $t = \cdot 00047$, $\phi = 4° \cdot 4$, and average $u = 1 \cdot 7 \times 10^{9}$, and N comes out to be 114.

Geiger showed that the most probable angle of deflexion for an atom was nearly proportional to its atomic weight. It consequently follows that the value of N for different atoms should be nearly proportional to their atomic weights, at any rate for atomic weights between gold and aluminium.

Since the atomic weight of platinum is nearly equal to that of gold, it follows from these considerations that the magnitude of the diffuse reflexion of α particles through more than 90° from gold and the magnitude of the average small angle scattering of a pencil of rays in passing through gold-foil are both explained on the hypothesis of single scattering by supposing the atom of gold has a central charge of about $100\ e$.

(d) *Experiments of Crowther on scattering of β rays.*— We shall now consider how far the experimental results of Crowther on scattering of β particles of different velocities by various materials can be explained on the general theory of single scattering. On this theory, the fraction of β particles p turned through an angle greater than ϕ is given by

$$p = \frac{\pi}{4} n \cdot t \cdot b^{2} \cot^{2} \phi/2.$$

In most of Crowther's experiments ϕ is sufficiently small that $\tan \phi/2$ may be put equal to $\phi/2$ without much error. Consequently

$$\phi^{2} = 2\pi\, n \cdot t \cdot b^{2} \quad \text{if } p = 1/2.$$

On the theory of compound scattering, we have already seen that the chance p_1 that the deflexion of the particles is greater than ϕ is given by

$$\phi^{2}/\log p_{1} = -\frac{9\pi^{3}}{64} n \cdot t \cdot b^{2}.$$

Since in the experiments of Crowther the thickness t of matter was determined for which $p_1 = 1/2$,

$$\phi^{2} = \cdot 96\pi\, n\, t\, b^{2}.$$

For a probability of $1/2$, the theories of single and compound

2 Y 2

scattering are thus identical in general form, but differ by a numerical constant. It is thus clear that the main relations on the theory of compound scattering of Sir J. J. Thomson, which were verified experimentally by Crowther, hold equally well on the theory of single scattering.

For example, if t_m be the thickness for which half the particles are scattered through an angle ϕ, Crowther showed that $\phi/\sqrt{t_m}$ and also $\dfrac{mu^2}{E} \cdot \sqrt{t_m}$ were constants for a given material when ϕ was fixed. These relations hold also on the theory of single scattering. Notwithstanding this apparent similarity in form, the two theories are fundamentally different. In one case, the effects observed are due to cumulative effects of small deflexions, while in the other the large deflexions are supposed to result from a single encounter. The distribution of scattered particles is entirely different on the two theories when the probability of deflexion greater than ϕ is small.

We have already seen that the distribution of scattered α particles at various angles has been found by Geiger to be in substantial agreement with the theory of single scattering, but cannot be explained on the theory of compound scattering alone. Since there is every reason to believe that the laws of scattering of α and β particles are very similar, the law of distribution of scattered β particles should be the same as for α particles for small thicknesses of matter. Since the value of mu^2/E for the β particles is in most cases much smaller than the corresponding value for the α particles, the chance of large single deflexions for β particles in passing through a given thickness of matter is much greater than for α particles. Since on the theory of single scattering the fraction of the number of particles which are deflected through a given angle is proportional to kt, where t is the thickness supposed small and k a constant, the number of particles which are undeflected through this angle is proportional to $1 - kt$. From considerations based on the theory of compound scattering, Sir J. J. Thomson deduced that the probability of deflexion less than ϕ is proportional to $1 - e^{-\mu/t}$ where μ is a constant for any given value of ϕ.

The correctness of this latter formula was tested by Crowther by measuring electrically the fraction I/I_0 of the scattered β particles which passed through a circular opening subtending an angle of $36°$ with the scattering material. If

$$I/I_0 = 1 - e^{-\mu/t},$$

the value of I should decrease very slowly at first with

increase of t. Crowther, using aluminium as scattering material, states that the variation of I/I_0 was in good accord with this theory for small values of t. On the other hand, if single scattering be present, as it undoubtedly is for α rays, the curve showing the relation between I/I_0 and t should be nearly linear in the initial stages. The experiments of Madsen * on scattering of β rays, although not made with quite so small a thickness of aluminium as that used by Crowther, certainly support such a conclusion. Considering the importance of the point at issue, further experiments on this question are desirable.

From the table given by Crowther of the value $\phi/\sqrt{t_m}$ for different elements for β rays of velocity $2\cdot68\times10^{10}$ cms. per second, the values of the central charge Ne can be calculated on the theory of single scattering. It is supposed, as in the case of the α rays, that for the given value of $\phi/\sqrt{t_m}$ the fraction of the β particles deflected by single scattering through an angle greater than ϕ is ·46 instead of ·5.

The values of N calculated from Crowther's data are given below.

Element.	Atomic weight.	$\phi/\sqrt{t_m}.$	N.
Aluminium	27	4·25	22
Copper	63·2	10·0	42
Silver.........................	108	15·4	78
Platinum	194	29·0	138

It will be remembered that the values of N for gold deduced from scattering of the α rays were in two calculations 97 and 114. These numbers are somewhat smaller than the values given above for platinum (viz. 138), whose atomic weight is not very different from gold. Taking into account the uncertainties involved in the calculation from the experimental data, the agreement is sufficiently close to indicate that the same general laws of scattering hold for the α and β particles, notwithstanding the wide differences in the relative velocity and mass of these particles.

As in the case of the α rays, the value of N should be most simply determined for any given element by measuring

* Phil. Mag. xviii. p. 909 (1909).

the small fraction of the incident β particles scattered through a large angle. In this way, possible errors due to small scattering will be avoided.

The scattering data for the β, rays, as well as for the α rays, indicate that the central charge in an atom is approximately proportional to its atomic weight. This falls in with the experimental deductions of Schmidt *. In his theory of absorption of β rays, he supposed that in traversing a thin sheet of matter, a small fraction α of the particles are stopped, and a small fraction β are reflected or scattered back in the direction of incidence. From comparison of the absorption curves of different elements, he deduced that the value of the constant β for different elements is proportional to nA^2 where n is the number of atoms per unit volume and A the atomic weight of the element. This is exactly the relation to be expected on the theory of single scattering if the central charge on an atom is proportional to its atomic weight.

§ 7. *General Considerations.*

In comparing the theory outlined in this paper with the experimental results, it has been supposed that the atom consists of a central charge supposed concentrated at a point, and that the large single deflexions of the α and β particles are mainly due to their passage through the strong central field. The effect of the equal and opposite compensating charge supposed distributed uniformly throughout a sphere has been neglected. Some of the evidence in support of these assumptions will now be briefly considered. For concreteness, consider the passage of a high speed α particle through an atom having a positive central charge Ne, and surrounded by a compensating charge of N electrons. Remembering that the mass, momentum, and kinetic energy of the α particle are very large compared with the corresponding values for an electron in rapid motion, it does not seem possible from dynamic considerations that an α particle can be deflected through a large angle by a close approach to an electron, even if the latter be in rapid motion and constrained by strong electrical forces. It seems reasonable to suppose that the chance of single deflexions through a large angle due to this cause, if not zero, must be exceedingly small compared with that due to the central charge.

It is of interest to examine how far the experimental evidence throws light on the question of the extent of the

* *Annal. d. Phys.* iv. 23. p. 671 (1907).

distribution of the central charge. Suppose, for example, the central charge to be composed of N unit charges distributed over such a volume that the large single deflexions are mainly due to the constituent charges and not to the external field produced by the distribution. It has been shown (§ 3) that the fraction of the α particles scattered through a large angle is proportional to $(NeE)^2$, where Ne is the central charge concentrated at a point and E the charge on the deflected particle. If, however, this charge is distributed in single units, the fraction of the α particles scattered through a given angle is proportional to Ne^2 instead of N^2e^2. In this calculation, the influence of mass of the constituent particle has been neglected, and account has only been taken of its electric field. Since it has been shown that the value of the central point charge for gold must be about 100, the value of the distributed charge required to produce the same proportion of single deflexions through a large angle should be at least 10,000. Under these conditions the mass of the constituent particle would be small compared with that of the α particle, and the difficulty arises of the production of large single deflexions at all. In addition, with such a large distributed charge, the effect of compound scattering is relatively more important than that of single scattering. For example, the probable small angle of deflexion of a pencil of α particles passing through a thin gold foil would be much greater than that experimentally observed by Geiger (§ *b–c*). The large and small angle scattering could not then be explained by the assumption of a central charge of the same value. Considering the evidence as a whole, it seems simplest to suppose that the atom contains a central charge distributed through a very small volume, and that the large single deflexions are due to the central charge as a whole, and not to its constituents. At the same time, the experimental evidence is not precise enough to negative the possibility that a small fraction of the positive charge may be carried by satellites extending some distance from the centre. Evidence on this point could be obtained by examining whether the same central charge is required to explain the large single deflexions of α and β particles; for the α particle must approach much closer to the centre of the atom than the β particle of average speed to suffer the same large deflexion.

The general data available indicate that the value of this central charge for different atoms is approximately proportional to their atomic weights, at any rate for atoms heavier than aluminium. It will be of great interest to examine experimentally whether such a simple relation holds also for the lighter atoms. In cases where the mass of the deflecting atom (for example, hydrogen, helium, lithium) is not very different from that of the α particle, the general theory of single scattering will require modification, for it is necessary to take into account the movements of the atom itself (see § 4).

It is of interest to note that Nagaoka * has mathematically considered the properties of a "Saturnian" atom which he supposed to consist of a central attracting mass surrounded by rings of rotating electrons. He showed that such a system was stable if the attractive force was large. From the point of view considered in this paper, the chance of large deflexion would practically be unaltered, whether the atom is considered to be a disk or a sphere. It may be remarked that the approximate value found for the central charge of the atom of gold (100 e) is about that to be expected if the atom of gold consisted of 49 atoms of helium, each carrying a charge 2 e. This may be only a coincidence, but it is certainly suggestive in view of the expulsion of helium atoms carrying two unit charges from radioactive matter.

The deductions from the theory so far considered are independent of the sign of the central charge, and it has not so far been found possible to obtain definite evidence to determine whether it be positive or negative. It may be possible to settle the question of sign by consideration of the difference of the laws of absorption of the β particle to be expected on the two hypotheses, for the effect of radiation in reducing the velocity of the β particle should be far more marked with a positive than with a negative centre. If the central charge be positive, it is easily seen that a positively charged mass if released from the centre of a heavy atom, would acquire a great velocity in moving through the electric field. It may be possible in this way to account for the high velocity of expulsion of α particles without supposing that they are initially in rapid motion within the atom.

Further consideration of the application of this theory to these and other questions will be reserved for a later paper, when the main deductions of the theory have been tested experimentally. Experiments in this direction are already in progress by Geiger and Marsden.

University of Manchester,
    April 1911.

* Nagaoka, Phil. Mag. vii. p. 445 (1904).

[581]

LIV. *Collision of α Particles with Light Atoms.* IV. *An Anomalous Effect in Nitrogen.* *By* Professor Sir E. RUTHERFORD, *F.R.S.**

IT has been shown in paper I. that a metal source, coated with a deposit of radium C, always gives rise to a number of scintillations on a zinc sulphide screen far beyond the range of the α particles. The swift atoms causing these scintillations carry a positive charge and are deflected by a magnetic field, and have about the same range and energy as the swift H atoms produced by the passage of α particles through hydrogen. These " natural " scintillations are believed to be due mainly to swift H atoms from the radioactive source, but it is difficult to decide whether they are expelled from the radioactive source itself or are due to the action of α particles on occluded hydrogen.

The apparatus employed to study these " natural " scintillations is the same as that described in paper I. The intense source of radium C was placed inside a metal box about 3 cm. from the end, and an opening in the end of the box was covered with a silver plate of stopping power equal to about 6 cm. of air. The zinc sulphide screen was mounted outside, about 1 mm. distant from the silver plate, to admit of the introduction of absorbing foils between them. The whole apparatus was placed in a strong magnetic field to deflect the β rays. The variation in the number of these " natural " scintillations with absorption in terms of cms. of air is shown in fig. 1, curve A. In this case, the air in the box was exhausted and absorbing foils of aluminium were used. When dried oxygen or carbon dioxide was admitted into the vessel, the number of scintillations diminished to about the amount to be expected from the stopping power of the column of gas.

A surprising effect was noticed, however, when dried air was introduced. Instead of diminishing, the number of scintillations was increased, and for an absorption corresponding to about 19 cm. of air the number was about twice that observed when the air was exhausted. It was clear from this experiment that the α particles in their passage through air gave rise to long-range scintillations which appeared to the eye to be about equal in brightness to H scintillations. A systematic series of observations was undertaken to account for the origin of these scintillations. In the first place we have seen that the passage of α particles through nitrogen and

* Communicated by the Author.

oxygen gives rise to numerous bright scintillations which have
a range of about 9 cm. in air. These scintillations have about
the range to be expected if they are due to swift N or O atoms,
carrying unit charge, produced by collision with α particles.

Fig. 1.

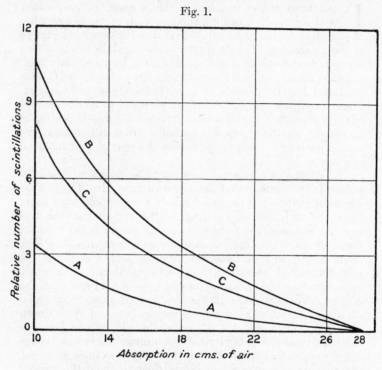

All experiments have consequently been made with an absorption greater than 9 cm. of air, so that these atoms are
completely stopped before reaching the zinc sulphide screen.

It was found that these long-range scintillations could not
be due to the presence of water vapour in the air ; for the
number was only slightly reduced by thoroughly drying
the air. This is to be expected, since on the average the
number of the additional scintillations due to air was equivalent to the number of H atoms produced by the mixture
of hydrogen at 6 cm. pressure with oxygen. Since on the
average the vapour pressure of water in air was not more
than 1 cm., the effects of complete drying would not reduce
the number by more than one sixth. Even when oxygen
and carbon dioxide saturated with water vapour at 20° C.

were introduced in place of dry air, the number of scintillations was much less than with dry air.

It is well known that the amount of hydrogen or gases containing hydrogen is normally very small in atmospheric air. No difference was observed whether the air was taken directly from the room or from outside the laboratory or was stored for some days over water.

There was the possibility that the effect in air might be due to liberation of H atoms from the dust nuclei in the air. No appreciable difference, however, was observed when the dried air was filtered through long plugs of cotton-wool, or by storage over water for some days to remove dust nuclei.

Since the anomalous effect was observed in air, but not in oxygen, or carbon dioxide, it must be due either to nitrogen or to one of the other gases present in atmospheric air. The latter possibility was excluded by comparing the effects produced in air and in chemically prepared nitrogen. The nitrogen was obtained by the well-known method of adding ammonium chloride to sodium nitrite, and stored over water. It was carefully dried before admission to the apparatus. With pure nitrogen, the number of long-range scintillations under similar conditions was greater than in air. As a result of careful experiments, the ratio was found to be 1·25, the value to be expected if the scintillations are due to nitrogen.

The results so far obtained show that the long-range scintillations obtained from air must be ascribed to nitrogen, but it is important, in addition, to show that they are due to collision of α particles with atoms of nitrogen through the volume of the gas. In the first place, it was found that the number of the scintillations varied with the pressure of the air in the way to be expected if they resulted from collision of α particles along the column of gas. In addition, when an absorbing screen of gold or aluminium was placed close to the source, the range of the scintillations was found to be reduced by the amount to be expected if the range of the expelled atom was proportional to the range of the colliding α particles. These results show that the scintillations arise from the volume of the gas and are not due to some surface effect in the radioactive source.

In fig. 1 curve A the results of a typical experiment are given showing the variation in the number of natural scintillations with the amount of absorbing matter in their path measured in terms of centimetres of air for α particles. In these experiments carbon dioxide was introduced at a pressure calculated to give the same absorption of the α rays as ordinary air. In curve B the corresponding curve is given when air at N.T.P. is introduced in place of carbon dioxide. The difference curve C shows the corresponding variation of the number of scintillations arising from the nitrogen in the air. It was generally observed that the ratio of the nitrogen effect to the natural effect was somewhat greater for 19 cm. than for 12 cm. absorption.

In order to estimate the magnitude of the effect, the space between the source and screen was filled with carbon dioxide at diminished pressure and a known pressure of hydrogen was added. The pressure of the carbon dioxide and of hydrogen were adjusted so that the total absorption of α particles in the mixed gas should be equal to that of the air. In this way it was found that the curve of absorption of H atoms produced under these conditions was somewhat steeper than curve C of fig. 1. As a consequence, the amount of hydrogen mixed with carbon dioxide required to produce a number of scintillations equal to that of air, increased with the increase of absorption. For example, the effect in air was equal to about 4 cm. of hydrogen at 12 cm. absorption, and about 8 cm. at 19 cm. absorption. For a mean value of the absorption, the effect was equal to about 6 cm. of hydrogen. This increased absorption of H atoms under similar conditions indicated either that (1) the swift atoms from air had a somewhat greater range than the H atoms, or (2) that the atoms from air were projected more in the line of flight of the α particles.

While the maximum range of the scintillations from air using radium C as a source of α rays appeared to be about the same, viz. 28 cm., as for H atoms produced from hydrogen, it was difficult to fix the end of the range with certainty on account of the smallness of the number and the weakness of the scintillations. Some special experiments were made to test whether, under favourable conditions, any scintillations due to nitrogen could be observed beyond 28 cm. of air absorption. For this purpose a strong source (about 60 mg. Ra activity) was brought within 2·5 cm. of the zinc sulphide screen, the space between containing dry air. On still further reducing the distance, the screen became too bright to detect very feeble scintillations. No certain evidence of scintillations was found beyond a range of 28 cm. It would therefore appear that (2) above is the more probable explanation.

In a previous paper (III.) we have seen that the number of swift atoms of nitrogen or oxygen produced per unit path by collision with α particles is about the same as the corresponding number of H atoms in hydrogen. Since the number of long-range scintillations in air is equivalent to that produced under similar conditions in a column of hydrogen at 6 cm.

pressure, we may consequently conclude that only one long-range atom is produced for every 12 close collisions giving rise to a swift nitrogen atom of maximum range 9 cm

It is of interest to give data showing the number of long-range scintillations produced in nitrogen at atmospheric pressure under definite conditions. For a column of nitrogen 3·3 cm. long, and for a total absorption of 19 cm. of air from the source, the number due to nitrogen per milligram of activity is ·6 per minute on a screen of 3·14 sq. mm. area.

Both as regards range and brightness of scintillations, the long-range atoms from nitrogen closely resemble H atoms, and in all probability are hydrogen atoms. In order, however, to settle this important point definitely, it is necessary to determine the deflexion of these atoms in a magnetic field. Some preliminary experiments have been made by a method similar to that employed in measuring the velocity of the H atom (see paper II.). The main difficulty is to obtain a sufficiently large deflexion of the stream of atoms and yet have a sufficient number of scintillations per minute for counting. The α rays from a strong source passed through dry air between two parallel horizontal plates 3 cm. long and 1·6 mm. apart, and the number of scintillations on the screen placed near the end of the plates was observed for different strengths of the magnetic field. Under these conditions, when the scintillations arise from the whole length of the column of air between the plates, the strongest magnetic field available reduced the number of scintillations by only 30 per cent. When the air was replaced by a mixture of carbon dioxide and hydrogen of the same stopping power for α rays, about an equal reduction was noted. As far as the experiment goes, this is an indication that the scintillations are due to H atoms ; but the actual number of scintillations and the amount of reduction was too small to place much reliance on the result. In order to settle this question definitely, it will probably prove necessary to employ a solid nitrogen compound, free from hydrogen, as a source, and to use much stronger sources of α rays. In such experiments, it will be of importance to discriminate between the deflexions due to H atoms and possible atoms of atomic weight 2. From the calculations given in paper III., it is seen that a collision of an α particle with a free atom of mass 2 should give rise to an atom of range about 32 cm. in air, and of initial energy about ·89 of that of the H atom produced under similar conditions. The deflexion of the pencil of these rays in a magnetic field should be about ·6 of that shown by a corresponding pencil of H atoms.

Discussion of results.

From the results so far obtained it is difficult to avoid the conclusion that the long-range atoms arising from collision of α particles with nitrogen are not nitrogen atoms but probably atoms of hydrogen, or atoms of mass 2.   If this be the case, we must conclude that the nitrogen atom is disintegrated under the intense forces developed in a close collision with a swift α particle, and that the hydrogen atom which is liberated formed a constituent part of the nitrogen nucleus.   We have drawn attention in paper III. to the rather surprising observation that the range of the nitrogen atoms in air is about the same as the oxygen atoms, although we should expect a difference of about 19 per cent.   If in collisions which give rise to swift nitrogen atoms, the hydrogen is at the same time disrupted, such a difference might be accounted for, for the energy is then shared between two systems.

It is of interest to note, that while the majority of the light atoms, as is well known, have atomic weights represented by $4n$ or $4n+3$ where n is a whole number, nitrogen is the only atom which is expressed by $4n+2$.   We should anticipate from radioactive data that the nitrogen nucleus consists of three helium nuclei each of atomic mass 4 and either two hydrogen nuclei or one of mass 2.   If the H nuclei were outriders of the main system of mass 12, the number of close collisions with the bound H nuclei would be less than if the latter were free, for the α particle in a collision comes under the combined field of the H nucleus and of the central mass.   Under such conditions, it is to be expected that the α particle would only occasionally approach close enough to the H nucleus to give it the maximum velocity, although in many cases it may give it sufficient energy to break its bond with the central mass.   Such a point of view would explain why the number of swift H atoms from nitrogen is less than the corresponding number in free hydrogen and less also than the number of swift nitrogen atoms.   The general results indicate that the H nuclei, which are released, are distant about twice the diameter of the electron (7×10^{-13} cm.) from the centre of the main atom.   Without a knowledge of the laws of force at such small distances, it is difficult to estimate the energy required to free the H nucleus or to calculate the maximum velocity that can be given to the escaping H atom. It is not to be expected, *a priori*, that the velocity or range of the H atom released from the nitrogen atom should be identical with that due to a collision in free hydrogen.

Taking into account the great energy of motion of the α particle expelled from radium C, the close collision of such an α particle with a light atom seems to be the most likely agency to promote the disruption of the latter ; for the forces on the nuclei arising from such collisions appear to be greater than can be produced by any other agency at present available. Considering the enormous intensity of the forces brought into play, it is not so much a matter of surprise that the nitrogen atom should suffer disintegration as that the α particle itself escapes disruption into its constituents. The results as a whole suggest that, if α particles—or similar projectiles—of still greater energy were available for experiment, we might expect to break down the nucleus structure of many of the lighter atoms.

I desire to express my thanks to Mr. William Kay for his invaluable assistance in counting scintillations.

University of Manchester,
    April 1919.

On the Interaction of Elementary Particles. I.

By Hideki YUKAWA.

(Read Nov. 17, 1934)

§ 1. Introduction

At the present stage of the quantum theory little is known about the nature of interaction of elementary particles. Heisenberg considered the interaction of " Platzwechsel " between the neutron and the proton to be of importance to the nuclear structure.[1]

Recently Fermi treated the problem of β-disintegration on the hypothesis of " neutrino ".[2] According to this theory, the neutron and the proton can interact by emitting and absorbing a pair of neutrino and electron. Unfortunately the interaction energy calculated on such assumption is much too small to account for the binding energies of neutrons and protons in the nucleus.[3]

To remove this defect, it seems natural to modify the theory of Heisenberg and Fermi in the following way. The transition of a heavy particle from neutron state to proton state is not always accompanied by the emission of light particles, i. e., a neutrino and an electron, but the energy liberated by the transition is taken up sometimes by another heavy particle, which in turn will be transformed from proton state into neutron state. If the probability of occurrence of the latter process is much larger than that of the former, the interaction between the neutron and the proton will be much larger than in the case of Fermi, whereas the probability of emission of light particles is not affected essentially.

Now such interaction between the elementary particles can be described by means of a field of force, just as the interaction between the charged particles is described by the electromagnetic field. The above considerations show that the interaction of heavy particles with this field is much larger than that of light particles with it.

---

(1) W. Heisenberg, Zeit f. Phys. **77,** 1 (1932); **78,** 156 (1932); **80,** 587 (1933). We shall denote the first of them by I.

(2) E. Fermi, ibid. **88,** 161 (1394).

(3) Ig. Tamm, Nature **133,** 981 (1934); D. Iwanenko, ibid. 981 (1934).

On the Interaction of Elementary Particles. I.

In the quantum theory this field should be accompanied by a new sort of quantum, just as the electromagnetic field is accompanied by the photon.

In this paper the possible natures of this field and the quantum accompanying it will be discussed briefly and also their bearing on the nuclear structure will be considered.

Besides such an exchange force and the oridinary electric and magnetic forces there may be other forces between the elementary particles, but we disregard the latter for the moment.

Fuller account will be made in the next paper.

§2. Field describing the interaction

In analogy with the scalar potential of the electromagnetic field, a function $U(x, y, z, t)$ is introducd to describe the field between the neutron and the proton. This function will satisfy an equation similar to the wave equation for the electromagnetic potential.

Now the eqnation

$$\left\{\Delta - \frac{1}{c^2}\frac{\partial^2}{\partial t^2}\right\} U = 0 \tag{1}$$

has only static solution with central symmetry $\frac{1}{r}$, except the additive and the multiplicative constants. The potential of force between the neutron and the proton should, however, not be of Coulomb type, but decrease more rapidly with distance. It can be expressed, for example, by

$$+ \text{ or } - g^2 \frac{e^{-\lambda r}}{r}, \tag{2}$$

where g is a constant with the dimension of electric charge, i. e., cm.$^{\frac{3}{2}}$ sec.$^{-1}$ gr.$^{\frac{1}{2}}$ and λ with the dimention cm.$^{-1}$

Since this function is a static solution with central symmetry of the wave equation

$$\left\{\Delta - \frac{1}{c^2}\frac{\partial^2}{\partial t^2} - \lambda^2\right\} U = 0, \tag{3}$$

let this equation be assumed to be the correct equation for U in vacuum. In the presence of the heavy particles, the U-field interacts with them and causes the transition from neutron state to proton state.

Now, if we introduce the matrices[4]

$$\tau_1 = \begin{pmatrix} 0 & 1 \\ 1 & 0 \end{pmatrix}, \quad \tau_2 = \begin{pmatrix} 0 & -i \\ i & 0 \end{pmatrix}, \quad \tau_3 = \begin{pmatrix} 1 & 0 \\ 0 & -1 \end{pmatrix}$$

and denote the neutron state and the proton state by $\tau_3 = 1$ and $\tau_3 = -1$ respectively, the wave equation is given by

$$\left\{ \Delta - \frac{1}{c^2} \frac{\partial^2}{\partial t^2} - \lambda^2 \right\} U = -4\pi g \, \tilde{\Psi} \frac{\tau_1 - i\tau_2}{2} \Psi, \qquad (4)$$

where Ψ denotes the wave function of the heavy particles, being a function of time, position, spin as well as τ_3', which takes the value either 1 or -1.

Next, the conjugate complex function $\tilde{U}(x, y, z, t)$, satisfying the equation

$$\left\{ \Delta - \frac{1}{c^2} \frac{\partial^2}{\partial t^2} - \lambda^2 \right\} \tilde{U} = -4\pi g \tilde{\Psi} \frac{\tau_1 + i\tau_2}{2} \Psi, \qquad (5)$$

is introduced, corresponding to the inverse transition from proton to neutron state.

Similar equation will hold for the vector function, which is the analogue of the vector potential of the electromagnetic field. However, we disregard it for the moment, as there's no correct relativistic theory for the heavy particles. Hence simple non-relativistic wave equation neglecting spin will be used for the heavy particle, in the following way

$$\left\{ \frac{h^2}{4} \left(\frac{1+\tau_3}{M_N} + \frac{1-\tau_3}{M_P} \right) \Delta + ih \frac{\partial}{\partial t} - \frac{1+\tau_3}{2} M_N c^2 - \frac{1-\tau_3}{2} M_P c^2 \right.$$
$$\left. -g \left(\tilde{U} \frac{\tau_1 - i\tau_2}{2} + U \frac{\tau_1 + i\tau_2}{2} \right) \right\} \Psi = 0, \qquad (6)$$

where h is Planck's constant divided by 2π and M_N, M_P are the masses of the neutron and the proton respectively. The reason for taking the negative sign in front of g will be mentioned later.

The equation (6) corresponds to the Hamiltonian

$$H = \left(\frac{1+\tau_3}{4M_N} + \frac{1-\tau_3}{4M_P} \right) p^2 + \frac{1+\tau_3}{2} M_N c^\circ + \frac{1-\tau_3}{2} M_P c^2$$
$$+ g \left(\tilde{U} \frac{\tau_1 - i\tau_2}{2} + U \frac{\tau_1 + i\tau_2}{2} \right) \qquad (7)$$

---

(4) Heisenberg, loc, cit. I.

where p is the momentum of the particle.   If we put $M_{NC}^2 - M_{PC}^2 = D$ and $M_N + M_P = 2M$, the equation (7) becomes approximately

$$H = \frac{p^2}{2M} + \frac{g}{2} \{ \tilde{U}(\tau_1 - i\tau_2) + U(\tau_1 + i\tau_2) \} + \frac{D}{2}\tau_3, \qquad (8)$$

where the constant term Mc^2 is omitted.

Now consider two heavy particles at points (x_1, y_1, z_1) and (x_2, y_2, z_2) respectively and assume their relative velocity to be small.   The fields at (x_1, y_1, z_1) due to the particle at $(x_2 y_2, z_2)$ are, from (4) and (5),

$$U(x_1, y_1, z_1) = g \frac{e^{-\lambda r_{12}}}{r_{12}} \frac{(\tau_1^{(2)} - i\tau_2^{(2)})}{2}$$

and
$$\tilde{U}(x, y_1, z_1) = g \frac{e^{-\lambda r_{12}}}{r_{12}} \frac{(\tau_1^{(2)} + i\tau_2^{(2)})}{2}, \qquad (9)$$

where $(\tau_1^{(1)}, \tau_2^{(1)}, \tau_3^{(1)})$ and $(\tau_1^{(2)}, \tau_2^{(2)}, \tau_3^{(2)})$ are the matrices relating to the first and the second particles respectively, and r_{12} is the distance between them.

Hence the Hamiltonian for the system is given, in the absence of the external fields, by

$$H = \frac{p_1^2}{2M} + \frac{p_2^2}{2M} + \frac{g^2}{4} \{ (\tau_1^{(1)} - i\tau_2^{(1)})(\tau_1^{(2)} + i\tau_2^{(2)}).$$

$$+ (\tau_1^{(1)} + i\tau_2^{(1)})(\tau_1^{(2)} - i\tau_2^{(2)}) \} \frac{e^{-\lambda r_{12}}}{r_{12}} + (\tau_3^{(1)} + \tau_3^{(2)})D$$

$$= \frac{p_1^2}{2M} + \frac{p_2^2}{2M} + \frac{g^2}{2}(\tau_1^{(1)}\tau_1^{(2)} + \tau_2^{(1)}\tau_2^{(2)}) \frac{e^{-\lambda r_{12}}}{r_{12}} + (\tau_3^{(1)} + \tau_3^{(2)})D, \quad (10)$$

where p_1, p_2 are the momenta of the particles.

This Hamiltonian is equivalent to Heisenberg's Hamiltonian (1),[5] if we take for " Platzwechselintegral "

$$J(r) = -g^2 \frac{e^{-\lambda r}}{r}, \qquad (11)$$

except that the interaction between the neutrons and the electrostatic repulsion between the protons are not taken into account.   Heisenberg took the positive sign for $J(r)$, so that the spin of the lowest energy state of H^2 was O, whereas in our case, owing to the negative sign in front of g^2, the lowest energy state has the spin 1, which is required

---

(5) Heisenberg, I.

from the experiment.

Two constants g and λ appearing in the above equations should be determined by comparison with experiment. For example, using the Hamiltonian (10) for heavy particles, we can calculate the mass defect of H^2 and the probability of scattering of a neutron by a proton provided that the relative velocity is small compared with the light velocity.[6]

Rough estimation shows that the calculated values agree with the experimental results, if we take for λ the value between $10^{12} \mathrm{cm}^{-1}$. and $10^{13} \mathrm{cm}^{-1}$. and for g a few times of the elementary charge e, although no direct relation between g and e was suggested in the above considerations.

§ 3. Nature of the quanta accompanying the field

The U-field above considered should be quantized according to the general method of the quantum theory. Since the neutron and the proton both obey Fermi's statistics, the quanta accompanying the U-field should obey Bose's statistics and the quantization can be carried out on the line similar to that of the electromagnetic field.

The law of conservation of the electric charge demands that the quantum should have the charge either $+e$ or $-e$. The field quantity U corresponds to the operator which increases the number of negatively charged quanta and decreases the number of positively charged quanta by one respectively. $\tilde{U}$, which is the complex conjugate of U, corresponds to the inverse operator.

Next, denoting

$$p_x = -ih\frac{\partial}{\partial x}, \quad \text{etc.,} \quad W = ih\frac{\partial}{\partial t},$$

$$m_U c = \lambda h,$$

the wave equation for U in free space can be written in the form

$$\left\{ p_x^2 + p_y^2 + p_z^2 - \frac{W^2}{c^2} + m_U c^2 \right\} U = 0, \tag{12}$$

so that the quantum accompanying the field has the proper mass $m_U = \dfrac{\lambda h}{c}$.

---

(6) These calculations were made previously, according to the theory of Heisenberg, by Mr. Tomonaga, to whom the writer owes much. A little modification is necessary in our case. Detailed accounts will be made in the next paper.

144

1935] *On the Interaction of Elementary Particles. I.* 53

Assuming $\lambda = 5 \times 10^{12} \mathrm{cm}^{-1}$., we obtain for m_U a value 2×10^2 times as large as the electron mass. As such a quantum with large mass and positive or negative charge has never been found by the experiment, the above theory seems to be on a wrong line. We can show, however, that, in the ordinary nuclear transformation, such a qnantum can not be emitted into outer space.

Let us consider, for example, the transition from a neutron state of energy W_N to a proton state of energy W_P, both of which include the proper energies. These states can be expressed by the wave functions

$$\Psi_N(x, y, z, t, 1) = u(x, y, z)e^{-iW_Nt/h}, \quad \Psi_N(x, y, z, t, -1) = 0$$

and

$$\Psi_P(x, y, z, t, 1) = 0, \quad \Psi_P(x, y, z, t, -1) = v(x, y, z)e^{-iW_Pt/h},$$

so that, on the right hand side of the equation (4), the term

$$-4\pi g\tilde{v}ue^{-it(W_N - W_P)/h}$$

appears.

Putting $U = U'(x, y, z)e^{i\omega t}$, we have from (4)

$$\left\{ \Delta - \left(\lambda^2 - \frac{\omega^2}{c^2} \right) \right\} U' = -4\pi g\tilde{v}u, \tag{13}$$

where $\omega = \dfrac{W_N - W_P}{h}$. Integrating this, we obtain a solution

$$U'(r) = g \iiint \frac{e^{-\mu|r-r'|}}{|r-r'|} \tilde{v}(r')u(r')dv', \tag{14}$$

where $\mu = \sqrt{\lambda^2 - \dfrac{\omega^2}{c^2}}$.

If $\lambda > \dfrac{|\omega|}{c}$ or $m_Uc^2 > |W_N - W_P|$, μ is real and the function $J(r)$ of Heisenberg has the form $-g^2\dfrac{e^{-\mu r}}{r}$, in which μ, however, depends on $|W_N - W_P|$, becoming smaller and smaller as the latter approaches m_Uc^2. This means that the range of interaction between a neutron and a proton increases as $|W_N - W_P|$ increases.

Now the scattering (elastic or inelastic) of a neutron by a nucleus can be considered as the result of the following double process: the neutron falls into a proton level in the nucleus and a proton in the latter jumps to a neutron state of positive kinetic energy, the total energy being conserved throughout the process. The above argument, then, shows that the probability of scattering may in some case increase with the velocity of the neutron.

According to the experiment of Bonner[7], the collision cross section of the neutron increases, in fact, with the velocity in the case of lead whereas it decreases in the case of carbon and hydrogen, the rate of decrease being slower in the former than in the latter. The origih of this effect is not clear, but the above considerations do not, at least, contradict it. For, if the binding energy of the proton in the nucleus becomes comparable with $m_U c^2$, the range of interaction of the neutron with the former will increase considerably with the velocity of the neutron, so that the cross section will decrease slower in such case than in the case of hydrogen, i. e., free proton. Now the binding energy of the proton in C^{12}, which is estimated from the difference of masses of C^{12} and B^{11}, is

$$12,0036 - 11,0110 = 0,9926.$$

This corresponds to a binding energy 0,0152 in mass unit, being thirty times the electron mass. Thus in the case of carbon we can expect the effect observed by Bonner. The arguments are only tentative, other explanations being, of course, not excluded.

Next if $\lambda < \dfrac{|\omega|}{c}$ or $m_U c^2 < |W_N - W_P|$, μ becomes pure imaginary and U expresses a spherial undamped wave, implying that a quantum with energy greater than $m_U c^2$ can be emitted in outer space by the transition of the heavy particle from neutron state to proton state, provided that $|W_N - W_P| > m_U c^2$.

The velocity of U-wave is greater but the group velocity is smaller than the light velocity c, as in the case of the electron wave.

The reason why such massive quanta, if they ever exist, are not yet discovered may be ascribed to the fact that the mass m_U is so large that condition $|W_N - W_P| > m_U c^2$ is not fulfilled in ordinary nuclear transformation.

§4. Theory of β-disintegration

Hitherto we have considered only the interaction of U-quanta with heavy particles. Now, according to our theory, the quantum emitted when a heavy particle jumps from a neutron state to a proton state, can be absorbed by a light particle which will then in consequence of energy absorption rise from a neutrino state of negative energy to an

---

(7) T. W. Bonner, Phys. Rev. **45**, 606 (1934).

146

1935] *On the Interaction of Elementary Particles. I.* 55

electron state of positive energy. Thus an anti-neutrino and an electron are emitted simultaneously from the nucleus. Such intervention of a massive quantum does not alter essentially the probability of β-disintegration, which has been calculated on the hypothesis of direct coupling of a heavy particle and a light particle, just as, in the theory of internal conversion of γ-ray, the intervation of the proton does not affect the final result.[8] Our theory, therefore, does not differ essentially from Fermi's thory.

Fermi considered that an electron and a neutrino are emitted simultaneously from the radioactive nucleus, but this is formally equivalent to the assumption that a light particle jumps from a neutrino state of negative energy to an electron state of positive energy.

For, if the eigenfunctions of the electron and the neutrino be ψ_k, φ_k respectively, where $k = 1, 2, 3, 4$, a term of the form

$$-4\pi g' \sum_{k=1}^{4} \tilde{\psi}_k \varphi_k \tag{15}$$

should be added to the right hand side of the equation (5) for $\tilde{U}$, where g' is a new constant with the same dimension as g.

Now the eigenfunctions of the neutrino state with energy and momentum just opposite to those of the state φ_k is given by $\varphi_k' = -\delta_{kl}\tilde{\varphi}_l$ and conversely $\varphi_k = \delta_{kl}\tilde{\varphi}_l'$, where

$$\delta = \begin{pmatrix} 0 & -1 & 0 & 0 \\ 1 & 0 & 0 & 0 \\ 0 & 0 & 0 & 1 \\ 0 & 0 & -1 & 0 \end{pmatrix},$$

so that (15) becomes

$$-4\pi g' \sum_{k,l=1}^{4} \tilde{\psi}_k \delta_{kl} \tilde{\varphi}_l'. \tag{16}$$

From equations (13) and (15), we obtain for the matrix element of the interaction energy of the heavy particle and the light particle an expression

$$gg' \int \cdots \int \tilde{v}(\boldsymbol{r}_1) u(\boldsymbol{r}_1) \sum_{k=1}^{4} \tilde{\psi}_k(\boldsymbol{r}_2) \varphi_k(\boldsymbol{r}_2) \frac{e^{-\lambda r_{12}}}{r_{12}} dv_1 dv_2, \tag{17}$$

corresponding to the following double process : a heavy particle falls

---

(8) H. A. Taylor and N. F Mott, Proc. Roy. Soc. A, **138**, 665 (1932).

from the neutron state with the eigenfunction $u(\boldsymbol{r})$ into the proton state with the eigenfunction $v(\boldsymbol{r})$ and simultaneously a light particle jumps from the neutrino state $\varphi_k(\boldsymbol{r})$ of negative energy to the electron state $\psi_k(\boldsymbol{r})$ of positive energy. In (17) λ is taken instead of μ, since the difference of energies of the neutron state and the proton state, which is equal to the sum of the upper limit of the energy spectrum of β-rays and the proper energies of the electron and the neutrino, is always small compared with $m_U c^2$.

As λ is much larger than the wave numbers of the electron state and the neutrino state, the function $\dfrac{e^{-\lambda r_{12}}}{r_{12}}$ can be regarded approximately as a δ-function multiplied by $\dfrac{4\pi}{\lambda^2}$ for the integrations with respect to x_2, y_2, z_2. The factor $\dfrac{4\pi}{\lambda^2}$ comes from

$$\iiint \frac{e^{-\lambda r_{12}}}{r_{12}} dv_2 = \frac{4\pi}{\lambda^2}.$$

Hence (17) becomes

$$\frac{4\pi gg'}{\lambda^2} \iiint \hat{v}(\boldsymbol{r}) u(\boldsymbol{r}) \sum_k \hat{\psi}_k(\boldsymbol{r}) \varphi_k(\boldsymbol{r}) dv \tag{18}$$

or by (16)

$$\frac{4\pi gg'}{\lambda^2} \iiint \check{v}(\boldsymbol{r}) u(\boldsymbol{r}) \sum_{k,\,l} \hat{\psi}(\boldsymbol{r}) \delta_{kl}' \check{\varphi}_l'(\boldsymbol{r}) dv, \tag{19}$$

which is the same as the expression (21) of Fermi, corresponding to the emission of a neutrino and an electron of positive energy states $\varphi_k'(\boldsymbol{r})$ and $\psi_k(\boldsymbol{r})$, except that the factor $\dfrac{4\pi gg'}{\lambda^2}$ is substituted for Fermi's g.

Thus the result is the same as that of Fermi's theory, in this approximation, if we take

$$\frac{4\pi gg'}{\lambda^2} = 4 \times 10^{-50} \text{cm}^3.\ \text{erg},$$

from which the constant g' can be determined. Taking, for example, $\lambda = 5 \times 10^{12}$ and $g = 2 \times 10^{-9}$, we obtain $g' \cong 4 \times 10^{-17}$, which is about 10^{-8} times as small as g.

This means that the interaction between the neutrino and the electron is much smaller than that between the neutron and the proton so that the neutrino will be far more penetrating than the neutron and consequently more difficult to observe. The difference of g and g' may be due to the difference of masses of heavy and light particles.

§5. Summary

The interaction of elementary particles are described by considering a hypothetical quantum which has the elementary charge and the proper mass and which obeys Bose's statistics. The interaction of such a quantum with the heavy particle should be far greater than that with the light particle in order to account for the large interaction of the neutron and the proton as well as the small probability of β-disintegration.

Such quanta, if they ever exist and approach the matter close enough to be absorbed, will deliver their charge and energy to the latter. If, then, the quanta with negative charge come out in excess, the matter will be charged to a negative potential.

These arguments, of course, of merely speculative character, agree with the view that the high speed positive particles in the cosmic rays are generated by the electrostatic field of the earth, which is charged to a negative potential.[9]

The massive quanta may also have some bearing on the shower produced by cosmic rays.

In conclusion the writer wishes to express his cordial thanks to Dr. Y. Nishina and Prof. S. Kikuchi for the encouragement throughout the course of the work.

Department of Physics,
Osaka Imperial University.

(Received Nov. 30, 1934)

(9) G. H. Huxley, Nature **134**, 418, 571 (1934) ; Johnson, Phys. Rev. **45**, 569 (1934).

BIBLIOGRAPHY

The List of Journals and their abbreviations used in this
BIBLIOGRAPHY *may be found at the end of this Section.*

CHAPTER ONE

Isotopes and Mass Measurements

Aldrich and Nier, *The abundance of He^3 in atmospheric and well helium,* Phys. Rev. *70,* 983 (1946).

Aldrich and Nier, *The abundance of He^3 in helium.* Phys. Rev. *71,* 143 (1947).

Allison, Graves, Skaggs and Smith, *A precise measurement of the mass difference $_4Be^9 - _4Be^8$. The stability of $_4Be^8$.* Phys. Rev. *55,* 107 (1939).

Allison, *The masses of Li^6, Li^7, Be^8, Be^9, Be^{10} and B^{11}.* Phys. Rev. *55,* 624 (1939).

Almy and Irwin, *Mass ratio of the lithium isotopes from the spectrum of Li_2.* Phys. Rev. *49,* 72 (1936).

Alvarez and Cornog, *He^3 in helium.* Phys. Rev. *56,* 379 (1939).

Alvarez and Cornog, *Helium and hydrogen of mass three.* Phys. Rev. *56,* 613 (1939).

Asada, Okuda, Ogata and Yoshimoto, *Preliminary report on the masses of ^{12}C and ^{14}N.* Nature. *143,* 797 (1939).

Asada, Okuda, Ogata and Yoshimoto, *Isotopic weights of C and N by doublet method.* Proc. Phys.-Math. Soc. Jap. (3) *22,* 41 (1940).

Aston, *Isotopic constitution and atomic weights of zinc, tin, chromium and molybdenum.* Proc. Roy. Soc. *130,* 302 (1931).

Aston, *Isotopic constitution and atomic weights of Se, Br, B, W, Sb, Os, Ru, Te, Ge, Re and Cl.* Proc. Roy. Soc. *130,* 302 (1931) and *132,* 487 (1931).

Aston, *Isotopes of the rare earth elements,* Proc. Roy. Soc. *146,* 46 (1934).

Aston, *Masses of some light atoms determined by a new method.* Nature *135,* 541 (1935).

Aston, *Isotopic constitution and atomic weights of Hf, Th, Rh, Ti, Zr, Ca, Ga, Ag, C, Ni, Cd, Fe and In.* Proc. Roy. Soc. *149,* 396 (1935).

Aston, *Masses of some light atoms measured by means of a new mass spectrograph.* Nature *137,* 357 (1936).

Aston, *New data on isotopes.* Nature *137,* 613 (1936).

Aston, *Isotopic weights by the doublet method.* Nature *138,* 1094 (1936).

Aston, *Isotopic weight of ^{12}C.* Nature *139,* 922 (1937).

Aston, *Packing fractions of bromine, chromium, nickel and titanium.* Nature *141,* 1096 (1938).

Bacher and Sawyer, *Isotope shift in Mg I.* Phys. Rev. *47,* 587 (1935).

Bainbridge, *The constitution of tellurium.* Phys. Rev. *39,* 1021 (1932).

Bainbridge, *Isotopic weight of H^2.* Phys. Rev. *42,* 1 (1932).

Bainbridge, *Comparison of the masses of He and H^1 on a mass spectrograph.* Phys. Rev. *43,* 103 (1933).

Bainbridge, *Measurement of the masses of He and H^1, Ne^{20}, Ne^{22}, B^{11}, Cl^{35} and Cl^{37} with a mass spectrograph.* Phys. Rev. *43,* 378 (1933).

Bainbridge, *The masses of Ne^{20} and B^{11}. The mass of Ne^{22} and the disintegration of F^{19}.* Phys. Rev. *43,* 424 (1933).

Bainbridge, *Atomic masses and the structure of atomic nuclei.* J. Franklin Inst. *215,* 509 (1933).

Bainbridge, *The isotopic constitution of krypton, mercury, selenium, cadmium and germanium.* Phys. Rev. *43,* 1056 (1933).

Bainbridge, *The masses of the lithium isotopes.* Phys. Rev. *44,* 56 (1933).

Bainbridge, *Comparison of the masses of H^2 and He.* Phys. Rev. *44,* 57 (1933).

Bainbridge and Jordan, *The mass spectrographic determination of mass changes in some carbon transmutations.* Phys. Rev. *49,* 883 (1936).

Bainbridge and Jordan, *Atomic masses of hydrogen, helium, carbon and nitrogen isotopes.* Phys. Rev. *51,* 384 (1937).

Bainbridge, *The isotopic weight of helium by direct comparison with oxygen.* Phys. Rev. *53,* 922 (1938).

Barkas, Creutz, Delsasso, Fox and White, *Carbon isotopes of mass 10 and 11.* Phys. Rev. *57,* 562 (1940).

Bartlett and Gibbons, *Isotope shift in neon.* Phys. Rev. *44,* 538 (1933).

Beck, *Remarks on the systematics of isotopes.* Phys. Rev. *48,* 47 (1935).

Bethe, *Masses of light atoms from transmutation data.* Phys. Rev. 47, 633 (1935).

Birge, *Further evidence of the carbon isotope, mass 13.* Phys. Rev. *34,* 379 (1929).

Birge, *The relative abundance of the oxygen isotopes and the bases of the atomic weight system.* Phys. Rev. *37,* 1669 (1931).

Bleakney, *Relative abundance of neon isotopes.* Phys. Rev. *43,* 1056 (1933).

Bleakney and Gould, *The relative abundance of hydrogen isotopes.* Phys. Rev. *44,* 265 (1933).

Bleakney, *Some properties of the hydrogen isotopes as revealed by the mass spectrograph.* Phys. Rev. *45,* 762 (1934).

Bleakney, Harnwell, Lozier, Smith and Smyth, *Production and identification of helium of mass three.* Phys. Rev. *46,* 81 (1934).

Bleuler and Zünti, *Isotopes S³⁷ and P³⁴ produced by irradiation of chlorine with rapid neutrons* (in German). Helv. Phys. Acta. *19,* 137 (1946).

Blewett and Sampson, *Isotopic constitution of strontium, barium and indium.* Phys. Rev. *49,* 778 (1936).

Blewett, *Mass spectrograph analysis of bromine.* Phys. Rev. *49,* 900 (1936).

Bothe and Maier-Leibnitz, *Relations between the masses of the light atoms.* Naturwiss. *25,* 25 (1937).

Bothe and Gentner, *Production of a new isotope from the nuclear photo effect.* Naturwiss. *25,* 126 (1937).

Brewer, *The abundance ratio of the isotopes of lithium.* Phys. Rev. *47,* 571 (1935).

Brewer, *Abundance ratio of potassium isotopes in mineral and plant sources.* J. Amer. Chem. Soc. *58,* 365 (1936).

Brosi and Harkins, *The abundance ratio of the isotopes in natural or isotopically separated carbon.* Phys. Rev. *52,* 472 (1937).

Cohen, *The isotopes of cerium and rhodium.* Phys. Rev. *63,* 219 (1943).

Corson, MacKenzie and Segrè, *Some chemical properties of element 85.* Phys. Rev. *57,* 1087 (1940).

Corson, MacKenzie and Segrè, *Astatine, the element of atomic number 85.* Nature *159,* 24 (1947).

Corvalen, *Concentration of plutonium in pitchblende.* Phys. Rev. *71,* 132 (1947).

Delsasso, White, Barkas and Creutz, *Carbon isotopes of mass ten and eleven.* Phys. Rev. *58,* 586 (1940).

Dempster, *Positive ray analysis of magnesium.* Phys. Rev. *17,* 427 (1921).

Dempster, *Positive ray analysis of lithium and magnesium.* Phys. Rev. *18,* 415 (1921).

Dempster, *Positive ray analysis of magnesium.* Proc. Nat., Acad. Sci. 7, 45 (1921).

Dempster, *Isotopic constitution of palladium and gold.* Nature *136,* 65 (1935).

Dempster, *Isotopic constitution of uranium.* Nature *136,* 180 (1935).

Dempster, *The isotopic constitution of barium and cerium.* Phys. Rev. *49,* 947 (1936).

Dempster, *The isotopic constitution of iron and nickel.* Phys. Rev. *50,* 98 (1936).

Dempster, *The isotopic constitution of strontium and tellurium.* Phys. Rev. *50,* 186 (1936).

Dempster, *The isotopic constitution of tungsten.* Phys. Rev. *52,* 1074 (1937).

Dempster, *The atomic masses of the heavy elements.* Phys. Rev. *53,* 64 (1938).

Dempster, *The isotopic constitution of gadolinium, dysprosium, erbium and ytterbium.* Phys. Rev. *53,* 727 (1938).

Dempster, *The energy content of the heavy nuclei.* Phys. Rev. *53,* 869 (1938).

Dempster, *Isotopic constitution of hafnium, yttrium, lutecium and tantalum.* Phys. Rev. *55,* 794 (1939).

Dempster, *Abnormal isotope abundances produced by neutron absorption in cadmium.* Phys. Rev. *71,* 144 (1947).

Duckworth and Hogg, *Relative abundance of the copper isotopes and the suitability of the photometric method for detecting small variations in isotopic abundance.* Phys. Rev. *71*, 212 (1947).

Eastman, *Indication of a genetic relation between indium and tin.* Phys. Rev. *52*, 1226 (1937).

Ewald, *The relative isotopic frequencies and the atomic weight of copper and ruthenium.* Z. Phys. *122*, 487 (1944).

Ewald, *Photometric determination of rare isotopes. Relative isotope abundances and atomic weight of nickel.* Z. Phys. *122*, 686 (1944).

Flügge and Mattauch, *Report on isotopes.* Phys. Z. *42*, 1 (1941).

Flügge and Mattauch, *Report on isotopes.* Phys. Z. *43*, 1 (1942).

Fretter, *The mass of cosmic-ray mesotrons.* Phys. Rev. *70*, 625 (1946).

Fuchs and Kopfermann, *On the isotopes of platinum.* Naturwiss. *23*, 372 (1935).

Giauque and Johnston, *An isotope of oxygen of mass 17 in the earth's atmosphere.* Nature *123*, 831 (1929).

Gibert, Roggen and Rossel, *Masses of Cl^{35} and Cl^{37}* (in French). Portugaliae Physica *1*, 43 (1944).

Gier and Zeeman, *Isotopes of nickel.* Proc. K. Ned. Akad. Wet. *38*, 810 (1935).

Götte, *New strontium, ytterbium isotopes from uranium fission.* Naturwiss. *29*, 496 (1941).

Hahn and Strassmann, *Problems on the existence of transuranic elements.* Naturwiss. *27*, 451 (1939).

Hahn, Flügge and Mattauch, *Chemical elements and natural atomic species according to results of isotopic and nuclear investigation.* Phys. Z. *41*, 1 (1940).

Harnwell and Bleakney, *Relative abundance of lithium isotopes.* Phys. Rev. *45*, 117 (1934).

Hayden and Lewis, *Assignments of mass to fission isotopes by mass spectrograph.* Phys. Rev. *70*, 111 (1946).

Hill, *Production of He^3.* Phys. Rev. *59*, 103 (1940).

Holleck, *Statistics of the isotope distribution in particles deposited by condensation.* Z. Phys. *116*, 624 (1940).

Huber, Huber and Scherrer, *Determination of the mass of $_8C^{14}$ from the nuclear reaction N (n, p) C* (in German). Helv. Phys. Acta *13*, 209 (1940).

Hughes, *The mass of the neutron.* Phys. Rev. *70*, 219 (1946).

Hughes, *Mass of the mesotron as determined by cosmic-ray measurements.* Phys. Rev. *71*, 387 (1947).

Hulubei and Cauchois, *New researches on element 93 in natural state.* C. R., Paris *209*, 476 (1939).

Inghram, *The isotopic constitution of tungsten, silicon and boron.* Phys. Rev. *70*, 653 (1946).

Jenkins and Woolridge, *Mass ratio of the carbon isotopes from the spectrum of CN.* Phys. Rev. *53*, 137 (1938).

Jensen, *Present state of systematics of stable atomic nuclei.* Naturwiss. *27*, 793 (1939).

Jordan and Bainbridge, *The mass spectrographic measurement of the mass separation of certain doublets.* Phys. Rev. *49*, 883 (1936).

Jordan and Bainbridge, *A mass spectrographic determination of the mass difference $N^{14} + H^1 - N^{15}$ and the nitrogen disintegration reactions.* Phys. Rev. *50*, 98 (1936).

Jordan and Bainbridge, *Atomic masses of beryllium, barium, neon and argon.* Phys. Rev. *51*, 385 (1937).

Jordan, *$C^{12}H_2 - N^{14}$ mass difference.* Phys. Rev. *58*, 1009 (1940).

Jordan, *The mass difference of the fundamental doublets used in the determination of the isotopic weights C^{12} and N^{14}.* Phys. Rev. *60*, 710 (1941).

Jordan and Young, *A short history of isotopes and the measurement of their abundances.* J. Appl. Phys. *13*, 526 (1942).

Kamen and Ruben, *Production and properties of carbon (14).* Phys. Rev. *58*, 194 (1940).

Kennedy, Seaborg, Segrè and Wahl, *Properties of 94 (239).* Phys. Rev. *70*, 555 (1946).

Kurbatov, *A classification of stable nuclei.* J. Phys. Chem. *49*, 110 (1945).

Kurbatov and Kurbatov, *Classification of stable nuclei by means of particles of mass numbers three and four.* Phys. Rev. *65,* 351 (1944).

Kurbatov, *Deviations in the classification of stable nuclei.* Phys. Rev. *65,* 351 (1944).

Laaff, *Beryllium isotope of mass eight.* Ann. Phys., Leipzig *32,* 743 (1938).

Lauritsen and Crane, *Transmutation of lithium by deutons and its bearing on the mass of the neutron.* Phys. Rev *45,* 550 (1934).

Lichtblau, *The isotope constitution and the atomic weights of europium.* Naturwiss. *27,* 260 (1939).

Livingston and Hoffman, *Slow neutron disintegration of Li⁶ and the disintegration mass scale.* Phys. Rev. *50,* 401 (1936).

Luhr and Harris, *Mass-spectrograph determination of the relative abundance of heavy hydrogen in a sample.* Phys. Rev. *45,* 843 (1934).

Massey and Mohr, *The masses of Be⁸ and C¹².* Nature *136,* 141 (1935).

Mattauch, *Isotope research.* Phys. Z. *35,* 567 (1934).

Mattauch, *Scheme of isotopes.* Z. Phys. *91,* 361 (1934).

Mattauch, *New mass-spectrograph measurements.* S.B. Akad. Wiss. Wien *145,* 461 (1936).

Mattauch, *The mass spectrographic measurement of nuclear binding energies.* Naturwiss. *25,* 156 (1937).

Mattauch, *The packing fraction of Sr⁸⁶ and Sr⁸⁷ by the doublet method.* Naturwiss. *25,* 170 (1937).

Mattauch, *The pair Rb⁸⁷ - Sr⁸⁷ and the isobaric rule.* Naturwiss *25,* 189 (1937).

Mattauch and Herzog, *The doublets of the C_1 group and the binding energy of the nuclei between C¹² and O¹⁶.* Naturwiss, *25,* 747 (1937).

Mattauch, *The substandard in mass spectroscopy.* Phys. Z. *39,* 892 (1938)

Mattauch, Lichtblau, Schüler and Gollnow, *Remarkable isotope of lutecium.* Z. Phys. *111,* 514 (1939).

Mattauch and Lichtblau, *The isotopic constitution and the atomic weight of molybdenum.* Z. Phys. Chem. *B42,* 288 (1939).

Mattauch, Ewald, Hahn and Strassmann, *Has a cesium isotope of long half life existed? A contribution to the explanation of unusual lines in mass spectroscopy.* Z. Phys. *120,* 598 (1943).

Mattauch and Ewald, *A new method for the measurement of relative frequency of isotopes. The isotopic constitution and the atomic weight of hafnium.* Z. Phys. *122,* 314 (1946).

McKellar, *Mass ratio of the lithium isotopes from the spectrum of Li₂.* Phys. Rev. *44,* 155 (1933).

McKellar, *On the relative abundance of silicon isotopes.* Phys. Rev. *45,* 761 (1934).

Mitchell, Brown and Fowler, *On the isotopic constitution of cobalt.* Phys. Rev. *60,* 359 (1941).

Murphey, *Relative abundances of the oxygen isotopes.* Phys. Rev. *59,* 320 (1941).

Murphey and Nier, *Variations in the relative abundance of the abundance of the carbon isotopes.* Phys. Rev. *59,* 771 (1941).

Naudé, *Isotopes of N¹⁵ and O¹⁸, O¹⁷ and their abundances.* Phys. Rev. *36,* 333 (1930).

Ney and McQueen, *A mass spectrographic study of the isotopes of silicon.* Phys. Rev. *69,* 41 (1946).

Nier, *Evidence for the existence of an isotope of potassium of mass 40.* Phys. Rev. *48,* 283 (1935).

Nier, *The isotopic constitution of rubidium, zinc and argon.* Phys. Rev. *49,* 272 (1936).

Nier and Hanson, *A mass spectrographic analysis of the ions produced in HCl under electron impact.* Phys. Rev. *50,* 722 (1936).

Nier, *A mass spectrographic study of the isotopes of argon, potassium, rubidium, zinc and cadmium.* Phys. Rev. *50,* 1041 (1936).

Nier, *The isotopic constitution of mercury and lead.* Nier, Phys. Rev. *51,* 1007 (1937).

Nier, *The isotopic constitution of osmium.* Phys. Rev. *52,* 885 (1937).

Nier, *A mass-spectrographic study of the isotopes of silver, xenon, krypton, beryllium, iodine, arsenic and cesium.* Phys. Rev. *52,* 933 (1937).

Nier, *Isotope abundance of common lead.* J. Amer. Chem. Soc. *60,* 1571 (1938).

Nier, *The isotopic constitution of calcium, titantium, sulphur and argon.* Phys. Rev. *53,* 282 (1938).

Nier, *The isotopic constitution of strontium, barium, bismuth, thorium and mercury.* Phys. Rev. *54,* 275 (1938).

Nier, *The isotopic constitution of uranium and the half-lives of the uranium isotopes. I.* Phys. Rev. *55,* 150 (1939).

Nier, *The isotopic constitution of radiogenic leads and the measurement of geological time. II.* Phys. Rev. *55,* 153 (1939).

Nier, *The isotopic constitution of iron and chromium.* Phys. Rev. *55,* 1143 (1939).

Ogata, *Isotopic weight of C^{12}.* Proc. Phys.-Math. Soc. Jap. (3) *22,* 486 (1940).

Okuda, Ogata, Aoki and Sugawara, *On the isotopic weights of chlorine, argon and iron by the doublet method.* Phys. Rev. *58,* 578 (1940).

Okuda, Ogata, Kuroda, Shima and Shindo *Isotopic weights of nickel isotopes by the doublet method.* Phys. Rev. *59,* 104 (1941).

Okuda and Ogata, *Isotopic weights of sulphur and titanium.* Phys. Rev. *60,* 690 (1941).

Okuda and Ogata, *Isotopic weight of sulphur and titanium.* Proc. Phys.-Math. Soc. Jap. (3) *25,* 371 (1943)

Paneth, *The making of the missing chemical elements.* Nature *159,* 8 (1947).

Pecher, *Long-lived isotope of yttrium.* Phys. Rev. *58,* 843 (1940).

Perey, *On element 87, derived from actinium.* C. R., Paris *208,* 97 (1939).

Perey, *Element 87, Ac K, a derivative of actinium.* J. Phys. Rad. *10,* 435 (1939).

Perrier and Segrè, *Technetium, the element of atomic number 43.* Nature *159,* 24 (1947).

Pierce and Brown, *Mass spectrographic confirmation of the isotopic assignment for the long-lived radioactivity in Be.* Phys. Rev. *70,* 779 (1946).

Pollard and Brasefield, *Alpha particle bombardment of neon, calcium and argon, and masses of light nuclei.* Phys. Rev. *51,* 8 (1937).

Pollard, *Mass and energy levels of S^{33}.* Phys. Rev. *56,* 961 (1939).

Pollard, *Mass and stability of C^{14}.* Phys. Rev. *56,* 1168 (1939).

Rall, *Mass assignments of some radioactive isotopes of palladium and iridium.* Phys. Rev. *70,* 112 (1946).

Ray, *Extension of periodic table and elements beyond uranium.* Science and Culture *4,* 167 (1938).

Ruark and Western, *Radium-uranium ratio and the number of actino-uranium isotopes.* Phys. Rev. *45,* 69 (1934).

Sachs, *On the level scheme of Mg^{24} and the mass of Na^{24}.* Phys. Rev. *70,* 572 (1946).

Sampson and Bleakney, *A mass spectrograph study of barium, strontium, indium, gallium, lithium and sodium.* Phys. Rev. *50,* 456 (1936).

Sampson and Bleakney, *The relative abundance of the isotopes in manganese, columbium, palladium, platinum, iridium, rhodium and cobalt.* Phys. Rev. *50,* 732 (1936).

Seaborg, *Table of isotopes.* Rev. Mod. Phys. *16,* 1 (1944).

Seaborg, *Discovery of elements 95 and 96 and the chemical properties of the transuranic elements.* Nature *157,* 307 (1946).

Segrè, *An unsuccessful search for transuranic elements.* Phys. Rev. *55,* 1104 (1939).

Sherr, Smith and Bleakney, *On the existence of H^3.* Phys. Rev. *54,* 388 (1938).

Smythe, *On the isotopic ratio in oxygen.* Phys. Rev. *45,* 299 (1934).

Stephens, *Neutron hydrogen mass difference and the neutron mass.* Rev. Mod. Phys. *19,* 19 (1947).

Strauss, *The abundance ratio $Ni^{61} : Ni^{64}$.* Phys. Rev. *59,* 102 (1941).

Swartout, *Protium-deuterim ratio and atomic weight of hydrogen.* J. Amer. Chem. Soc. *61,* 2025 (1939).

Thode and Graham, *A mass spectrometer investigation of the isotopes of xenon and krypton resulting from the fission of U²³⁵ by thermal neutrons.* Canad. J. Res. *25*, 1 (1947).

Turner, *Nonexistence of transuranic elements.* Phys. Rev. *57*, 157 (1940).

Turner, *Missing heavy nuclei.* Phys. Rev. *57*, 950 (1940).

Tuve, Hafstad and Dahl, *Stable hydrogen isotope of mass three.* Phys. Rev. *45*, 840 (1934).

Tuve and Hafstad, *The carbon reactions and the corrected mass scale.* Phys. Rev. *48*, 106 (1935).

Urey, Brickwedde and Murphy, *A hydrogen isotope of mass two and its concentration.* Phys. Rev. *40*, 1 (1932).

Urey and Teal, *Hydrogen isotope of atomic weight two.* Rev. Mod. Phys. *7*, 34 (1935).

Urey, *Separation and use of stable isotopes,* J. Appl. Phys. *12*, 270 (1941).

Valley and Anderson, *The relative abundance of the stable isotopes of terrestrial and meteoric iron.* Phys. Rev. *59*, 113 (1941).

Valley, *The stable isotopes of nickel.* Phys. Rev. *59*, 836 (1941).

Vaughan, Williams, and Tate, *Isotopic abundance ratios of carbon, nitrogen, argon, neon and helium.* Phys. Rev. *46*, 327 (1934).

Venkatesaschar and Sibaiya, *Isotope abundance in platinum.* Proc. Indian Acad. Sci. (A) *2*, 101 (1935).

Watson, *Mass ratio of hydrogen and deuterium from band spectra.* Phys. Rev. *49*, 70 (1936).

Wiens and Alvarez, *Spectroscopically pure mercury (198).* Phys. Rev. *58*, 1005 (1940).

Williams, Haxby and Shepherd, *The disintegration of beryllium and the masses of beryllium isotopes.* Phys. Rev. *52*, 1031 (1937).

Williams and Yuster, *Isotopic constitution of tellurium, silicon, tungsten, molybdenum and bromine.* Phys. Rev. *69*, 556 (1946).

Williams and Yuster, *Isotopic constitution of tellurium and tungsten.* Phys. Rev. *70*, 118 (1946).

Yamaguchi, *On the mass spectrographic determination of mass number of artificially produced radioactive atoms.* Proc. Phys.-Math. Soc. Jap. (3) *23*, 264 (1941).

Zingg, *The isobaric pairs cadmium-indium, indium-tin, antimony-tellurium, rhenium-osmium* (In German). Helv. Phys. Acta *13*, 219 (1940).

CHAPTER TWO

Hyperfine Structures, Nuclear Moments, and Spin

Alvarez and Bloch, *Quantitative determination of the neutron moment in absolute nuclear magnetons.* Phys. Rev. *57*, 111 (1940).

Anderson, *Nuclear mechanical moment of lanthanum from hyperfine structure.* Phys. Rev. *45*, 685 (1934).

Anderson, *Nuclear magnetic moment of lanthanum.* Phys. Rev. *46*, 473 (1934).

Anderson and Novick, *Magnetic moment of the triton.* Phys. Rev. *71*, 372 (1947).

Arnold and Roberts, *The magnetic moments of the neutron and the deuteron.* Phys. Rev. *70*, 766 (1946).

Arvidsson, *Hyperfine structure in some spectral lines from highly ionized atoms of thallium and bismuth.* Nature *126*, 565 (1931).

Ashley, *Nuclear spin of phosphorus from band spectrum analysis.* Phys. Rev. *44*, 919 (1933).

Atherton, *Determination of nuclear moment of rubidium by resonance radiation.* Phys. Rev. *64*, 43 (1943).

Bacher, *Note on the magnetic moment of the nitrogen nucleus.* Phys. Rev. *43*, 1001 (1933).

Bacher and Tomboulian, *In 115 hyperfine structure deviations and electric quadrupole moment.* Phys. Rev. *50*, 1096 (1936).

Bacher and Tomboulian, *The electric quadrupole moment of In¹¹⁵.* Phys. Rev. *52*, 836 (1937).

Back and Goudsmit, *Nuclear moment and Zeeman effect for bismuth.* Z. Phys. *47*, 174 (1928).

Badami, *Hyperfine structures in the antimony spark spectrum and the nuclear moments of antimony isotopes.* Z. Phys. *79*, 206 (1932).

Badami, *Hyperfine sructures in the antimony arc spectrum.* Z. Phys. *79*, 224 (1932).

Ballard, *Nuclear spin of columbium.* Phys. Rev. *46*, 233 (1934).

Ballard, *Nuclear moments of columbium from hyperfine structure.* Phys. Rev. *46*, 806 (1934).

Bartlett, *Isotopic displacement in hyperfine sructure.* Nature *128*, 408 (1931).

Bartlett, Gibbons and Watson, *The magnetic moment of the Li^7 nucleus. II.* Phys. Rev. *50*, 315 (1936).

Benson and Sawyer, *The nuclear moment of barium.* Phys. Rev. *52*, 1127 (1937).

Bethe, *Ionization power of a neutrino with magnetic moment.* Proc. Camb. Phil. Soc. *31*, 108 (1935).

Bethe, *Magnetic moment of Li^7 in the alpha particle model.* Phys. Rev. *53*, 842 (1938).

Bloch and Rabi, *Atoms in variable magnetic fields.* Rev. Mod. Phys. *17*, 237 (1945).

Bloch, Hansen and Packard, *Nuclear induction.* Phys. Rev. *69*, 127 (1946).

Bloch, Hansen and Packard, *Nuclear induction.* Phys. Rev. *69*, 680 (1946).

Bloch, *Nuclear induction.* Phys. Rev. *70*, 460 (1946).

Bloch, Hansen and Packard, *The nuclear induction experiment.* Phys. Rev. *70*, 474 (1946).

Bloch, Graves, Packard and Spence, *Spin and magnetic moment of tritium.* Phys. Rev. *71*, 373 (1947).

Bloch, Graves, Packard and Spence, *Relative moments of H^1 and H^3.* Phys. Rev. *71*, 551 (1947).

Bloembergen, Pound and Purcell, *The width of the nuclear magnetic resonance absorption line in gases, liquids, and solids.* Phys. Rev. *71*, 466 (1947).

Boggs and Webb, *The hyperfine structure of the mercury triplet 6^3P_{012}-7^3S_1 in optical excitation.* Phys. Rev. *48*, 226 (1935).

Boyd and Sawyer, *The hyperfine structure of Cs II.* Phys. Rev. *61*, 601 (1942).

Breit, *Possible effects of nuclear spin on x-ray terms.* Phys. Rev. *35*, 1447 (1930).

Breit, *Derivation of hyperfine structure formulas for one electron spectra.* Phys. Rev. *37*, 51 (1931).

Breit, *On the hyperfine structure of heavy elements.* Phys. Rev. *38*, 463 (1931).

Breit and Rabi, *Measurement of nuclear spin.* Phys. Rev. *38*, 2082 (1931).

Breit, *The isotope displacement in the hyperfine structure.* Phys. Rev. *42*, 348 (1932).

Breit and Wills, *Hyperfine structure in intermediate coupling.* Phys. Rev. *44*, 470 (1933).

Breit and Rabi, *On the interpretation of present values of nuclear moments.* Phys. Rev. *46*, 230 (1934).

Breit and Stehn, *The fine structure of the nuclear ground level of Li^7.* Phys. Rev. *53*, 459 (1938).

Breit, *Relativistic corrections to magnetic moments of nuclear particles.* Phys. Rev. *71*, 400 (1947).

Brooks, *Interaction between nuclear spin and molecular rotation.* Phys. Rev. *60*, 168 (1941).

Brown and Bartlett, *Hyperfine structure of fluorine.* Phys. Rev. *45*, 527 (1934).

Brown and Cook, *Nuclear magnetic moment of aluminum.* Phys. Rev. *45*, 731 (1934).

Bryden, *The structure of the nucleus and its total moment of momentum.* Phys. Rev. *38*, 1989 (1931).

Brylinski, *Magnetic moment of the electron in the atom.* Rev. Gén. Élect. *52*, 249 (1943).

Caldirola, *Relativistic correction in calculating the magnetic moment of the deuteron.* Phys. Rev. *69*, 608 (1946).

Campbell, *Nuclear moments of the gallium isotopes 69 and 71.* Nature *131*, 204 (1933).

Campbell, *Hyperfine structure in the arc spectrum of fluorine.* Z. Phys. *84*, 393 (1933).

Casimir, *Hyperfine structure of europium* (In German). Physica 2, 719 (1935).

Chakrabarty and Majumdar, *On the spin of the meson*. Phys. Rev. *65*, 206 (1945).

Christy and Kusaka, *Electric quadrupole moment of the deuteron*. Phys. Rev. *55*, 665 (1939).

Christy and Keller, *Precise determination of the fine structure constant from x-ray spin doublet splitting*. Phys. Rev. *61*, 147 (1942).

Cohen, *Nuclear spin of cesium*. Phys. Rev. *46*, 713 (1934).

Crawford and Crooker, *Nuclear moment of arsenic*. Nature *131*, 655 (1933).

Crawford and Bateson, *Nuclear moments of antimony isotopes; discussion of Landé's theory*. Canad. J. Res. *10*, 693 (1934).

Crawford and Grace, *Hyperfine structures in La III. Nuclear magnetic moment of lanthanum*. Phys. Rev. *47*, 536 (1935).

Crawford, *Hyperfine structure formulas for the configuration d^2S. Application to the $5d^26s\ {}^4F$ states of La I*. Phys. Rev. *47*, 768 (1935).

Crawford and Wills, *Hyperfine structure. Formulas for the configuration p^3s*. Phys. Rev. *48*, 69 (1935).

Dickinson, *Specific isotope effect in the hyperfine spectrum of the lead atom*. Phys. Rev. *46*, 598 (1934).

Ellet and Heydenburg, *Nuclear magnetic moments from the polarization of resonance radiation: Sodium, $3^2S_{1/2}\text{-}4^2P_{3/2, 1/2}$*. Phys. Rev. *46*, 583 (1934).

Elliott, *Absorption band spectrum of Chlorine. Part II*. Proc. Roy. Soc. *127*, 638 (1930).

Elliott and Wulff, *Hyperfine structure of gold*. Phys. Rev. *55*, 170 (1939).

Estermann, Frisch and Stern. *Magnetic moment of the proton*. Nature *132*, 169 (1933).

Eastermann and Stern, *Magnetic moment of the deuton*. Nature *133*, 911 (1934).

Estermann and Stern, *Magnetic moment of the deuton*. Phys. Rev. *45*, 761 (1934).

Estermann, Simpson and Stern, *The magnetic moment of the proton*. Phys. Rev. *52*, 535 (1937).

Farkas and Farkas, *Ratio of the magnetic moments of the proton and deuton*. Proc. Roy. Soc. *152*, 152 (1935).

Feld and Lamb, *Effect of nuclear electric quadrupole moment on the energy levels of a diatomic molecule in a magnetic field. Part I. Heteronuclear molecules*. Phys. Rev. *67*, 15, 59 (1945).

Feld, *Nuclear electric quadrupole moment and the radio frequency spectra of homonuclear diatomic molecules*. Phys. Rev. *70*, 112 (1946).

Fermi, *Magnetic moments of atomic nuclei*. Z. Phys. *60*, 320 (1930).

Fermi and Segrè, *Theory of hyperfine structure*. Z. Phys. *82*, 729 (1933).

Fisher and Peck, *Hyperfine structure of manganese I and nuclear magnetic moment*. Phys. Rev. *55*, 270 (1939).

Flügge, *Quadrupole moment of the deuteron and nuclear forces*. Z. Phys. *113*, 587 (1939).

Fox and Rabi, *On the nuclear moments of lithium, potassium and sodium*. Phys. Rev. *48*, 746 (1935).

Frisch and Stern, *Magnetic deviation of hydrogen molecules and the magnetic moment of the proton*. Z. Phys. *85*, 4 (1933).

Frisch, von Halban jr. and Koch, *A method of measuring the magnetic moment of free neutrons*. Nature *139*, 756 (1937).

Frisch, von Halban jr. and Koch, *Sign of the magnetic moment of free neutrons*. Nature *139*, 1021 (1937).

Fröhlich, Heitler and Kemmer, *Nuclear forces and the magnetic moments of the neutron and the proton*. Proc. Roy. Soc. *166*, 154 (1938).

Galanin, *Properties of electrons and meson spin in the classical approximation*. J. Phys., USSR, *6*, 35 (1942).

Gale and Monk, *Band spectrum of fluorine*. Astrophys. J. *69*, 77 (1929).

Gibbons and Bartlett, *The magnetic moment of the K^{39} nucleus*. Phys. Rev *47*, 692 (1935).

Ginsburg, *The relativistic theory of excited spin states of the proton and the neutron*. Phys. Rev. *63*, 1 (1943).

Gollnow, *Second isotope of cassiopeium and magnetic moment and the quadrupole moment of the ${}^{175}_{71}Cp$ nucleus*. Z. Phys. *103*, 443 (1936).

Gorham, *The signs of the nuclear magnetic moments of Li6 and K41*. Phys. Rev. *53*, 563 (1938).

Gorter and Broer, *Negative result of an attempt to observe nuclear magnetic resonance in solids* (in English). Physica *9*, 591 (1942).

Goudsmit and Back, *Fine structure and arrangement of terms of the bismuth spectrum*. Z. Phys. *43*, 321 (1927).

Goudsmit and Bacher, *Separations in hyperfine structure*. Phys. Rev. *34*, 1501 (1929).

Goudsmit, *Theory of hyperfine structure separations*. Phys. Rev. *37*, 663 (1931).

Goudsmit, *Nuclear magnetic moments*. Phys. Rev. *43*, 636 (1933).

Goudsmit and Bacher, *Anomalies in hyperfine structure*. Phys. Rev. *43*, 894 (1933).

Grace, *Nuclear moments and their dependence upon atomic number and mass number*. Phys. Rev. *44*, 361 (1944).

Granath, *The nuclear spin and magnetic moment of Li7*. Phys. Rev. *42*, 44 (1932).

Granath and Van Atta, *Nuclear spin and magnetic moment of sodium from hyperfine structure*. Phys. Rev. *44*, 935 (1934).

Granath and Stranathan, *Magnetic moment of cesium determined from the hyperfine structure of the $6p2P_{1/2}$ state*. Phys. Rev. *46*, 317 (1934).

Granath and Stranathan, *The hyperfine structure and nuclear magnetic moment of cesium I*. Phys. Rev. *48*, 725 (1935).

Gray, *The nuclear spin of Li7 from hyperfine structure data*. Phys. Rev. *44*, 570 (1933).

Güttinger, *Hyperfine structure of the Li II spectrum*. Z. Phys. *64*, 749 (1930).

Güttinger and Pauli, *On the hyperfine structure of Li+*. Z. Phys. *67*, 743 (1931).

Halpern and Gwathmey, *Effect of the similarity of particles on gas kinetic quantities with application to nuclear spin*. Phys. Rev. *52*, 944 (1937).

Halpern and Johnson, *On the neutron's magnetic moment*. Phys. Rev. *57*, 160 (1940).

Hamilton, *An atomic beam study of the hyperfine structure of the metastable $2P_{3/2}$ state of In115. I. The electric quadrupole moment of In115*. Phys. Rev. *56*, 30 (1939).

Hamilton, *Molecular beams and nuclear moments*. Amer. J. Phys. *9*, 319 (1941).

Hardy, *The nuclear magnetic moment of In115*. Phys. Rev. *59*, 686 (1941).

Hardy and Millman, *Nuclear spin and magnetic moments of In113*. Phys. Rev. *60*, 167 (1941).

Hardy and Millman, *Radiofrequency spectrum of indium. Nuclear spin of In113*. Phys. Rev. *61*, 459 (1942).

Hartree and Black, *Theoretical investigation of the oxygen atom in various states of ionization*. Proc. Roy. Soc. *139*, 311 (1933).

Hay, *Nuclear magnetic moment of C13*. Phys. Rev. *58*, 180 (1940).

Hay, *The nuclear magnetic moments of Ba135 and Ba137*. Phys. Rev. *59*, 686 (1941).

Hay, *The nuclear magnetic moments of C13, Ba135 and Ba137*. Phys. Rev. *60*, 75 (1941).

Heitler, *Scattering of mesons and the magnetic moments of proton and neutron*. Nature *145*, 29 (1940).

Heyden, *Spin change in the radioactive beta decay of Rb87 into Sr87*. Z. Phys. *108*, 232 (1938).

Heyden and Ritschl, *Nuclear moment of aluminum*. Z. Phys. *108*, 739 (1938).

Heydenburg, *Paschen-Back effect of hyperfine structure and polarization of resonance radiation. Cadmium $61P_1 - 51S_0$*. Phys. Rev. *43*, 640 (1933).

Heydenburg, *Nuclear magnetic moment of cesium from the polarization of resonance radiation*. Phys. Rev. *46*, 802 (1934).

Hill, *Relative intensities in nuclear spin multiplets*. Proc. Nat. Acad. Sci. *15*, 779 (1929).

Hill, *Hyperfine structure in silver*. Phys. Rev. *48*, 233 (1935).

Hoffman, Livingston and Bethe, *Some direct evidence of the magnetic moment of the neutron*. Phys. Rev. *51*, 214 (1937).

Hollenberg, *Hyperfine structure of the resonance lines of rubidium*. Phys. Rev. *52*, 139 (1937).

Hori, *Analysis of the hydrogen band spectrum in the extreme ultra-violet*. Z. Phys. *44*, 834 (1927).

Inglis and Landé, *Magnetic moment of the neutron*. Phys. *45*, 842 (1934).

Inglis, *On nuclear moments*. Phys. Rev. *47*, 84 (1935).

Inglis, *On nuclear magnetic moments*. Phys. Rev. *53*, 470 (1938).

Inglis, *On interpreting related magnetic moments of light nuclei*. Phys. Rev. *55*, 329 (1939).

Inglis, *Nuclear spin of C^{13}*. Phys. Rev. *58*, 577 (1940).

Inglis, *On the triplets of helium*. Phys. Rev. *61*, 297 (1942).

Jackson, *Nuclear moment of gallium*. Z. Phys. *75*, 229 (1932).

Jackson, *Hyperfine structure of the lines of the arc spectrum of rubidium*. Proc. Roy. Soc. *139*, 673 (1933).

Jackson, *Nuclear moment of indium*. Z. Phys. *80*, 59 (1933).

Jackson and Kuhn, *Hyperfine structure of the resonance lines of potassium*. Proc. Roy. Soc. *148*, 335 (1935).

Jackson and Kuhn, *Hyperfine structure and Zeeman effect of resonance lines of silver*. Proc. Roy. Soc. *158*, 372 (1937).

Jackson and Kuhn, *Hyperfine structure and Zeeman effect of the resonance lines of lithium*. Proc. Roy. Soc. *173*, 278 (1939).

Jaeckel, *Hyperfine structure of platinum isotope*. Z. Phys. *100*, 513 (1936).

Jauch, *Meson theory of the magnetic moment of proton and neutron*. Phys. Rev. *63*, 334 (1943).

Jenkins and Ashley, *Nuclear spin of phosphorus from the band spectrum*. Phys. Rev. *39*, 552 (1932).

Joffe and Urey, *The spin of the sodium nucleus*. Phys. Rev. *43*, 761 (1933).

Joffe, *Nuclear spin of sodium*. Phys. Rev. *45*, 468 (1934).

Jones, *Hyperfine structure in the spark spectrum of cadmium*. Proc. Phys. Soc., Lond. *45*, 625 (1933).

Jones, *Hyperfine structure in the arc spectrum of xenon*. Proc. Roy. Soc. *144*, 587 (1934).

Kalckar and Teller, *Theory of the catalysis of the ortho-para transformation by paramagnetic gases*. Proc. Roy. Soc. *150*, 520 (1935).

Kellogg, Rabi and Zacharias, *Magnetic moment of the deuteron*. Phys. Rev. *49*, 867 (1936).

Kellogg, Rabi and Zacharias, *The gyromagnetic properties of the hydrogens*. Phys. Rev. *50*, 472 (1936).

Kellogg, Rabi, Ramsey and Zacharias, *An electrical quadrupole moment of the deuteron*. Phys. Rev. *55*, 318 (1939).

Kellogg, Rabi, Ramsey and Zacharias, *The magnetic moments of the proton and the deuteron. The radiofrequency spectrum of H_2 in various magnetic fields*. Phys. Rev. *56*, 728 (1939).

Kellogg, Rabi, Ramsey and Zacharias, *Electrical quadrupole moment of the deuteron. The radiofrequency spectra of HD and D_2 modecules in a magnetic field*. Phys. Rev. *57*, 677 (1940).

Kofink, *Magnetic and electric moment of the electron in Dirac's theory*. Ann. Phys., Leipzig *30*, 91 (1937).

Kopfermann, *Determination of the mechanical moment of the cesium nucleus from the hyperfine structure of some Cs lines*. Z. Phys. *73*, 437 (1931).

Kopfermann, *Nuclear moments of three lead isotopes*. Z. Phys. *75*, 363 (1932).

Kopfermann, *Hyperfine structure and nuclear moment of rubidium*. Z. Phys. *83*, 417 (1933).

Kopfermann and Wieth-Knudsen, *Hyperfine structure and nuclear moment of krypton*. Z. Phys. *85*, 353 (1933).

Kopfermann and Rasmussen, *The mechanical moment of the cobalt nucleus*. Naturwiss. *22*, 291 (1934).

Kopfermann and Rindal, *Nuclear moments of xenon*. Z. Phys. *87*, 460 (1934).

Kopfermann and Rasmussen, *Nuclear moment of scandium*. Z. Phys. *92*, 82 (1934).

Kopfermann and Rasmussen, *On the mechanical moment of the cobalt nucleus*. Z. Phys. *94*, 58 (1935).

Kopfermann and Rasmussen, *Hyperfine structure of vanadium multiplets.* Z. Phys. *98*, 624 (1936).

Kopfermann and Wittke, *Magnetic moment of the scandium nucleus.* Z. Phys. *105*, 16 (1937).

Korsching, *Quadrupole moments of $^{83}_{36}Kr$ and the mechanical moment of $^{131}_{54}Xe$.* Z. Phys. *109*, 349 (1938).

Krasnikov, *Hyperfine structure of secondary x-ray spectra* (in English). C. R., URSS *49*, 337 (1945).

Kruger, Gibbs and Williams, *The nuclear moment of barium as determined from the hyperfine structure of the Ba II lines.* Phys. Rev. *41*, 322 (1932).

Kruger, *Nuclear shells, angular and magnetic moments of nuclei.* Phys. Rev. *47*, 605 (1935).

Kusch, Millman and Rabi, *On the nuclear magnetic moment of beryllium.* Phys. Rev. *55*, 666 (1939).

Kusch, Millman and Rabi, *The nuclear magnetic moments of N^{14}, Na^{23}, K^{39} and Cs^{133}.* Phys. Rev. *55*, 1176 (1939).

Kusch and Millman, *On the nuclear magnetic moments of the isotopes of rubidium and chlorine.* Phys. Rev. *56*, 527 (1939).

Kusch, Millman and Rabi, *Radiofrequency spectra of atoms. Hyperfine and Zeeman effect in the ground state of Li^6, Li^7, K^{39} and K^{41}.* Phys. Rev. *57*, 765 (1940).

Landé, *The absolute intervals of the optical doublets and triplets.* Z. Phys. *25*, 46 (1924).

Landé, *Magnetic moment of the neutron.* Phys. Rev. *46*, 477 (1934).

Lapp, *On the spin of the mesotron from burst measurements.* Phys. Rev. *64*, 255 (1943).

Larrick, *Nuclear magnetic moments from the polarization of resonance radiation. Sodium, $3^2S_{1/2} - 3^2P_{3/2, 1/2}$.* Phys. Rev. *46*, 581 (1934).

Lopes, *The influence of the recoil of heavy particles on the nuclear potential energy.* Phys. Rev. *67*, 60 (1945).

Lyshede and Rasmussen, *Nuclear moment of Zn^{67}.* Z. Phys. *104*, 434 (1937).

Ma, *Electrostatic dipole moment of a nucleus in the meson theory.* Proc. Camb. Phil. Soc. *36*, 438 (1940).

Manley, *The nuclear spin and magnetic moment of potassium 41.* Phys. Rev. *49*, 921 (1936).

Manley and Millman, *The nuclear spin and magnetic moment of Li^6.* Phys. Rev. *51*, 19 (1937).

Margenau, *Relativistic magnetic moment of a charged particle.* Phys. Rev. *57*, 383 (1940).

Markov, *Proper masses and magnetic moments of elementary particles and the Wentzel-Dirac λ process* (in English). C. R., URSS *47*, 177 (1945).

McLay and Crawford, *Multiplet and hyperfine structure analysis of Bi IV. Discussion of perturbation effects.* Phys. Rev. *44*, 986 (1933).

McLennan and Allin, *Spectrum of thallium III.* Proc. Roy. Soc. *129*, 43 (1930).

McLennan and Allin, *Hyperfine structure of lines in the arc and first spectrum of indium.* Proc. Roy. Soc. *129*, 208 (1930).

McLennan, Crawford and Leppard, *Nuclear moments of the isotopes of lead. Relative values of the g (I) factors of Pb^{207} and Tl.* Nature *128*, 301 (1931).

McLennan and Crawford, *Hyperfine structure of Tl II.* Proc. Roy. Soc. *132*, 10 (1931).

McMillan and Grace, *Hyperfine structure in the tantalum arc spectrum.* Phys. Rev. *44*, 949 (1933).

Meggers, *The optical spectra of rhenium.* Phys. Rev. *37*, 219 (1931).

Meggers, King and Bacher, *Hyperfine structure and nuclear moment of rhenium.* Phys. Rev. *38*, 1258 (1931).

Meissner, *Application of atomic beams in spectroscopy.* Rev. Mod. Phys. *14*, 68 (1942).

Millman, *On the nuclear spins and magnetic moments of the principal isotopes of potassium.* Phys. Rev. *47*, 739 (1935).

Millman and Fox, *Nuclear spins and magnetic moments of Rb^{85} and Rb^{87}.* Phys. Rev. *50*, 220 (1936).

Millman, Rabi and Zacharias, *On the nuclear moments of indium.* Phys. Rev. *53*, 384 (1938).

Millman, Kusch and Rabi, *The nuclear magnetic moment of N^{14}*. Phys. Rev. *54*, 968 (1938).

Millman, *On the determination of the signs of nuclear magnetic moments by the molecular beam method of magnetic resonance*. Phys. Rev. *55*, 628 (1939).

Millman, Kusch and Rabi, *On the nuclear magnetic moments of the boron isotopes*. Phys. Rev. *56*, 165 (1939).

Millman and Kusch, *Nuclear spin and magnetic moment of $_{13}Al^{27}$*. Phys. Rev. *56*, 303 (1939).

Millman, *On the precise measurement of nuclear magnetic moments by the molecular beam magnetic resonance method. The moments of H^1, Li^7, F^{19} and Na^{23}*. Phys. Rev. *60*, 91 (1941).

Mintz and Granath, *A test of the interval rule in the $^2D_{3/2}$ state of Bi I*. Phys. Rev. *49*, 196 (1936).

More, *Nuclear mechanical moment of cobalt*. Phys. Rev. *46*, 470 (1934).

Mrozowski, *Hyperfine structure on the quadrupole line 2815 A and some other lines of ionized mercury*. Phys. Rev. *57*, 207 (1940).

Mukerji, *On the hyperfine structure and analysis of the complex line λ 3842.82 $sp^{33}D_3$ - $5p$. $^3P_2$ in the first spark spectrum of arsenic in the ultra-violet region*. Indian J. Phys. *19*, 180 (1945).

Murakawa, *Hyperfine structure of arc and spark spectra of barium*. Tokyo Inst. Phys. Chem. Res. *18*, 304 (1932).

Murakawa, *Spectra of I II, I I and Cl II*. Z. Phys. *109*, 162 (1938).

Murakawa, *Electrical quadrupole moment of the iodine nucleus*. Z. Phys. *112*, 234 (1939).

Murakawa, *Electric quadrupole moment of the iodine nucleus*. Z. Phys. *114*, 651 (1939).

Murphy and Johnston, *Nuclear spin of deuterium*. Phys. Rev. *45*, 550 (1934).

Murphy and Johnston, *Nuclear spin of deuterium*. Phys. Rev. *46*, 95 (1934).

Nierenberg, Ramsey and Brody, *Measurements of nuclear quadrupole moment interaction*. Phys. Rev. *70*, 773 (1946).

Nierenberg, Ramsey and Brody, *Measurements of nuclear quadrupole interactions*. Phys. Rev. *71*, 466 (1947).

Nikolsky, *On isotopic spin*. Phys. Rev *67*, 366 (1945).

Ornstein and Van Wijk, *Negative band spectrum of nitrogen*. Z. Phys. *49*, 315 (1928).

Packard, Hansen and Bloch, *Nuclear induction*. Phys. Rev. *69*, 680 (1946).

Paschen and Ritschl, *Infra-red grating spectra*. Ann. Phys., Leipzig *18*, 867 (1933).

Paschen and Campbell, *The nuclear moment of indium*. Naturwiss. 22, 136 (1934).

Paschen, *Line groups and fine structure. Part II*. S. B. preuss. Akad. 24, 430 (1935).

Payne, *Interpretation of the spinning electron with bipolar coordinates*. Phys. Rev. *65*, 39 (1944).

Phillips, *On the magnetic moments of light nuclei*. Phys. Rev. *57*, 160 (1940).

Pound, Purcell and Torrey, *Measurement of magnetic resonance absorption by nuclear moments in a solid*. Phys. Rev. *69*, 681 (1946).

Powers, H. Beyer and Dunning, *Experiments on the magnetic moment of the neutron*. Phys. Rev. *51*, 371 (1937).

Powers, Carrol and Dunning, *Experiments on the magnetic moment of the neutron*. Phys. Rev. *51*, 1112 (1937).

Powers, Carroll, H. Beyer and Dunning, *The sign of the magnetic moment of the neutron*. Phys. Rev. *52*, 38 (1937).

Purcell, Torrey and Pound, *Resonance absorption by nuclear magnetic moments in a solid*. Phys. Rev. *69*, 37 (1946).

Purcell, *Spontaneous emission probabilities at radio frequencies*. Phys. Rev. *69*, 681 (1946).

Purcell, Pound and Bloembergen, *Nuclear magnetic resonance absorption in hydrogen gas*. Phys. Rev. *70*, 986 (1946).

Purcell, Bloembergen and Pound. *Resonance absorption by nuclear magnetic moments in a single crystal of CaF_2*. Phys. Rev. *70*, 988 (1946).

Rabi, Kellogg and Zacharias, *Magnetic moment of the proton*. Phys. Rev. *45*, 761 (1934).

Rabi, Kellogg and Zacharias, *Magnetic moment of the deuton*. Phys. Rev. *45*, 769 (1934).

Rabi, Kellogg and Zacharias, *Magnetic moment of the proton*. Phys. Rev. *46*, 157 (1934).

Rabi, Kellogg and Zacharias, *Magnetic moment of the deuton*. Phys. Rev. *46*, 163 (1934).

Rabi and Cohen, *Measurment of nuclear spin by the method of molecular beams. The nuclear spin of the deuton*. Phys. Rev. *46*, 707 (1934).

Rabi, *On the process of space quantization*. Phys. Rev. *49*, 324 (1936).

Rabi, Kellogg and Zacharias, *The sign of the magnetic moment of the proton*. Phys. Rev. *49*, 421 (1936).

Rabi, *Space quantization in a gyrating magnetic field*. Phys. Rev. *51*, 652 (1937).

Rabi, Zacharias, Millman and Kusch, *A new method of measuring nuclear magnetic moment*. Phys. Rev. *53*, 318 (1938).

Rabi, Millman, Kusch and Zacharias, *Magnetic moments of $_3Li^6$, $_3Li^7$ and $_9F^{19}$*. Phys. Rev. *53*, 495 (1938).

Rabi, Millman, Kusch and Zacharias, *The molecular beam resonance method for measuring nuclear magnetic moments. The magnetic moments of $_3Li^6$, $_3Li^7$ and $_9F^{19}$*. Phys. Rev. *55*, 526 (1939).

Racah, *Theory of hyperfine structure*. Z. Phys. *71*, 431 (1931).

Racah, *Isotopic displacement and hyperfine structure*. Nature *129*, 723 (1932).

Rasetti, *Raman spectra of nitrogen and oxygen*. Z. Phys. *61*, 598 (1930).

Rasmussen, *Hyperfine structure in the hafnium spectrum*. Naturwiss. *23*, 69 (1935).

Rasmussen, *On the nuclear moment of the zinc isotope Zn^{67}*. Z. Phys. *104*, 434 (1936).

Renzetti, *Electric quadrupole moments of Ga^{69} and Ga^{71}. An atomic beam study of the hyperfine structures of the $^2P_{1/2}$ and $^2P_{3/2}$ states of Ga^{69} and Ga^{71}*. Phys. Rev. *57*, 753 (1940).

Ritschl, *The hyperfine structures in copper and gold*. Naturwiss. *19*, 690 (1931).

Ritschl and Sawyer, *Hyperfine structure and Zeeman effect of the resonance lines of Ba II*. Z. Phys. *72*, 36 (1931).

Ritschl, *Hyperfine structure of the arc lines and the nuclear moment of copper*. Z. Phys. *79*, 1 (1932).

Ritschl, *Hyperfine structure in aluminum*. Nature *131*, 58 (1933).

Roberts, Beers and Hill, *The measurement of nuclear spin, magnetic moment and hyperfine structure separation by a microwave frequency modulation method*. Phys. Rev. *70*, 112 (1946).

Robertson, *Magnetic moments and the vibratory electron*. Phil. Mag. *34*, 182 (1943).

Rollin, *Nuclear magnetic resonance and spin lattice equilibrium*. Nature *158*, 669 (1946).

Rose, *Hyperfine structure of singly ionized lead*. Phys. Rev. *47*, 122 (1935).

Rose and Bethe, *Nuclear spins and magnetic moments in the Hartree model*. Phys. Rev. *51*, 205 (1937). Errata, *51*, 993 (1937).

Rose, *The electrical quadrupole and magnetic dipole moments of Li^6 and N^{14}*. Phys. Rev. *56*, 1064 (1939).

Rosenthal and Breit, *The isotope shift in hyperfine structure*. Phys. Rev. *41*, 459 (1932).

Sachs, *Nuclear spins and magnetic moments by the alpha particle model*. Phys. Rev. *55*, 825 (1939).

Sachs, *The magnetic moments of light nuclei*. Phys. Rev. *69*, 611 (1946).

Sachs and Schwinger, *The magnetic moments of H^3 and He^3*. Phys. Rev. *70*, 41 (1946).

Sachs, *On the magnetic moment of the triton*. Phys. Rev. *71*, 457 (1947).

Sakata and Yukawa, *Dirac's generalized wave equation*. Proc. Phys.-Math. Soc. Jap. (3) *19*, 91 (1937).

Schmidt, *Nuclear magnetic moment of $_{78}^{195}Pt$*. Z. Phys. *101*, 486 (1936).

Schmidt, *Magnetic moments of $_{63}^{151}Eu$, $_{75}^{153}Eu$, $_{63}^{185}Re$, $_{75}^{187}Re$, $_{35}^{79}Br$, $_{35}^{81}Br$*. Z. Phys. *108*, 408 (1938).

Schmidt, *Quadrupole moment and magnetic moment of the* $^{127}_{53}I$ *nucleus.* Z. Phys. *112*, 199 (1939).

Schmidt, *Electrical quadrupole moment of the I nucleus.* Z. Phys. *113*, 140 (1939).

Schmidt, *A note on the quadrupole moments of the atomic nuclei.* Naturwiss. *28*, 565 (1940).

Schmidt, *On the quadrupole moment of the atomic nucleus* $^{181}_{73}Ta$. Z. Phys. *121*, 63 (1943).

Schüler and Bruck, *Hyperfine structure in triplet spectra and its meaning in connection with the determination of nuclear moments.* Z. Phys. *56*, 291 (1929).

Schüler, *Nuclear moments of* Li^6 *and* Li^7. Z. Phys. *66*, 431 (1930).

Schüler, *Hyperfine structures and nuclear moments.* Phys. Z. *32*, 667 (1931).

Schüler and Keyston, *Measurements of intensity in certain hyperfine spectral structures of Cd I in relation to nuclear moment and ratio between isotopes.* Z. Phys. *67*, 433 (1938).

Schüler and Keyston, *On an isotopic shift of the hyperfine structure of thallium.* Z. Phys. *70*, 1 (1931).

Schüler and Keyston, *Intensity changes of hyperfine structure lines.* Z. Phys. *71*, 413 (1931).

Schüler and Keyston, *Hyperfine structures and nuclear moments of mercury.* Z. Phys. *72*, 423 (1931).

Schüler and Jones. *Hyperfine structure and nuclear moments of mercury.* Z. Phys. *74*, 631 (1932).

Schüler and Jones, *Hyperfine structures of lead lines in the wave-length interval* $\lambda = 5000$ *to* $\lambda = 8000$. *Confirmation of lead isotope 204.* Z. Phys. *75*, 563 (1932).

Schüler and Westmeyer, *Hyperfine structure of the resonance lines of Sr II.* Naturwiss. *21*, 561 (1933).

Schüler and Westmeyer, *The nuclear moment of tin.* Naturwiss. *21*, 660 (1933).

Schüler and Westmeyer, *Isotope displacement and the nuclear moments of zinc.* Z. Phys. *81*, 565 (1933).

Schüler and Westmeyer, *Isotope displacement effect.* Z. Phys. *82*, 685 (1933).

Schüler and Gollnow, *Atomic weight and the mechanical nuclear moment of protactinium.* Naturwiss. *22*, 511 (1934).

Schüler and Schmidt, *The nuclear moment of cassiopeium (lutecium).* Naturwiss. *22*, 714 (1934).

Schüler and Gollnow, *The nuclear moment of terbium (element number 65).* Naturwiss. *22*, 730 (1934) .

Schüler and Schmidt, *The nuclear moment of scandium,* Sc^{45}. Naturwiss. *22*, 758 (1934).

Schüler and Schmidt, *On the nuclear moments of thulium* (Tu^{169}), *yttrium* (Y^{89}) *and rhodium* (Rh^{103}). Naturwiss. 22, 838 (1934).

Schüler, *Expression of the nuclear moments by nucleus vectors.* Z. Phys. *88*, 323 (1934).

Schüler and Schmidt, *Phenomenon exhibited by the isotopes of samarium.* Z. Phys. *92*, 148 (1934).

Schüler and Schmidt, *The nuclear moment of holmium* (Ho^{165}:). Naturwiss. *23*, 69 (1935).

Schüler and Schmidt, *Deviations of the atomic nucleus from spherical symmetry.* Z. Phys. *94*, 457 (1935).

Schüler and Schmidt, *Deviation of the lutecium nucleus from spherical symmetry.* Z. Phys. *95*, 265 (1935).

Schüler and Schmidt, *On the asymmetry of the electrical charge distribution of the* $^{201}_{80}Hg$ *nucleus.* Z. Phys. *98*, 239 (1935).

Schüler and Schmidt, *Observations on the electrical quadrupole moments of some atomic nuclei and on the magnetic moment of the proton.* Z. Phys. *98*, 430 (1936).

Schüler and Schmidt, *Electrical quadrupole moment of the* $^{209}_{83}Bi$ *nucleus.* Z. Phys. *99*, 717 (1936).

Schüler and Schmidt, *Electric quadrupole moment and magnetic moment of* $^{63}_{29}Cu$ *and* $^{65}_{29}Cu$. Z. Phys. *100*, 113 (1936).

Schüler and Marketu, *Quadrupole moment and magnetic moment of* $^{75}_{33}As$. Z. Phys. *102*, 703 (1936).

Schüler and Korsching, *Quadrupole moment and magnetic moment of* $^{69}_{31}Ga$ *and* $^{71}_{31}Ga$. Z. Phys. *103*, 434 (1936).

Schüler and Schmidt, *Quadrupole moment and magnetic moment of* $^{115}_{49}In$. Z. Phys. *104*, 468 (1937).

Schüler and Korsching, *Nuclear structure and quadrupole moment of* $^{187}_{75}Re$ *and* $^{185}_{75}Re$. Z. Phys. *105*, 168 (1937).

Schüler, Roig and Korsching, *Mechanical moments of* $^{171, 173}Yb$, *quadrupole moment of* ^{173}Yb *and abundance ratio* $^{173}Yb/^{171}Yb$. Z. Phys. *111*, 165 (1938).

Schüler and Korsching, *Magnetic moment of* $^{171, 173}Yb$ *and isotope shifting in Yb I.* Z. Phys. *111*, 386 (1938).

Schüler and Gollnow, *Mechanical and magnetic moments and the quadrupole moment of the rare* ^{176}Cp *nucleus.* Z. Phys. *113*, 1 (1939).

Schwinger, *On the non-adiabatic processes in inhomogeneous fields.* Phys. Rev. *51*, 648 (1937).

Schwinger, *On the spin of the neutron.* Phys. Rev. *52*, 1250 (1937).

Schwinger, *The quadrupole moment of the deuteron and the range of nuclear forces.* Phys. Rev. *60*, 164 (1941).

Shoupp, Bartlett and Dunn, *The magnetic moment of the* Na^{23} *nucleus.* Phys. Rev. *47*, 705 (1935).

Shrader, *Nuclear spin of* Cl^{37}. Phys. Rev. *58*, 475 (1940).

Shrader, Millman and Kusch. *On the nuclear gyromagnetic ratios of the chlorine isotopes.* Phys. Rev. *58*, 925 (1940).

Shrader, *Determination of the nuclear spin of* Cl^{37}. Phys. Rev. *64*, 57 (1943).

Sibaiya, *Structure of iridium lines.* Phys. Rev. *56*, 768 (1939).

Smith, *Determination of nuclear spin from the Stark effect of microwave rotational spectra.* Phys. Rev. *71*, 126 (1947).

Solomon, *Magnetic moment of proton,* C. R., Paris *208*, 1795 (1939).

Tolansky, *Fine structure in the arc spectra of bromine and iodine.* Proc. Roy. Soc. *136*, 585 (1932).

Tolansky, *Nuclear spin of arsenic.* Proc. Roy. Soc. *137*, 541 (1932).

Tolansky, *Nuclear moments of the lighter elements.* Z. Phys. *78*, 71 (1932).

Tolansky, *Nuclear spin of tin.* Proc. Roy. Soc. *144*, 574 (1934).

Tolansky, *Nuclear spins and magnetic moments of the isotopes of antimony.* Proc. Roy. Soc. *146*, 182 (1934).

Tolansky, *Nuclear spin of iodine. Part I. Fine structure in the first spark spectrum.* Proc. Roy. Soc. *149*, 269 (1935).

Tolansky, *Nuclear spin of iodine. Part II. Fine structure in arc spectrum and fine structure perturbation effect.* Proc. Roy. Soc. *152*, 663 (1935).

Tolansky, *Fine structure in the limit terms of the* (^2D) *series of the I* $^+$ *spectrum.* Proc. Phys. Soc., Lond. *48*, 49 (1936).

Tolansky and Lee, *Fine structure in the arc spectrum of platinum.* Proc. Roy. Soc. *158*, 110 (1937).

Tolansky and Forester, *Nuclear spin of iodine. Part III.* Proc. Roy. Soc. *162*, 78 (1938).

Tolansky, *Laws of nuclear structure derived from nuclear spins.* Nature *147*, 269 (1941).

Tolansky and Forester, *Hyperfine structure in the arc spectrum of tin.* Phil. Mag. *32*, 315 (1941).

Tomboulian and Bacher, *On the hyperfine structure interval rule in iridium.* Phys. Rev. *54*, 446 (1938).

Tomboulian and Bacher, *Evidence for a nuclear electric quadrupole moment for* Sb^{123}. Phys. Rev. *58*, 52 (1940).

Torrey, *The sign of the magnetic moment of the potassium 39 nucleus.* Phys. Rev. *51*, 501 (1937).

Torrey, Purcell and Pound, *Theory of magnetic resonance absorption by nuclear moments in solids.* Phys. Rev. *69*, 680 (1946).

Townes and Smythe, *The spin of carbon 13.* Phys. Rev. *56*, 1210 (1939).

Townes, Holden and Merritt, *Rotational spectra of some linear molecules near 1 cm wave-length and nuclear quadrupole moments of bromine and chlorine.* Phys. Rev. *71*, 479 (1947).

Van Vleck, *Formula for the coupling of nuclear quadrupole moments in symmetrical polyatomic molecules.* Phys. Rev. *71*, 468 (1947).

Venkatesaschar and Sibaiya, *Platinum isotopes and their nuclear spin.* Nature *136*, 65 (1935).

Venkatesaschar and Sibaiya, *Iridium isotopes and their nuclear spin*. Nature *136*, 437 (1935).

Venkatesaschar and Sibaiya, *Platinum isotopes and their nuclear spin*. Proc. Indian Acad. Sci. (A) *1*, 955 (1935).

Venkatesaschar and Sibaiya, *Iridium isotopes and their nuclear spins*. Proc. Indian Acad. Sci. *2*, 203 (1935).

Way, *The liquid drop model and nuclear moments*. Phys. Rev. *55*, 963 (1939).

Welles, *Electric quadrupole moments of light and heavy nuclei*. Phys. Rev. *62*, 197 (1942).

Westmeyer, *Hyperfine structure of the red cadmium line 6438 A and of strontium, tin and magnesium*. Z. Phys. *94*, 590 (1935).

White, *Hyperfine structure in singly ionized praseodymium*. Phys. Rev. *34*, 1397 (1929).

White and Ritschl, *Hyperfine structure in the spectrum of neutral manganese*. Phys. Rev. *35*, 1146 (1930).

White and Eliason, *Relative intensity tables for spectrum lines*. Phys. Rev. *44*, 753 (1933).

Wick, *Theory of beta rays and the magnetic moment of the proton*. Atti Accad. Lincei *21*, 170 (1935).

Williams and Granath, *Hyperfine structure of boron, yttrium, rhodium and palladium*. Phys. Rev. *54*, 338 (1938).

Wills, *Nuclear moment of thallium*. Phys. Rev. *45*, 883 (1934).

Wills and Breit, *Nuclear magnetic moment of Na^{23}*. Phys. Rev. *47*, 704 (1935).

Wolff, *The hyperfine structure and nuclear moment of gold*. Phys. Rev. *44*, 512 (1933).

Wood and Dieke, *The nuclear spin of N^{15}*. J. Chem. Phys. *6*, 908 (1938).

Wood and Dieke, *Negative bands of heavy nitrogen molecules*. J. Chem. Phys. *8*, 351 (1940).

Wurm, *The nuclear moment of Se^{80}*. Naturwiss. *20*, 85 (1932).

Young, *The proton's magnetic moment*. Phys. Rev. *52*, 138 (1937).

Zacharias and Kellogg, *The nuclear magnetic moment of N^{15}*. Phys. Rev. *57*, 570 (1940).

Zacharias, *The nuclear spin and magnetic moment of K^{40}*. Phys. Rev. *60*, 168 (1941).

Zacharias, *The nuclear spin and magnetic moment of K^{40}*. Phys. Rev. *61*, 270 (1942).

Zeeman, Gisolf and de Bruin, *Magnetic resolution and nuclear moment of rhenium*. Nature *128*, 637 (1931).

CHAPTER THREE

Scattering and Collision Processes

Adler, *On the slowing down of neutrons by elastic collisions*. Phys. Rev. *60*, 279 (1941).

Ageno, *The albedo of thermal neutrons*. Phys. Rev. *69*, 241 (1946).

Ageno, *The albedo of slow neutrons*. Nuovo Cim. (9) *3*, 13 (1946).

Ageno, Amaldi, Bocciarelli and Trabacchi, *On the scattering of fast neutrons by protons and deuterons*. Phys. Rev. *71*, 20 (1947).

Akhieser and Pomerantschuk, *Coherent scattering of gamma rays of nuclei*. Phys. Z. Sowjet, *11*, 478 (1937).

Akhieser and Pomerantschuk, *On the scattering of low energy neutrons in helium*. J. Phys., USSR *9*, 461 (1945).

Akhieser and Pomerantschuk, *On the elastic scattering of fast charged particles by nuclei*. J. Phys., USSR *9*, 471 (1945).

Alichanian, Alichanow and Weissenberg, *Scattering of relativistic electrons through large angles*. J. Phys., USSR *9*, 280 (1945).

Amaldi, Bocciarelli, Rasetti and Trabacchi, *On the scattering of neutrons from the $C + D$ reaction*. Phys. Rev. *56*, 881 (1939).

Amaldi, Bocciarelli, Caccipuoti and Trabacchi, *Diffraction effects in the scattering of fast neutrons*. Nuovo Cim. (9) *3*, 15 (1946).

Amaldi, Bocciarelli and Trabacchi, *Diffraction phenomena in angular distribution of fast neutrons scattered by lead nuclei. Phys. Rev. *70*, 103 (1946).

Aoki, *On the scattering of fast neutrons of different energy.* Phys. Rev. *55*, 795 (1939).

Aoki, *On the scattering of fast neutrons.* Proc. Phys.-Math. Soc. Jap. (3) *21*, 232 (1939).

Ashkin and Marshak, *Neutron-proton and proton-proton scattering at 200-Mev.* Phys. Rev. *71*, 467 (1947).

Badarau, *Passage of particles through spherical potential barriers.* C. R., Paris *207*, 39 (1938).

Bailey, Phillips and Williams, *The yield of neutrons from deuterons on carbon.* Phys. Rev. *62*, 80 (1942).

Bailey, Bennett, Bergstrahl, Nuckolls, Richards and Williams, *The neutron-proton and neutron-carbon scattering cross sections for fast neutrons.* Phys. Rev. *70*, 583 (1946).

Barbre and Goldhaber, *Resonance scattering of neutrons by manganese.* Phys. Rev. *71*, 141 (1947).

Barschall and Ladenburg, *Elastic and inelastic scattering of fast neutrons.* Phys. Rev. *61*, 129 (1942).

Barschall, Battat and Bright, *Scattering of fast neutrons by boron.* Phys. Rev. *70*, 458 (1946).

Beck and Rodrigues Martins, *Spin inversion processes and nuclear spectroscopy.* Phys. Rev. *62*, 554 (1942).

Bennett, *Stopping power of mica for alpha particles.* Proc. Roy. Soc. *155*, 419 (1936).

Bethe, *Theory of the passage of swift corpuscular rays through matter.* Ann. Phys., Leipzig *5*, 325 (1930).

Bethe, *Scattering of electrons.* Z. Phys. *76*, 293 (1932).

Bethe and Peierls, *Scattering of neutrons by protons.* Proc. Roy. Soc. *149*, 176 (1935).

Bethe, Rose and Smith, *Multiple scattering of electrons.* Proc. Amer. Phil. Soc. *78*, 573 (1938).

Bethe and Oppenheimer, *Reaction of radiation of electron scattering and Heitler's theory of radiation damping.* Phys. Rev. *70*, 451 (1946).

Bethe, *Multiple scattering and the mass of the meson.* Phys. Rev. *70*, 821 (1946).

Beyer, H. G., and Whitaker, *Interference phenomena in the scattering of slow neutrons.* Phys. Rev. *57*, 976 (1940).

Bhabha, *Scattering of charged mesons.* Proc. Indian Acad. Sci. (A) *13*, 9 (1941).

Blackett and Champion, *Scattering of slow alpha particles by helium.* Proc. Roy. Soc. *130*, 380 (1931).

Blackett and Lees, *Further investigations with a Wilson chamber. II. The range and velocity of recoil atoms.* Proc. Roy. Soc. *134*, 658 (1932).

Blackett, *Loss of energy of alpha particles and H— particles.* Proc. Roy. Soc. *135*, 132 (1932).

Blatt, *On the Heitler theory of radiation damping.* Phys. Rev. *71*, 468 (1947).

Bleuler, Scherrer and Zünti, *Single scattering of fast electrons.* Phys. Rev. *61*, 95 (1942).

Bloch, *Stopping power of matter for swiftly moving charged particles.* Ann. Phys., Leipzig *16*, 285 (1933).

Bloch, *Stopping power of atoms with several electrons.* Z. Phys. *81*, 363 (1933).

Bloch and Hamermesh, *Further results on magnetic scattering of neutrons.* Phys. Rev. *61*, 203 (1942).

Blokhintzev, *An equation for the scattering of particles, taking into account the reaction of emission* (in English). C. R., URSS, *53*, 201 (1946).

Bohm and Thomas, *High energy scattering of deuterons by protons.* Phys. Rev. *61*, 203 (1942).

Bohr, *On the theory of the decrease of velocity of moving electrified particles on passing through matter.* Phil. Mag. (6) *25*, 10 (1913).

Bohr, *On the decrease of velocity of swiftly moving electrified particles on passing through matter.* Phil. Mag. (6) *30*, 581 (1915).

Bonner, *Neutron-proton scattering and the disintegration of deuterium by deuterons.* Phys. Rev. *52*, 685 (1937).

Booth and Wilson, *Scattering of neutral mesons.* Proc. Camb. Phil. Soc. *36*, 446 (1940).

Bradbury, Bloch, Tatel and Ross, *The scattering and absorption cross section of neutrons in cobalt.* Phys. Rev. *52,* 1023 (1937).

Breit, Condon and Present, *Theory of scattering of protons by protons.* Phys. Rev. *50,* 825 (1936).

Breit, *The scattering of slow neutrons by bound protons. I. Methods of calculation.* Phys. Rev. *71,* 215 (1937).

Breit, Thaxton and Eisenbud, *Analysis of experiments on the scattering of protons.* Phys. Rev. *55,* 1018 (1939).

Breit, Kittel and Thaxton, *Note on proton wave anomalies in proton-proton scattering.* Phys. Rev. *57,* 255 (1946).

Breit and Zilsel, *The scattering of slow neutrons by bound protons. II. Harmonic binding—neutrons of zero energy.* Phys. Rev. *71,* 232 (1947).

Brickwedde, Dunning, Hoge and Manley, *Neutron scattering cross sections of para and ortho-hydrogen, and of N_2, O_2 and H_2O.* Phys. Rev. *54,* 266 (1935).

Brown and Plesset, *On the equality of the proton-proton and proton-neutron interactions.* Phys. Rev. *56,* 841 (1939).

Brubaker, *The scattering of alpha particles by argon, oxygen and neon.* Phys. Rev. *54,* 1011 (1938).

Brubaker, *The scattering of alpha particles by nitrogen.* Phys. Rev. *56,* 1181 (1939).

Buckingham and Massey, *The scattering of neutrons by deuterons and the nature of nuclear forces.* Proc. Roy. Soc. *179,* 123 (1941).

Buechner, Burrill, Sperduto, Van de Graaff and Feshbach, *Further experiments on the elastic single scattering of electrons by nuclei.* Phys. Rev. *71,* 142 (1947).

Bunge, *Neutron-proton scattering at 8.8 and 13 Mev.* Nature *156,* 301 (1945).

Busshard and Scherrer, *Dependence of scattering of rapid electrons at the nitrogen nucleus on energy* (in German). Helv. Phys. Acta *14,* 85 (1941).

Carroll, *The interaction of slow neutrons with nuclei.* Phys. Rev. *60,* 702 (1942).

Chadwick, *Charge on the atomic nucleus and the law of force; valiidty of the inverse square law for the platinum atom.* Phil. Mag. (6) *40,* 734 (1920).

Chadwick and Bieler, *Collisions of alpha particles with hydrogen nuclei.* Phil. Mag. (6) *42,* 923 (1921).

Chadwick, *Scattering of alpha particles in helium.* Proc. Roy. Soc. *128,* 114 (1930).

Champion and Barber, *Elastic scattering of fast beta particles by atomic nuclei.* Phys. Rev. *55,* 111 (1939).

Champion, *Elastic scattering of fast electrons by nitrogen nuclei.* Nature *148,* 727 (1941).

Champion and Powell, *Applications of the photographic method to problems in nuclear physics. II. The scattering of 8.8 and 13 Mev neutrons by protons.* Proc. Roy. Soc. *183,* 64 (1944).

Cohen, Goldsmith and Schwinger, *The neutron-proton scattering cross section.* Phys. Rev. *55,* 106 (1939).

Coon, Davis and Barschall, *Angular distribution of neutron-deuteron scattering for 2.5 Mev neutrons.* Phys. Rev. *70,* 104 (1946).

Cooper and Morrison, *Internal scattering of gamma rays.* Phys. Rev. *57,* 862 (1940).

Coster, de Vries and Diemer, *The resonance levels for neutron capture. III. Scattering of resonance neutrons* (in English). Physica *10,* 299 (1943).

Creutz, *The resonance scattering of protons by lithium.* Phys. Rev. *55,* 819 (1939).

Creutz, *Analysis of proton-proton scattering data.* Phys. Rev. *56,* 893 (1939).

Creutz and Wilson, *Evidence of P— wave proton-proton scattering.* Phys. Rev. *61,* 388 (1942).

Dicke and Marshall, *Inelastic scattering of protons.* Phys. Rev. *63,* 86 (1943).

Diemer, *Further experiments upon the scattering of silver 22 sec. resonance neutrons* (in English). Physica *11,* 481 (1946).

Dmitrieff, *Scattering of fast neutrons by protons.* Phys. Z. Sowjet. *11,* 225 (1937).

Drăganu, *Passage of very fast protons through matter.* C. R., Paris *205,* 897 (1937).

Duncanson, *Calculations on the range-velocity relations for alpha particles and protons.* Proc. Camb. Phil. Soc. *30,* 102 (1934).

Dunlap and Little, *The scattering of fast neutrons by lead.* Phys. Rev. *60,* 693 (1941).

Dunning, *Emission and scattering of neutrons.* Phys. Rev. *45,* 586 (1934).

Eaton, *Collision of alpha particles with neon nuclei.* Phys. Rev. *48,* 921 (1935).

Eddington, *Theory of scattering of protons by protons.* Proc. Roy. Soc. *162,* 155 (1937).

Fay, *The scattering of fast neutrons by heavy nuclei.* Phys. Rev. *50,* 560 (1936).

Feather, *Collisions of alpha particles with fluorine nuclei.* Proc. Roy. Soc. *141,* 194 (1933).

Feather, *Collisions of neutrons with light nuclei.* Proc. Roy. Soc. *142,* 689 (1933).

Fermi, *On the recombination of neutrons and protons.* Phys. Rev. *48,* 570 (1938).

Fermi and Marshall, *Phase of neutron scattering.* Phys. Rev. *70,* 103 (1946).

Feshbach, Peaslee and Weisskopf, *On the scattering and absorption of particles by atomic nuclei.* Phys. Rev. *71,* 145 (1947).

Fisk and Morse, *The elastic scattering of neutrons by protons.* Phys. Rev. *51,* 54 (1937).

Flügge, *Cross sections for reactions between very light nuclei.* Z. Phys. *108,* 545 (1938).

Flügge, *On the derivation of the Breit-Wigner formula.* Z. Naturforsch. *1,* 121 (1946).

Fowler and Oppenheimer, *Scattering and loss of energy of fast electrons and positrons in lead.* Phys. Rev. *54,* 320 (1938).

Fowler, *Scattering of fast electrons.* Phys. Rev. *54,* 773 (1938).

Géhéniau and van Isacker, *Effective cross section of scattering of particles with spin n/2.* C. R., Paris *222,* 484 (1946).

Geiger, *The scattering of the alpha particles by matter.* Proc. Roy. Soc. *83,* 492 (1910).

Geiger, *The ionization produced by an alpha particle. Part II. Connection between ionization and absorption.* Proc. Roy. Soc. *83,* 505 (1910).

Geiger, *Range measurements with alpha rays.* Z. Phys. *8,* 45 (1921).

Gentner, *Absorption and scattering of hard gamma rays.* Phys. Z. *38,* 836 (1937).

Gibert and Rossel, *Effect of temperature on neutron-proton scattering* (in French). Helv. Phys. Acta *19,* 285 (1946).

Ginsburg, *On nuclear scattering of mesotrons.* J. Phys., USSR *10,* 298 (1946).

Goldhaber and Yalow, *Resonance scattering of group neutrons.* Phys. Rev. *69,* 47 (1946).

Goldsmith, Cohen and Dunning, *Scattering of slow neutrons by uranium.* Phys. Rev. *55,* 1124 (1939).

Goldstein, *Nuclear collisions.* J. Phys. Rad. *9,* 96 (1938).

Goldstein, *Nuclear collisions.* J. Phys. Rad. *10,* 23 (1939).

Goloborodko and Leipunski, *Scattering of photoneutrons from deuterium by the nuclei of atoms of light elements.* Phys. Rev. *56,* 891 (1939).

Goloborodko, *Scattering of neutrons by protons.* J. Phys., USSR *8,* 13 (1944).

Goloborodko, *Anomalous scattering of photoneutrons by nuclei of heavy elements.* J. Phys. USSR *8,* 106 (1944).

Grahame, Seaborg and Gibson, *Inelastic scattering of fast neutrons.* Phys. Rev. *51,* 590 (1937).

Grahame and Seaborg, *Elastic and inelastic scattering of fast neutrons.* Phys. Rev. *53,* 795 (1938).

Grahame, *Resonance scattering of fast neutrons.* Phys. Rev. *69,* 369 (1946).

Gupta and Majumdar, *On the collision between meson and electron.* Proc. Nat. Inst. Sci. India *8,* 199 (1942).

Gupta, *On the elastic scattering of the fast mesons.* Proc. Nat. Inst. Sci. India *8,* 369 (1942).

Guth, *Nuclear spectroscopy and inelastic scattering of particles by nuclei.* Phys. Rev. *68,* 279 (1945).

Guth, *Nuclear spectroscopy and energy distribution of charged particles inelastically scattered by nuclei.* Phys. *68,* 280 (1945).

Guth, *Nuclear processes by polarization scattering of deuterons.* Phys. Rev. *69,* 47 (1946).

Hafstad, Heydenburg and Tuve, *Scattering of protons by protons*. Phys. Rev. *49*, 402 (1936).

Halpern, Lueneberg and Clark, *On multiple scattering of neutrons. I. Theory of the albedo of a plane boundary*. Phys. Rev. *53*, 173 (1938).

Halpern and Johnson, *On the magnetic scattering of neutrons*. Phys. Rev. *55*, 898 (1939).

Hamermesh, *Magnetic scattering of neutrons*. Phys. Rev. *61*, 17 (1942).

Hamermesh and Schwinger, *The scattering of slow neutrons by ortho- and para-deuterium*. Phys. Rev. *69*, 145 (1946).

Hamilton and Peng, *On the production of mesons by light quanta and related processes*. Proc. R. Irish Acad. *49*, 197 (1944).

Hansen and Wrenshall, *Collisions of alpha particles with chlorine nuclei*. Phys. Rev. *57*, 750 (1940).

Harish-Chandra, *On the scattering of scalar mesons*. Proc. Indian Acad. Sci. (A) *21*, 135 (1945).

Harkins, Gans, Kamen and Newson, *The scattering of protons in collision with neutrons*. Phys. Rev. *47*, 511 (1935).

Harkins, Kamen, Newson and Gans, *Neutron-proton interaction: the scattering of neutrons by protons*. Phys. Rev. *50*, 980 (1936).

Heisenberg, *Passage of very high energy particles through atomic nuclei*. Nuovo Cim. *15*, 31 (1938).

Heitler, *Influence of radiation damping on the scattering of light and mesons by free particles*. Proc. Camb. Phil. Soc. *37*, 291 (1941).

Heitler and Peng, *Anomalous scattering of mesons*. Phys. Rev. *62*, 81 (1942).

Heitler, *Quantum theory of damping*. Phys. Rev. *70*, 795 (1946).

Hellund, *Phase series*. Phys. Rev. *59*, 395 (1941).

Henneberg, *Electron scattering by heavy atoms*. Z. Phys. *83*, 555 (1933).

Herb, Kerst, Parkinson and Plain, *The scattering of protons by protons*. Phys. Rev. *55*, 998 (1939).

Heydenburg, Hafstad and Tuve, *The scattering of protons by protons*. Phys. Rev. *56*, 1078 (1939).

Heydenburg and Roberts, *Deuteron-deuteron, proton-helium, and deuteron-helium scattering*. Phys. Rev. *56*, 1092 (1939).

Heydenburg and Ramsey, *The scattering of 1-3 Mev protons by helium*. Phys. Rev. *60*, 42 (1941).

Höcker, *Effective cross section of reactions between neutrons and deuterons*. Phys. Z. *43*, 236 (1942).

Hoisington, Share and Breit, *Effects of shape of potential energy wells detectable by experiments on proton-proton scattering*. Phys. Rev. *56*, 884 (1939).

Horsley, *Schematic representation of the anomalous scattering by suitable nuclear fields*. Phys. Rev. *48*, 1 (1935).

Hsüeh and Ma, *Approximate solutions of the integral equations in scattering problems*. Phys. Rev. *67*, 303 (1945).

Huber, Huber and Scherrer, *Effective cross section for elastic scattering and nuclear reactions of fast neutrons in N_2 (in German)*. Helv. Phys. Acta *13*, 212 (1940).

Hudspeth and Dunlap, *Anomalous scattering of neutrons by helium and the d-d neutron spectrum*. Phys. Rev. *57*, 971 (1940).

Hulthén, *On the meson field theory of nuclear forces and the scattering of fast neutrons by protons (in English)*. Ark. Mat. Astr. Fys. *29A*, No. 33 (1943).

Huthén, *On the meson field theory of nuclear forces and the scattering of fast neutrons by protons, II (in English)*. Ark. Mat. Astr. Fys. *30A*, No. 9 (1943).

Hulthén, *On the meson field theory of nuclear forces and the scattering of fast neutrons by protons, III (in English)*. Ark. Mat. Astr. Fys. *31A*, No. 15 (1944).

Hulthén, *On the scattering of neutrons by protons*. Phys. Rev. *63*, 383 (1943).

Hulthén, *High energy neutron-proton scattering and the saturation problem*. Phys. Rev. *67*, 193 (1945).

Jauch, *Scattering problems in the electron pair theory* (in German). Helv. Phys. Acta *15*, 221 (1942).

Jauch, *Neutron-proton scattering and the meson theory of nuclear forces*. Phys. Rev. *71*, 125 (1945).

Kahan, *Collisions of fast neutrons with protons*. C. R., Paris *206*, 742 (1938).

Kar and Basu, *Neutron-proton scattering*. Phil. Mag. (7) *27*, 76 (1939).

Kar and Basu, *Proton-proton scattering*. Phil. Mag. (7) *29*, 200 (1940).

Kar, *On nuclear scattering*. Indian J. Phys. *15*, 113 (1941).

Kar, *Proton-proton interaction and the Yukawa particle*. Indian J. Phys. *16*, 187 (1942).

Kar and Roy, *The theory of neutron-proton scattering*. Indian J. Phys. *17*, 321 (1943).

Kar, *The distortion of plane x wave and its effect on elastic scattering*. Indian J. Phys. *18*, 144 (1944).

Kar and Basu, *The scattering of fast beta particles by electrons*. Indian J. Phys. *18*, 223 (1944).

Kar and Mitra, *Proton-proton scattering at low velocity*. Indian J. Phys. *18*, 303 (1944).

Kikuchi and Aoki, *The scattering of the D-D neutrons*. Phys. Rev. *55*, 108 (1939).

Kikuchi, Aoki and Wakatuki, *On the angular distribution of the fast neutrons scattered by atoms*. Phys. Rev. *55*, 1264 (1939).

Kikuchi and Aoki, *The scattering of fast neutrons by atoms*. Proc. Phys.-Math. Soc. Jap. (3) *21*, 75 (1939).

Kikuchi, Aoki and Wakatuki, *On the angular distribution of the fast neutrons scattered by atoms*. Proc. Phys. Math. Soc. Jap. (3) *21*, 410, 656 (1939).

Kimura, *Scattering of slow neutrons by some elements*. Proc. Phys.-Math. Soc. Jap. (3) *22*, 391 (1940).

Kimura, *On the resonance level of Hg at negative energy*. Phys. Rev. *60*, 688 (1941).

Kimura, *On the variation in scattering and absorption cross sections with neutron energy*. Proc. Phys.-Math. Soc. Jap. (3) *24*, 569 (1942).

Kimura, *On the variation in scattering and absorption cross sections with resonance neutron energy. II*. Proc. Phys.-Math. Soc. Jap. (3) *25*, 481 (1943).

Kimura and Hasiguti, *On the scattering of thermal neutrons by solids*. Proc. Phys.-Math. Soc. Jap. (3) *25*, 530 (1943).

Kittel and Breit, *Note on the scattering of neutrons by protons*. Phys. Rev. *56*, 744 (1939).

Klein and Nishina, *Scattering of radiation by free electrons in the new relativistic quantum dynamics of Dirac*. Z. Phys. *52*, 853 (1928).

Kobayasi and Utiyama, *Pair creation of mesons by gamma rays and the "bremsstrahlung" of mesons in the nuclear field*. Tokyo Inst. Phys. Chem. Res. *37*, 221 (1940).

Kruger, Shoupp and Stallmann, *Scattering of neutrons of homogeneous energy by protons*. Phys. Rev. *51*, 1021 (1937).

Kruger, Shoupp and Stallmann, *The scattering of protons by neutrons*. Phys. Rev. *52*, 678 (1937).

Kruger, Shoupp, Watson and Stallmann, *The scattering of neutrons by deuterons*. Phys. Rev. *53*, 1014 (1938).

Kulchitsky and Latyshev, *The scattering of fast electrons*. J. Phys., USSR *5*, 249 (1941).

Kulchitsky and Latyshev, *The multiple scattering of fast electrons*. Phys. Rev. *61*, 254 (1941).

Kurie, *The collisions of neutrons with protons*. Phys. Rev. *44*, 463 (1933).

Kuronuma, Sueoka and Toya, *Phase shift in deuteron-deuteron scattering*. Proc. Phys.-Math. Soc. Jap. (3) *22*, 862 (1940).

Landau and Smorodinsky, *On the theory of scattering of protons by protons*. J. Phys., USSR *8*, 154 (1944).

Langer and Daly, *The scattering of thermal neutrons in polycrystalline materials*. Phys. Rev. *71*, 464 (1947).

Langevin, *Collisions between fast neutrons and nuclei of any mass*. Ann. Phys., Paris *17*, 303 (1943).

Laporte, *Elastic scattering of Yukawa particles. I*. Phys. Rev. *54*, 905 (1938).

Laslett, *The magnetic scattering of slow neutrons.* Phys. Rev. *51,* 22 (1937).

Libby and Long, *The scattering of slow neutrons by gaseous ortho- and para-hydrogen: spin dependence on the n-p force.* Phys. Rev. *55,* 339 (1939).

Lifshitz, *Transfer of neutrons between nuclei in heavy nuclei in colllisions* (in Russian). J. Exp. Theor. Phys., USSR *9,* 237 (1939).

Little, Long and Mandeville, *The scattering of neutrons by magnesium.* Phys. Rev. *69,* 414 (1946).

Lopes, *High energy neutron-proton scattering and the meson theory of nuclear forces with a strong coupling.* Phys. Rev. *69,* 252 (1946).

Lopes, *High energy neutron-proton scattering and the meson theory of nuclear forces with strong coupling.* Phys. Rev. *70,* 5 (1946).

MacPhail, *Anomalous scattering of fast neutrons.* Phys. Rev. *57,* 669 (1940).

Ma, *Calculations of the scattering of mesons by the matrix* (in English). Sci. Rec. Acad. Sinica. *1,* 123 (1942).

Ma, *Relativistic treatment of the scattering of charged mesons under influence of radiation damping* (in English). Sci. Rec. Acad. Sinica *1,* 381 (1945).

Ma, *On a general condition of Heisenberg for the S matrix.* Phys. Rev. *71,* 195 (1947).

Manley, Agnew, Barschall, Bright, Coon, Graves, Jorgensen and Waldman, *Elastic back scattering of d-d neutrons.* Phys. Rev. *70,* 602 (1946).

Mano, *Retardation of alpha particles in hydrogen.* C. R., Paris *197,* 47 (1933).

Mano, *Absorption of alpha particles.* Ann. Phys., Paris *1,* 407 (1934).

Mano, *Absorption of alpha rays and H particles by matter.* J. Phys. Rad. *5,* 628 (1934).

Manu, *Retardation of alpha rays by magnesium, zinc and molybdenum.* Bull. Soc. Roumaine de Phys. *41,* 33 (1940).

Marsden and Taylor, *The decrease in velocity of alpha particles in passing through matter.* Proc. Roy. Soc. *88,* 443 (1913).

Marshak and Weisskopf, *On the scattering of mesons of spin* $\frac{1}{2}$ $\hbar$ *by atomic nuclei.* Phys. Rev. *59,* 130 (1941).

Massey and Burhop, *The probability of K shell ionization of silver by cathode rays.* Phys. Rev. *48,* 468 (1935).

Massey and Mohr, *Interaction of light nuclei. Part I.* Proc. Roy. Soc. *148,* 206 (1935).

Massey and Buckingham, *Nature of interaction between neutron and proton from scattering experiments.* Proc. Roy. Soc. *163,* 281 (1937).

Massey and Corben, *Elastic collisions of mesons with electrons and protons.* Proc. Camb. Phil. Soc. *35,* 463 (1939).

Massey and Buckingham, *The collision of neutrons with deuterons and the reality of exchange forces.* Phys. Rev. *71,* 558 (1947).

Members of the Nuclear Physics Laboratory of the Physical Institute of Osaka Imperial University, *The scattering of neutrons by protons.* Proc. Phys.-Math. Soc. Jap. (3) *23,* 855 (1941).

Mitchell and Murphy, *Scattering of neutrons.* Phys. Rev. *47,* 881 (1935).

Mitchell and Murphy, *Scattering of slow neutrons.* Phys. Rev. *48,* 653 (1935).

Mitchell, Murphy and Langer, *Continuation of work on scattering of slow neutrons.* Phys. Rev. *49,* 400 (1936).

Mitchell, Murphy and Whitaker, *Scattering of slow neutrons. I.* Phys. Rev. *50,* 133 (1936).

Mitchell and Varney, *Neutron scattering cross section as a function of energy.* Phys. Rev. *52,* 282 (1937).

Miyazima, *Spin orbit interaction between elementary particles and angular asymmetry in neutron-proton scattering.* Proc. Phys.-Math. Soc. Jap. (3) *22,* 188 (1940).

Miller, *Passage of hard beta rays through matter.* Ann. Phys., Leipzig *14,* 531 (1932).

Mohr and Pringle, *Collision forces between light nuclei.* Nature *137,* 865 (1936).

Morse, Fisk and Schiff, *Collision of neutron and proton.* Phys. Rev. *50,* 748 (1936).

Morse, Fisk and Schiff, *Collision of neutron and proton. II.* Phys. Rev. *51,* 706 (1937).

Mott, *The exclusion principle and aperiodic systems.* Proc. Roy. Soc. *125,* 222 (1929).

Mott, *Collision between two electrons.* Proc. Roy. Soc. *126,* 259 (1930).

Motz and Schwinger, *Neutron-deuteron scattering cross section.* Phys. Rev. *57,* 162 (1940).

Muehlhause and Goldhaber, *Capture cross sections for slow neutrons. II. Small capture cross sections.* Phys. Rev. *70,* 85 (1946).

Muto, *Radiative collisions of neutrons with protons.* Phys. Rev. *59,* 837 (1941).

Nelson and Oppenheimer, *Pair theories of meson scattering.* Phys. Rev. *61,* 202 (1942).

Nix and Clement, *Thermal neutrons scattering studies in metals.* Phys. Rev. *68,* 159 (1945).

Nogami, *Elastic collisions by protons and neutrons.* Tokyo Inst. Phys. Chem. Res. *36,* 244 (1939).

Nogami, *On the elastic collisions of neutrons and alpha particles.* Proc. Phys.-Math. Soc. Jap. (3) *24,* 26 (1942).

Nonaka, *On the cross sections of inelastic scattering of d-d neutrons.* Proc. Phys.-Math.. Soc. Jap. (3) *25,* 227 (1943).

Nuckolls, Bailey, Bennett, Bergstrahl, Richards and Williams, *The total scattering cross sections of deuterium and oxygen for fast neutrons.* Phys. Rev. *70,* 805 (1946).

Nuckolls, Bailey, Bennett, Bergstrahl, Richards and Williams, *The total scattering cross sections of deuterium and oxygen for fast neutrons.* Phys. Rev. *71,* 140 (1947).

O'Ceallaigh and MacCarthaigh, *A convenient method of measuring the intensity of nuclear scattering of fast beta particles in the cloud chamber.* Proc. R. Irish Acad. *50A,* 13 (1944).

Oleson, Chao and Crane, *The multiple scattering of fast electrons.* Phys. Rev. *60,* 378 (1941).

Ornstein, *Scattering of neutrons in matter. Part V.* Proc. K. Akad. Amsterdam *40,* 464 (1937).

Pais, *On the scattering of fast neutrons by protons.* Proc. Camb. Phil. Soc. *42,* 45 (1946).

Parkinson, Herb, Bellamy and Hudson, *The range of protons in aluminum and air.* Phys. Rev. *52,* 75 (1937).

Petukhov and Vyshinsky, *Nuclear scattering of electrons in thin metallic films. I.* J. Phys., USSR *4,* 235 (1941).

Petukhov and Vyshinsky, *The nuclear scattering of electrons in thin metallic films. II.* J. Phys., USSR *5,* 137 (1941).

Philipp, *Angular scattering of neutrons by protons.* Z. Phys. *107,* 683 (1937).

Plesset and Brown, *Scattering of slow neutrons by protons.* Proc. Nat. Acad. Sci. *25,* 600 (1939).

Pollard and Margenau, *Collisions of alpha particles in deuterium.* Phys. Rev. *47,* 833 (1935).

Pollard, *Large angle scattering of gamma radiation.* Phys. Rev. *71,* 134 (1947).

Pomerantschuk, *Scattering of slow neutrons in a crystal lattice.* Phys. Z. Sowjet. *13,* 65 (1938).

Pose and Diebner, *Scattering of alpha particles by H nuclei.* Z. Phys. *90,* 773 (1934).

Powell, Heitler and Champion, *Neutron-proton scattering at high energies.* Nature *146,* 716 (1940).

Powers, *The magnetic scattering of neutrons.* Phys. Rev. *54,* 827 (1938).

Primakoff, *On the relation of proton-deuteron scattering.* Phys. Rev. *52,* 1000 (1937).

Ragan, Kanne, Taschek, *The scattering of protons by protons from 200 to 300 Kev.* Phys. Rev. *60,* 628 (1941).

Ray and Champion, *Single scattering of fast beta particles by protons.* Nature *158,* 753 (1946).

Riezler, *Scattering of alpha particles by light elements.* Proc. Roy. Soc. *134,* 154 (1932).

Riezler, *Scattering of alpha particles from polonium by O_2 and N_2.* Ann. Phys., Leipzig *23,* 198 (1935).

Riezler, *Scattering of alpha particles on carbon.* Ann. Phys., Leipzig *38,* 304 (1940).

Rose, *The multiple scattering and stopping of electrons.* Phys. Rev. *51*, 1024 (1937).

Rose, *On the resonance scattering of alpha particles.* Phys. Rev. *57*, 958 (1940).

Rosenfeld, *Penetration of fast nucleons into heavy atomic nuclei.* Nature *156*, 141 (1945).

Ruark, *Multiple scattering of charged particles.* Phys. Rev. *57*, 62 (1940).

Rutherford, *The scattering of alpha and beta particles by matter and the structure of the atom.* Phil. Mag. (6) *21*, 669 (1911).

Rutherford, *The capture and loss of electrons by alpha particles.* Phil. Mag. (6) *47*, 277 (1924).

Rutherford and Chadwick, *Scattering of alpha particles by helium.* Phil. Mag. (7) *4*, 605 (1927).

Sachs and Teller, *The scattering of slow neutrons by molecular gases.* Phys. Rev. *60*, 18 (1941).

Sakata and Tanikawa, *On the capture of the mesotron by the atomic nucleus.* Proc. Phys.-Math. Soc. Jap. (3) *21*, 58 (1939).

Scherrer and Zünti, *Nuclear scattering of rapid electrons in argon* (in German). Helv. Phys. Acta *14*, 111 (1941).

Schiff, *Inelastic collision of deuteron and deuteron.* Phys. Rev. *51*, 783 (1937).

Schiff, *Scattering of neutrons by deuterons.* Phys. Rev. *52*, 149 (1937).

Schwinger, *On the magnetic scattering of neutrons.* Phys. Rev. *51*, 544 (1937).

Schwinger and Teller, *The scattering of neutrons by ortho- and para-hydrogen.* Phys. Rev. *51*, 775 (1937).

Schwinger and Teller, *The scattering of neutrons by ortho- and para-hydrogen.* Phys. Rev. *52*, 286 (1937).

Schwinger, *Polarization of neutrons by resonance scattering in helium.* Phys. Rev. *69*, 681 (1946).

Seaborg, Gibson and Grahame, *Inelastic scattering of fast neutrons.* Phys. Rev. *52*, 408 (1937).

Seeger and Teller, *On the inelastic scattering of neutrons by crystal lattices.* Phys. Rev. *62*, 37 (1942).

Shutt, *On the electrical and anomalous scattering of mesotrons.* Phys. Rev. *61*, 6 (1942).

Shutt, *Some experimental results concerning mesotrons.* Phys. Rev. *69*, 261 (1946).

Sigrist, *Scattering of rapid electrons by iodine nuclei* (in German). Helv. Phys. Acta *16*, 471 (1943).

Simons, *The neutron-proton scattering cross section.* Phys. Rev. *55*, 792 (1939).

Sleator, *Fast neutron collision cross sections of carbon and hydrogen.* Phys. Rev. *69*, 681 (1946)

Smirnov and Vonsovsky, *The effect of long-range order in alloys on the scattering of slow neutrons.* J. Phys., USSR *5*, 263 (1941).

Smith, *Theoretical range-energy values for protons in air and aluminum.* Phys. Rev. *71*, 32 (1946).

Smorodinsky, *On the scattering of neutrons by protons.* J. Phys., USSR *8*, 219 (1944).

Snyder, *Theory of multiple scattering.* Phys. Rev. *71*, 478 (1947).

Soden, *Ionization of the K level by electron collision.* Ann. Phys., Leipzig *19*, 409 (1934).

Sokolow, *Scattering of mesons taking damping into account.* J. Phys., USSR *5*, 231 (1941).

Staub and Stephens, *Anomalous scattering of neutrons by helium.* Phys. Rev. *55*, 131 (1939).

Staub and Tatel, *Resonance scattering of neutrons in helium.* Phys. Rev. *57*, 936 (1940).

Stückelberg, *Radiation damping in the quantum theory* (in French). Helv. Phys. Acta *16*, 427 (1943).

Sundarachar and Streib, *The elastic scattering of neutrons in deuterium.* J. Mysore Univ. *3*, 55 (1942).

Sandarachar and Streib, *Scattering of neutrons in deuterium.* Nature *51*, 149 (1942).

Tanikawa and Yukawa, *On the scattering of mesons by nuclear particles.* Proc. Phys.-Math. Soc. Jap. (3) *23*, 445 (1941).

Taschek, *Scattering of protons by deuterium.* Phys. Rev. *61*, 13 (1942).

Tatel, *The angular distribution of protons scattered by high energy neutrons.* Phys. Rev. *61*, 450 (1942).

Taylor, *Interaction energy of two alpha particles at close distances determined from the anomalous scattering in helium.* Proc. Roy. Soc. *134*, 103 (1931).

Taylor, *Anomalous scattering of alpha particles by hydrogen and helium.* Proc. Soc. *136*, 605 (1932).

Teller, *Interference of neutron waves in ortho- and para-hydrogen.* Phys. Rev. *49*, 420 (1936).

Thaxton and Monroe, *Phase shifts (K_1) for the square well.* Phys. Rev. *56*, 616 (1939).

Thaxton and Hoisington, *Phase shift calculations for p-p scattering at high energies.* Phys. Rev. *56*, 1194 (1939).

Thaxton, *Scattering of protons from 200 to 300 Kev.* Phys. Rev. *60*, 173 (1941).

Thaxton, *Note on the low energy scattering of protons by protons.* Indian J. Phys. *16*, 133 (1942).

Tuve, Heydenburg and Hafstad, *The scattering of protons by protons.* Phys. Rev. *49*, 432 (1936).

Urban, *The scattering of fast electrons by nuclei.* Ann. Phys., Leipzig *43*, 557 (1943).

Van de Graaff, Buechner and Feshbach, *Experiments on the elastic single scattering of electrons by nuclei.* Phys. Rev. *69*, 453 (1946).

Van Isacker, *Certain formulas associated with plane waves of particles with spin $nh/2\pi$.* C. R., Paris 222, 375 (1946).

Van Vleck, *On the theory of the forward scattering of neutrons by paramagnetic media.* Phys. Rev. *55*, 924 (1939).

Wakatuki, *Angular distribution of fast neutrons, scattered by the atoms.* Proc. Phys.-Math. Soc. Jap. (3) *22*, 430 (1940).

Wang, *On the deceleration of high energy protons and neutrons by emanation of mesotrons.* Z. Phys. *115*, 431 1940).

Weinberg, *Scattering in the pair theory of nuclear forces.* Phys. Rev. *59*, 776 (1941).

Weinstock, *Inelastic scattering of slow neutrons.* Phys. Rev. *63*, 139 (1943).

Weinstock, *Inelastic scattering of slow neutrons.* Phys. Rev.. *65*, 1 (1944).

Wells, *The scattering of protons on protons.* Phys. Rev. *47*, 591 (1935).

Wentzel, *Theory of anomalous alpha ray scattering.* Z. Phys. Chem. *90*, 754 (1934).

Wentzel, *Anistropy of proton-neutron scattering and symmetrical meson theory* (in German). Helv. Phys. Acta *18*, 430 (1945).

Wheeler, *Scattering of alpha particles in helium.* Phys. Rev. *59*, 16 (1941).

Wheeler, *Mechanism of capture of slow mesons.* Phys. Rev. *71*, 320 (1947).

Wheeler, *Mechanism of absorption of negative mesons.* Phys. Rev. *71*, 462 (1947).

Whitaker, *Magnetic scattering of neutrons.* Phys. Rev. *52*, 384 (1937).

Whitaker, H. G. Beyer and Dunning, *Scattering of slow neutrons by paramagnetic salts.* Phys. Rev. *54*, 771 (1938).

Whitaker and Bright, *Total and scattering cross sections for slow neutrons.* Phys. Rev. *60*, 155 (1941).

Whitaker and Bright, *On the forward scattering of neutrons by paramagnetic media.* Phys. Rev. *60*, 280 (1941).

White, *Scattering of high energy protons in hydrogen.* Phys. Rev. *49*, 309 (1936).

Wick, *Scattering of slow neutrons by atomic lattices of crystals.* Phys. Z. *38*, 403 (1937).

Wigner, *Scattering of neutrons and protons.* Z. Phys. *83*, 253 (1933).

Wigner, *Resonance reactions and anomalous scattering.* Phys. Rev. *70*, 15 (1946).

Wigner, *Resonance reactions.* Phys. Rev. *70*, 606 (1946).

Wilkins, *Scattering of protons by magnesium and aluminum.* Phys. Rev. *60*, 365 (1941).

Williams, *Passage of alpha and beta particles through matter and Born's theory of collisions.* Proc. Roy. Soc. *135*, 108 (1932).

Williams, *Application of ordinary space-time concepts in collision problems* and relation of classical theory to Born's approximation. Rev. Mod. Phys. *17*, 217 (1945).

Wilson, *Quantum theory of radiation damping.* Proc. Camb. Phil. Soc. *37*, 301 (1941).

Wilson, *Range and ionization measurements on high speed protons.* Phys. Rev. *60*, 749 (1941).

Wilson and Creutz, *Proton-proton scattering at 8 Mev.* Phys. Rev. *71* 339 (1947).

Wilson, *Proton-proton scattering at 10 Mev.* Phys. Rev. *71*, 384 (1947).

Wilson, *Range, straggling and multiple scattering of fast protons.* Phys. Rev. *71*, 385 (1947).

Wilson, Lofgren, Richardson, Wright and Shankland, *Proton-proton scattering at 14.5 Mev.* Phys. Rev. *71*, 560 (1947).

Wrenshall, *Collision of alpha particles with carbon nuclei.* Phys. Rev. *57*, 1095 (1940).

Wrenshall, *Scattering of protons by magnesium nuclei.* Phys. Rev. *63*, 56 (1943).

Wright, *Scattering of alpha particles at small angles by helium.* Proc. Roy. Soc. *137*, 677 (1932).

Wu, Rainwater, Havens and Dunning, *Neutron scattering in ortho- and parahydrogen and the range of nuclear forces.* Phys. Rev. *69*, 236 (1946).

Yamasita and Nogami, *On the virtual P doublet of the compound nucleus* $_2He^5$. Proc. Phys.-Math. Soc. (3) *25*, 354 (1943).

Yukawa and Sakata, *Theory of collision of neutrons with deuterons.* Proc. Phys.-Math. Soc. Jap (3) *19*, 542 (1937).

Zah-Wei, *Collisions of positrons with negatons.* C. R., Paris 222, 1168 (1946).

CHAPTER FOUR

Various Papers on Disintegraion Processes (Experimental)

Allan and Clavier, *Disintegration of magnesium and aluminum by deuterons.* Nature *158*, 832 (1946).

Allen, *The disintegration of beryllium by protons.* Phys. Rev. *51*, 182 (1937).

Allen, *Experimental evidence for the existence of a neutrino.* Phys. Rev. *61*, 692 (1942).

Allison, *Experiments on the efficiencies of production, and half lives of radio carbon and radio nitrogen.* Phys. Rev. *49*, 420 (1936).

Allison, Skaggs and Smith, *A precise measurment of the energy change in the transmutation of beryllium into lithium by proton bombardment.* Phys. Rev. *54*, 171 (1938).

Allison, Skaggs and Smith, *Re-measurement of the energies released in the reactions* Li^7 *(p,* α*)* He^4 *and* Li^6 *(d,* α*)* He^4. Phys. Rev. *56*, 288 (1939).

Allison, Graves and Skaggs, *Alpha particle groups from the disintegration of beryllium by neutrons.* Phys. Rev. *57*, 158 (1940).

Allison, Miller, Perlow, Skaggs and Smith. *Energy released in the reaction* Li^6 *(p.* α*)* He^3 *and the mass of* He^3. Phys. Rev. *58*, 178 (1940).

Allison, Miller, Skaggs and Smith. *Electrostatic deflection studies of alpha particles. The alpha's from* Li^6 *(p,* α*)* He^3. Phys. Rev. *59*, 108 (1941).

Alvarez, *Nuclear K electron capture.* Phys. Rev. *52*, 134 (1937).

Alvarez, *The capture of orbital electrons by nuclei.* Phys. Rev. *54*, 486 (1938).

Alvarez, Helmholz and Nelson, *Isomeric silver and the Weizsäcker theory.* Phys. Rev. *57*, 660 (1940).

Alvarez, Helmholz and Wright, *Recoil from K capture.* Phys. Rev. *60*, 160 (1941).

Amaldi, *Disintegration of boron and other light elements by slow neutrons.* Nuovo. Cim. *12*, 223 (1935).

Amaki and Sugimoto, *On the relative cross sections of the (n,* α*) and (n, p) reactions produced by fast neutrons.* Tokyo Inst. Phys. Chem. Res. *38*, 377 (1941).

Anderson, *The positive electron.* Phys. Rev. *43*, 491 (1933).

Anderson and Neddermeyer, *Cloud chamber observations of cosmic rays at 4300* meters elevation and near sea level. Phys. Rev. *50*, 263 (1936).

Anthony, *Collisions of alpha particles with sulphur nuclei*. Phys. Rev. *50*, 726 (1936).

Aoki, *Disintegration of carbon by fast neutrons*. Proc. Phys.-Math. Soc. Jap. (3) *20*, 755 (1938).

Arakatsu, Sonada, Uemura, Shimizu and Kimura, *A type of nuclear photo-disintegration. The expulsion of alpha particles from various substances irradiated by the gamma rays of lithium and fluorine bombarded by high speed protons*. Proc. Phys.-Math. Soc. Jap. (3) *25*, 173 (1943).

Argo, Hemmendinger, Kratz, Perry, Sherr, Taschek and Williams, *Reaction constant Li7 (p, n) Be7*. Phys. Rev. *70*, 102 (1946).

Arnett, Sharff-Goldhaber and Klaiber, *Search for (n, α) reaction in rare earth elements with slow neutrons*. Phys. Rev. *68*, 100 (1945).

Asano, *Nuclear reaction F (p, α) O and the new effect on the angular distribution of the emitted alpha particles* (in German). Proc. Phys.-Math. Soc. Jap. (3) *25*, 432 (1943).

Bachelet and Bouissières, *Measurement of the paths of alpha rays of protactinium*. J. Phys. Rad. *7*, 151 (1946).

Bagge, *The disintegration of the deuteron by fast neutrons*. Phys. Z. *43*, 226 (1942).

Bailey and Williams, *Thick target yield of Na24 under deuteron bombardment*. Phys. Rev. *61*, 539 (1942).

Bainbridge, *The equivalence of mass and energy*. Phys. Rev. *44*, 123 (1933).

Baldinger and Huber, *Nuclear transformations of nitrogen with rapid neutrons* (in German). Helv. Phys. Acta *12*, 330 (1939).

Baldwin and Klaiber, *Multiple nuclear disintegrations induced by 100 Mev. x-rays*. Phys. Rev. *70*, 259 (1946).

Baldwin and Koch, *Threshold for the nuclear photo-effect*. Phys. Rev. *63*, 462 (1943).

Baldwin and Koch, *Threshold measurements on the nuclear photo-effect*. Phys. Rev. *67*, 1, 1945).

Barkas, *Some new reactions in light nuclei with high energy protons*. Phys. Rev. *56*, 287 (1939).

Barkas and White, *Disintegration of deuterium by protons and p-n reactions in light gaseous elements*. Phys. Rev. *56*, 288 (1939).

Barkas, *Helium of mass three as an agent in nuclear reactions*. Phys. Rev. *56*, 1242 (1939).

Barkas, Creutz, Delsasso and Sutton, *Processes induced in the nuclei of iodine and bromine by high energy protons*. Phys. Rev. *57*, 1087 (1940).

Barker, *Thick target yields with deuteron-proton reactions*. Phys. Rev. *70*, 101 (1946).

Barker, *The total disintegration energy of Na24*. Phys. Rev. *71*, 453 (1947).

Barnes, *Proton reactions of $_{48}Cd$ and $_{49}In$*. Phys. Rev. *55*, 241 (1939).

Barschall and Battat, *On the disintegration of nitrogen by fast neutrons*. Phys. Rev. *70*, 245 (1946).

Barton and Mueller, *Detection of nuclear disintegration products*. Phys. Rev. *45*, 650 (1934).

Becker, Fowler and Lauritsen, *The short range alpha particles from F19 and H1*. Phys. Rev. *59*, 217 (1941).

Becker, Fowler and Lauritesen, *Short range alpha particles from fluorine bombarded with protons*. Phys. Rev. *62*, 186 (1942).

Becker, Hanson and Diven, *Thresholds for the photo-disintegration of Li7, Mg24, Al27, Si28, S32 and Ca40*. Phys. Rev. *71*, 466 (1947).

Bennett, Bonner, Hudspeth, Richards and Watt, *Protons from C13 and H2*. Phys. Rev. *58*, 478 (1940).

Bennett, Bonner, Hudspeth, Richards and Watt, *The disintegration of carbon by deuterons*. Phys. Rev. *59*, 781 (1941).

Bennett, Bonner, Mandeville and Watt, *Resonances in the disintegration of fluorine by protons*. Phys. Rev. *70*, 882 (1946).

Bennett, Bonner, Richards and Watt, *The disintegration of Li7 by deuterons*. Phys. Rev. *71*, 11 (1947).

Beringer and Montgomery, *The angular distribution of positron annihilation radiation.* Phys. Rev. *61*, 222 (1942).

Bernardini, *Neutron disintegration of beryllium and boron by bombardment with alpha particles from polonium.* Nuovo. Cim. *15*, 220 (1938).

Bhabha, *Experimental test of the proton-neutron exchange interaction.* Nature *139*, 1021 (1937).

Bhattacharya, *A search for the double proton.* Proc. Nat. Inst. Sci. India 7, 275 (1941).

Bittencourt and Goldhaber, *The "two step" isomeric transition in Te^{121}.* Phys. Rev. *70*, 780 (1946).

Blackett and Lees, *Photography of artificial disintegration collisions and accuracy of angle determinations.* Proc. Roy. Soc. *134*, 658 (1932).

Blau, *Photographic investigation of artificial disintegration.* J. Phys. Rad. *5*, 61 (1934).

Bleuler, Scherrer and Zünti, *Determination of nuclear masses with the aid of beta-gamma coincidences* (in German). Helv. Phys. Acta *18*, 262 (1945).

Bleuler and Zünti, *The absorption method determining beta and gamma energies* (in German). Helv. Phys. Acta *19*, 375 (1946).

Bleuler and Zünti, *Decay energy of O^{19}, Na^{25} and K^{42}* (in German). Helv. Phys. Acta *19*, 421 (1946).

Bonner and Brubaker, *The disintegration of nitrogen by slow neutrons.* Phys. Rev. *48*, 469 (1935).

Bonner and Brubaker, *Disintegration of lithium by deuterons.* Phys. Rev. *48*, 742 (1935).

Bonner and Brubaker, *The disintegration of nitrogen by neutrons.* Phys. Rev. *49*, 223 (1936).

Bonner and Brubaker, *The disintegration of nitrogen by slow neutrons.* Phys. Rev. *49*, 778 (1936).

Bonner and Brubaker, *Disintegration of beryllium, boron and carbon by deuterons.* Phys. Rev. *50*, 308 (1936).

Bonner and Brubaker, *The disintegration of nitrogen by fast neutrons.* Phys. Rev. *50*, 781 (1936).

Bonner, *Formation of an excited He^3 in the disintegration of deuterium by deuterons.* Phys. Rev. *53*, 711 (1938).

Bonner and Hudspeth, *Resonance in the disintegration of carbon by deuterons.* Phys. Rev. *57*, 1075 (1940).

Bonner, *Neutrons from the disintegration of fluorine by deuterons.* Proc. Roy. Soc. *174*, 339 (1940).

Bonner, Freidlander, Pepkowitz and Perlman, *Search for the transmutation products of gamma-meson reactions induced by 100 Mev. x-rays.* Phys. Rev. *71*, 511 (1947).

Borst and Harkins, *Search for a neutron-deuteron reaction.* Phys. Rev. *57*, 659 (1940).

Borst, *Deuteron-tritium reaction in nitrogen.* Phys. Rev. *59*, 941 (1941).

Borst, *Deuteron-tritium reaction in nitrogen and fluorine.* Phys. Rev. *61*, 106 (1942).

Bothe and Fränz, *Atomic disintegration of alpha rays from polonium.* Z. Phys. *43*, 456 (1927).

Bothe and Fränz, *Products of atomic disintegration, reflected alpha particles and x-rays excited by alpha particles.* Z. Phys. *49*, 1 (1928).

Bothe, *Disintegration experiments on boron with polonium alpha rays.* Z. Phys. *63*, 381 (1930).

Bothe, *Artificial nuclear disintegration.* Phys. Z. *36*, 776 (1935).

Bothe, *Nuclear spectra of light atoms.* Z. Phys. *100*, 273 (1936).

Bothe and Gentner, *Further atomic transmutations by gamma rays.* Naturwiss. *25*, 191 (1937).

Bothe and Gentner, *Nuclear isomerism in bromine.* Naturwiss. *25*, 284 (1937).

Bothe and Gentner, *Atomic transformations by gamma rays.* Z. Phys. *106*, 236 (1937).

Bothe and Maier-Leibnitz, *Transmutation of B^{10} into C^{13}.* Z. Phys. *107*, 513 (1937).

Bothe and Gentner, *Nuclear photoeffect with gamma rays from B^{11} (p, ∂).* Naturwiss. *26*, 497 (1938).

Bower and Burcham, *Transmutation of fluorine by deuterons.* Proc. Roy. Soc. *173*, 379 (1939).

Bowersox, *Efficiency of production of short range particles from lithium and of 4.4cm alpha particles from boron under proton bombardment.* Phys. Rev. *55,* 323 (1939).

Bradt, Guillissen and Scherrer, *Pair production by fast beta rays* (in German). Helv. Phys. Acta *13,* 259 (1940).

Bradt, *Pair production by electron and positron emission from beta radiators* (in German). Helv. Phys. Acta *17,* 59 (1944).

Bradt, Gugelot, Huber, Medicus, Preiswerk and Scherrer, *K capture and positron emission of 6.7 hr. Cd*$^{107, 109}$ (in German). Experienta *1,* 119 (1945).

Bradt, Gugelot, Huber, Medicus, Preiswerk and Scherrer, *Taking up of K-electron and positron emission of Cu*^{61} *and Cu*^{64} (in German). Helv. Phys. Acta *18,* 252 (1945).

Bradt, and Scherrer, *Disintegration of UZ and the UX$_2$ - UZ isomerism* (in German). Helv. Phys. Acta *18,* 405 1945).

Bradt, Halter, Heine and Scherrer, *Pair emission of Th C''* (in German). Helv. Phys. Acta *18,* 457 (1945).

Bradt, Gugelot, Huber, Medicus, Preiswerk, Scherrer and Steffen, *The silver isomers Ag* and Ag**^{109} (in German). Helv. Phys. Acta *19,* 218 (1946).

Bradt, Gugelot, Huber, Medicus, Preiswerk, Scherrer and Steffen, *The dual radioactive transformation of Cu*^{64} (in German). Helv. Phys. Acta *19,* 219 (1946).

Bradt, Gugelot, Huber, Medicus, Preiswerk, Scherrer and Steffen, *The transformation scheme of Zn*^{63} (in German). Helv. Phys. Acta *19,* 221 (1946).

Bradt, Gugelot, Huber, Medicus, Preiswerk, Scherrer and Steffen, *K capture of Fe*^{55} (in German). Helv. Phys. Acta *19,* 222 (1946).

Bradt, Halter, Heine and Scherrer, *Pair emission of Th C''* (in German). Helv. Phys. Acta *19,* 431 (1946).

Braga, *Investigation of the transmutation Ra D -> Ra E by magnetic spectrography of the beta radiation of internal conversion* (in French). Portugaliae Physica *1,* 159 (1944).

Brasch, *Production and application of fast corpuscular rays.* Naturwiss. *21,* 82 (1933).

Brasefield and Pollard, *The transmutation of sulphur by Th C'' alpha particles.* Phys. Rev. *50,* 296 (1936).

Bretscher and Cook, *Production of actinum C'' from lead.* Nature *146,* 430 (1940).

Briggs, *The straggling of alpha particles from Ra C.* Proc. Roy. Soc. *114,* 313 (1927).

Briggs, *The decrease in velocity of alpha particles from Ra C.* Proc. Roy. Soc. *114,* 341 (1927).

Briggs, *Redetermination of the velocities of alpha particles from Ra C, Th C and C'.* Proc. Roy. Soc. *118,* 549 (1928).

Briggs, *Relative velocities of the alpha particles from Th X and its products and from Ra C'.* Proc. Roy. Soc. *139,* 638 (1933).

Briggs, *Relative velocities of alpha particles from radon, Ra A and Ra C'.* Proc. Roy. Soc. *143,* 604 (1934).

Briggs, *Absolute velocity of alpha particles from radium C'.* Proc. Roy. Soc. *157,* 183 (1936).

Brubaker and Pollard, *Maximum energy of the protons emitted by boron under alpha particle bombardment.* Phys. Rev. *51,* 1013 (1937).

Buck, Strain and Valley, *Resonance transmutation of Si (34) by protons.* Phys. Rev. *51,* 1012 (1937).

Burcham and Goldhaber, *Disintegration of nitrogen by slow neutrons.* Proc. Camb. Phil. Soc. *32,* 632 (1936).

Burcham and Smith, *Transmutation of fluorine by protons and deuterons.* Proc. Roy. Soc. *168,* 176 (1938).

Burcham and Smith, *Short range alpha particles from oxygen, nitrogen and fluorine bombarded with protrons.* Nature *143,* 795 (1939).

Burcham and Devons, *Resonance phenomena in disintegration of fluorine by protons.* Proc. Roy. Soc. *173,* 555 (1939).

Burhop, *Atomic disintegration by particles of low energy.* Proc. Camb. Phil. Soc. *32,* 643 (1936).

Burson, Bittencourt, Duffield and Goldhaber, *Decay scheme for* Te^{121}. Phys. Rev. *70*, 566 (1946).

Carlson and Henderson, *Photonuclear disintegration of iron and manganese.* Phys. Rev. *58*, 193 (1940).

Chadwick, *Artificial disintegration of elements.* Phil. Mag. (7) *2*, 1056 (1926).

Chadwick, Constable and Pollard, *Artificial disintegration by alpha particles.* Proc. Roy. Soc. *130*, 463 (1931).

Chadwick and Constable, *Artificial disintegration by alpha particles.* Proc. Roy. Soc. *135*, 48 (1932).

Chadwick and Feather, *Nuclear transformations with alpha particles and neutrons.* Int. Conf. Phys., London *1*, 95 (1934).

Chadwick and Goldhaber, *Disintegration of diplons.* Nature *134*, 237 (1934).

Chadwick, Feather and *Davies, New disintegration produced by neutrons.* Proc. Camb. Phil. Soc. *30*, 357 (1934).

Chadwick, Blackett and Occhialini, *Some experiments on the production of positive electrons.* Proc. Roy. Soc. *144*, 235 (1934).

Chadwick and Goldhaber, *Disintegration by slow neutrons.* Nature *135*, 65 (1935).

Chadwick and Goldhaber, *Disintegration by slow neutrons.* Proc. Camb. Phil. Soc. *31*, 612 (1935).

Chadwick and Goldhaber, *Nuclear photoelectric effect.* Proc. Roy. Soc. *151*, 479 (1935).

Champion, *Nuclear energy levels.* Nature *153*, 720 (1944).

Champion and Ray, *Angular distribution of protons ejected in the disintegration of nitrogen by alpha particles.* Nature *159*, 127 (1947).

Chang, *Short range alpha particles from polonium.* Phys. Rev. *67*, 267 (1945).

Chang, *Low energy alpha ray spectra and mechanism of alpha decay.* Phys. Rev. *69*, 254 (1946).

Chang, *Low energy alpha particles from radium.* Phys. Rev. *70*, 632 (1946).

Chatterjee, *Disintegration of boron by slow neutrons.* Indian J. Phys. *18*, 269 (1944).

Christy and Rubin, *Angular distribution of alpha particles from* $Li^7 + p$. Phys. Rev. *71*, 275 (1947).

Clarke and Irvine, *Excitation functions obtained by bombardment of NaBr with 14 Mev deuterons.* Phys. Rev. *65*, 352 (1944).

Clarke and Irvine, *Nuclear excitation functions. I.* Na^{23} *(d, p)* Na^{24}, Br^{81} *(d, p)* Br^{82} *and Br (d, 2n) Kr (34 hrs).* Phys. Rev. *66*, 231 (1944).

Clarke and Irvine, *Nuclear excitation functions: 14 Mev deuteron reactions on copper, magnesium and aluminum.* Phys. Rev. *69*, 680 (1946).

Clarke and Irvine, *Experimental yields with 14 Mev deuterons.* Phys. Rev. *70*, 893 (1946).

Clarke, *Nuclear excitation functions. II.* Al^{27} *(d; p, α)* Na^{24}. Phys. Rev. *71*, 187 (1947).

Cockcroft and Walton, *Experiments with high velocity positive protons.* Proc. Roy. Soc. *137*, 229 (1932).

Cockcroft and Walton, *Disintegration of light elements by fast protons.* Nature *131*, 23 (1933).

Cockcroft, *Transmutations with protons and diplons.* Int. Conf. Phys. London *1*, 112 (1934).

Cockcroft and Walton, *High velocity positive ions. Part III. Disintegration of lithium, boron and carbon by diplons.* Proc. Roy. Soc. *144*, 704 (1934).

Cockcroft and Lewis, *Experiments with high velocity positive ions. Part V. Further experiments on the disintegration of boron. Part VI. The disintegration of carbon, nitrogen and oxygen by deuterons.* Proc. Roy. Soc. *154*, 46 (1936).

Collins, Waldman and Polye, *Electrodisintegration of beryllium.* Phys. Rev. *55*, 412 (1939).

Collins, Waldman, Stubblefield and Goldhaber, *Nuclear excitation of indium by x-rays.* Phys. Rev. *55*, 507 (1939).

Collins, Waldman and Guth, *Disintegration of beryllium by electrons.* Phys. Rev. *56*, 876 (1939).

Collins and Waldman, *Nuclear excitation of indium by electrons.* Phys. Rev. *57*, 1088 (1940).

Collins and Waldman, *Energy distribution of continuous x-rays from nuclear excitation.* Phys. Rev. *59,* 109 (1941).

Conversi, Pancini and Piccioni, *On the disintegration of negative mesons.* Phys. Rev. *71,* 209 (1947).

Cork and Lawrence, *Transmutation of platinum by deuterons, a resonance phenomenon.* Phys. Rev. *49,* 205 (1936).

Cork and Lawrence, *The transmutation of platinum by deuterons.* Phys. Rev. *49,* 788 (1936).

Cork and Thornton, *The bombardment of gold with deuterons.* Phys. Rev. *51,* 59 (1937).

Cork and Thornton, *The disintegration of cadmium with deuterons.* Phys. Rev. *51,* 608 (1937).

Cork and Curtis, *Excitation function for iron by energetic deuterons.* Phys. Rev. *55,* 1264 (1939).

Cork, Halpern and Tatel, *The production of radium E and radium F from bismuth.* Phys. Rev. *57,* 348 (1940).

Cork, Halpern and Tatel, *Production of radium E and radium F (polonium) from bismuth.* Phys. Rev. *57,* 371 (1940).

Cork and Halpern, *Relative probability for the loss of neutrons and alpha particles from an excited nucleus.* Phys. Rev. *57,* 667 (1940).

Cork and Middleton, *Simultaneous emission of three particles from an excited nucleus.* Phys. Rev. *58,* 474 (1940).

Cork, Hadley and Kent, *On the relative probability of the d, 2n and the d, n reactions in nickel and zinc.* Phys. Rev. *61,* 388 (1942).

Cork, *The relative probability of the (d, p) and the (d, n) reactions in bombarded bismuth.* Phys. Rev. *70,* 563 (1946).

Cork, *The escape probability of protons and neutrons from bombarded bismuth.* Phys. Rev. *71,* 144 (1947).

Corson, MacKenzie and Segrè, *Artificially produced alpha particle emitter.* Phys. Rev. *57,* 250 (1940).

Craggs and Smee, *Action of fast hydrogen ions on lithium chloride.* Nature *148,* 531 (1941).

Crane and Lauritsen, *Disintegration of beryllium by deutons.* Phys. Rev. *45,* 226 (1934).

Crane, Delsasso, Fowler and Lauritsen, *The emission of negative electrons from boron bombarded by deuterons.* Phys. Rev. *47,* 887 (1935).

Crane and Halpern, *New experimental evidence for the existence of a neutrino.* Phys. Rev. *53,* 789 (1938).

Crane and Halpern, *A reply to the above note by Wertenstein* (concerning the neutrino). Phys. Rev. *54,* 306 (1938).

Crane, *An attempt to observe the absorption of neutrons.* Phys. Rev. *55,* 501 (1939).

Crane and Halpern, *On the creation of pairs or positrons by fast electrons.* Phys. Rev. *55,* 838 (1939).

Crenshaw, Young, Manning, *Difference in loss of energy of deuterons in D_2 and H_2.* Phys. Rev. *61,* 388 (1942).

Curie and Joliot, *Emission of protons of great speed by hydrogenic substances under the influence of very penetrating gamma rays.* C. R., Paris *194,* 273 (1932).

Curie and Joliot, *Effect of absorption of gamma rays of high frequency by projection of heavy nuclei.* C. R., Paris *194,* 708 (1932).

Curie and Joliot, *On the origin of the positive electron.* C. R., Paris *196,* 1581 (1933).

Curie and Joliot, *Electrons produced by artificial disintegration.* J. Phys. Rad. *4,* 494 (1933).

Curie and Tsien, *Paths of alpha rays of ionium.* J. Phys. Rad. *6,* 162 (1945).

Curran and Strothers, *Bombardment of nitrogen and oxygen with protons.* Nature *145,* 224 (1940).

Darrow, *Transmutation by alpha particles.* Rev. Sci. Instrum. *5,* 66 (1934).

Davidson and Pollard, *The transmutation of titanium by Th C' alpha particles.* Phys. Rev. *54,* 408 (1938).

Davidson, *A study of the protons from calcium under deuteron bombardment.* Phys. Rev. *56,* 1061 (1939).

Davidson, *A study of the protons from vanadium, copper, manganese and scandium under bombardment by deuterons.* Phys. Rev. *56*, 1061 (1939).

Davidson, *Energy levels near the ground state in Co60 and As76.* Phys. Rev. *57*, 568 (1940).

Dee and Walton, *Transmutation of lithium and boron.* Proc. Roy. Soc. *141*, 733 (1933).

Dee, *Disintegration of the diplon.* Nature *133*, 564 (1934).

Dee, *Some experiments upon artificial transmutations, using the cloud track method.* Proc. Roy. Soc. *148*, 623 (1935).

Dee and Gilbert, *Transmutation of heavy hydrogen investigated by the cloud track method.* Proc. Roy. Soc. *149*, 200 (1935).

Dee and Gilbert, *Disintegration of boron into three alpha particles.* Proc. Roy. Soc. *154*, 279 (1936).

Dee, Curran and Petržilka, *Resonance transmutations of carbon by protons.* Nature *141*, 642 (1938).

Delsasso, Fowler and Lauritsen, *Protons from the disintegration of lithium by deuterons.* Phys. Rev. *48*, 848 (1935).

Delsasso, Fowler and Lauritsen, *Production of pairs and Compton electrons by gamma radiation from the bombardment of lithium by protons.* Phys. Rev. *50*, 389 (1936).

Deutsch, *Radiations from $I^{131} \rightarrow Xe^{131}$.* Phys. Rev. *59*, 940 (1941).

Deutsch and Elliott, *The disintegration of Co60.* Phys. Rev. *62*, 558 (1942).

Diebner and Pose, *Resonance penetration of alpha particles into the aluminum nucleus.* Z. Phys. *75*, 753 (1932).

Döpel, *Atomic disintegration by H and He particles.* Z. Phys. *91*, 796 (1934).

Doolittle, *The disintegration of Li7 bombarded by slow protons.* Phys. Rev. *49*, 779 (1936).

Doran and Henderson, *New reactions in nickel.* Phys. Rev. *60*, 411 (1941).

DuBridge, *Nuclear reactions produced by 6.5 Mev protons.* Phys. Rev. *55*, 603 (1939).

Duncanson and Miller, *Artificial disintegration by Ra C' alpha particles from aluminum and magnesium.* Proc. Roy. Soc. *146*, 396 (1934).

Dunworth and Pontecorvo, *Excitation of indium 113 by x-rays.* Proc. Camb. Phil. Soc. *43*, 123 (1947).

Eggen and Pool, *X-radiation and internal conversion from Ra + d.* Phys. Rev. *70*, 446 (1946).

Ehrenberg, *Excitation of nuclei by neutrons.* Nature *136*, 870 (1935).

Eklund and Hole, *On neutron induced activity in aluminum* (in English). Ark. Mat. Astr. Fys. *29A*, No. 26 (1943).

Eklund, *An attempt to obtain nuclear excitation by means of x-rays.* Nature *156*, 690 (1945).

Eklund, *Studies in nuclear physics. Excitation by means of x-rays. Activity of 87Rb* (in English). Ark. Mat. Astr. Fys. *33A*, No. 14 (1946).

Ellett and Huntoon, *Asymmetric distribution of protons from deuteron-deuteron reaction.* Phys. Rev. *54*, 87 (1938).

Ellett, McLean, Young and Plain, *Distribution in angle of alpha particles from F19 and H1.* Phys. Rev. *57*, 1083 (1940).

Elliott and Deutsch, *Nuclear energy levels in Fe56 from decay of Mn56 and Co56.* Phys. Rev. *63*, 321 (1943).

Ellis and Wooster, *The average energy of disintegration of radium E.* Proc. Roy. Soc. *117*, 109 (1927).

Evans and Livingston, *Correlation of nuclear disintegration processes.* Rev. Mod. Phys. *7*, 229 (1935).

Fea, *Bibliography and tables of artificial transmutation.* Nuovo Cim. *12*, 368 (1935).

Feather, *Artificial disintegration by neutrons.* Nature *130*, 257 (1932).

Feather, *Collisions of neutrons with nitrogen nuclei.* Proc. Roy. Soc. *136*, 709 (1932).

Feather and Dunworth, *Production of positron-electron pairs during the passage of beta particles through matter.* Proc. Camb. Phil. Soc. *34*, 435 (1938).

Feather and Dunworth, *Absorption and coincidence experiments on the radiation from the radioactive sodium Na24.* Proc. Camb. Phil. Soc. *34*, 442 (1938).

Feather and Bretscher, *Uranium Z and nuclear isomerism.* Proc. Roy. Soc. *165*, 530 (1938).

Feather and Dunworth, *Nuclear isomerism*. Proc. Roy. Soc. *168*, 566 (1938).

Feldmeier and Collins, *Excitation of nuclei by x-rays*. Phys. Rev. *59*, 937 (1941).

Fischer, *Nuclear transformations of nitrogen by means of fast neutrons*. Phys. Z. *43*, 507 (1942).

Fischer-Colbrie, *Nuclear disintegration using Ra (B + C) as source of radiation*. Akad. Wiss. Wien *145*, 283 (1936).

Fisk, *Transmutation of boron by slow neutrons with the emission of alpha particles and protons*. Phys. Rev. *55*, 1117 (1939).

Flammersfeld, *Isomers to stable nuclei by neutron capture in dysprosium and hafnium*. Z. Naturforsch. *1*, 190 (1946).

Flügge and Krebs, *Nuclear chemistry*. Phys. Z. *36*, 466 (1935).

Fowler and Lauritsen, *Pair emission from fluorine bombarded with protons*. Phys. Rev. *56*, 840 (1939).

Fränz, *Disintegration experiments on boron with alpha rays from Ra C'*. Z. Phys. *63*, 370 (1930).

Friedlander, Perlman and Pepkowitz, *Radiochemical search for gamma-meson reactions induced by 100 Mev x-rays*. Phys. Rev. *70*, 790 (1946).

Fünfer, *Transmutation of boron by slow neutrons, measured by a proportional counter*. Ann. Phys., Leipzig *29*, 1 (1937).

Gentner, *Processes of addition of rapid protons to nuclei*. Z. Phys. *107*, 354 (1937).

Gentner, *Nuclear photoeffect under simultaneous emission of two neutrons*. Naturwiss. *26*, 109 (1938).

Giarratana and Brennecke, *Angular distribution of the products of artificial nuclear disintegration*. Phys. Rev. *49*, 35 (1936).

Gibert, Roggen and Rossel, *Nuclear reactions of chlorine with neutrons* (in German). Helv. Phys. Acta *17*, 97 (1944).

Glückauf and Paneth, *Identification and measurement of He formed in Be by gamma rays*. Proc. Roy. Soc. *165*, 229 (1938).

Goldhaber, Muehlhause and Turkel, *L-converted isomeric transition*. Phys. Rev. *71*, 372 (1947).

Goldhaber, *L-converted isomeric transition*. Phys. Rev. *71*, 467 (1947).

Good and Hill, *Investigation of boron by slow neutrons*. Phys. Rev. *56*, 288 (1939).

Good, Peaslee and Deutsch, *Experimental test of beta ray theory for the positron emitters Na^{22}, V^{48}, Mn^{52}, Co^{58}*. Phys. Rev. *69*, 313 (1946). Erratum, *69*, 673 (1946).

Good and Peacock, *Positron emission, electron capture competition in Zn^{65}*. Phys. Rev. *69*, 680 (1946).

Graves, *Energy release from Be^9 (d, α) Li^7 and the production of Li^7*. Phys. Rev. *57*, 855 (1940).

Graves, Graves, Coon and Manley, *Cross section of D (d, p) H^3 reaction*. Phys. Rev. *70*, 101 (1946).

Graves and Coon, *Disintegrations of neon and argon by d-d neutrons*. Phys. Rev. *70*, 101 (1946).

Grinberg and Roussinow, *Structure of the lower excited levels of the Br^{80} nucleus*. Phys. Rev. *58*, 181 (1940).

Groetzinger, Kruger and Smith, *Evidence for the production of a nonionizing radiation other than neutrons and gamma rays by 10 Mev deuterons*. Phys. Rev. *67*, 202 (1945).

Groetzinger, Kruger and Lowen, *Evidence for the artificial production of a new neutral radiation*. Phys. Rev. *67*, 52 (1945).

Groetzinger and Smith, *Further work on the artificial production of mesotrons*. Phys. Rev. *68*, 55 (1945).

Günther, *Formation of helium from alpha ray emission*. Z. phys. Chem. *185*, 367 (1939).

Hacman and Haxel, *Energy and range of slow alpha rays*. Z. Phys. *120*, 486 (1943).

Hafstad and Tuve, *Resonance transmutation by protons*. Phys. Rev. *47*, 506 (1935).

Hafstad, Heydenburg and Tuve, *Widths of nuclear resonance levels and the calibration of ion beam energy levels*. Phys. Rev. *49*, 866 (1936).

Hafstad, Heydenburg and Tuve, *Excitation curves for fluorine and lithium.* Phys. Rev. *50,* 504 (1936).

Hahn and Strassmann, *Formation of zirconium and protactinium by bombardment of thorium with neutrons.* Naturwiss. *29,* 285 (1941).

Hall, *Disintegration of the deuteron by gamma rays.* Phys. Rev. *49,* 401 (1936).

Hamilton, *Electron-neutrino angular correlation in beta decay.* Phys. Rev. *71,* 456 (1947).

Hansen, Disintegration of nitrogen and oxygen by neutrons. Phys. Rev. *59,* 941 (1941).

Hanson and Benedict, *An independent determination of fixed points on the high voltage scale.* Phys. Rev. *65,* 33 (1944).

Harkins, Gans and Newson, *Disintegration of the nuclei of nitrogen and other light atoms by neutrons. I.* Phys. Rev. *44,* 529 (1933).

Harkins, Gans and Newson, *The disintegration of the nuclei of light atoms by neutrons. II. Neon, fluorine and carbon.* Phys. Rev. *47,* 52 (1935).

Harnwell, Smyth, van Voorhis and Kuper, *Production of $_1H^3$ by a canal ray discharge in deuterium.* Phys. Rev. *45,* 655 (1934).

Hatch, *The yield of alpha particles from beryllium bombarded by protons.* Phys. Rev. *54,* 165 (1938).

Haworth and King, *The excitation functions for the disintegration of Li^7 under bombardment by low energy protons.* Phys. Rev. *54,* 38 (1938).

Haworth and King, *The stopping power of lithium for low energy protons.* Phys. Rev. *54,* 48 (1938).

Haxby, Allen and Williams, *The angular distribution of the disintegration products of light elements.* Phys. Rev. *55,* 140 (1939).

Haxby, Shoupp, Stephens and Wells, *Lithium and carbon (p, n) threshholds.* Phys. Rev. *57,* 348 (1940).

Haxby, Shoupp, Stephens and Wells, *Beryllium (p, n) threshold.* Phys. Rev. *57,* 567 (1940).

Haxel, *Artificial disintegration of aluminum.* Z. Phys. *83,* 323 (1933).

Haxel, *The proton spectra of the elements magnesium, silicon and sulphur by irradiation with fast alpha rays.* Phys. Z. *36,* 804 (1935).

Haxel, *Nuclear transformations of nitrogen produced by fast alpha rays.* Z. Phys. *93,* 400 (1935).

Haxel, *Nuclear transformations of boron by slow neutrons.* Z. Phys. *104,* 540 (1937).

Henderson, *The disintegration of lithium by protons of high energy.* Phys. Rev. *43,* 98 (1935).

Henderson, Livingston and Lawrence, *Transmutation of fluorine by proton bombardment and the mass of fluorine 19.* Phys. Rev. *46,* 38 (1934).

Henneberg, *Excitation of atoms in inner shells by slow protons and alpha particles.* Z. Phys. *86,* 592 (1933).

Herb, Parkinson and Kerst, *Yield of alpha particles from lithium films bombarded by protons.* Phys. Rev. *48,* 118 (1935).

Hevesy and Levi, *The action of neutrons on the rare earth elements* (in English). Kgl. Danske Acad. *14,* No. 5 (1936).

Hevesy and Levi, *Action of slow neutrons on rare earth elements.* Nature *137,* 185 (1936).

Heyn, *Evidence for the expulsion of two neutrons from copper and zinc by one fast neutron.* Nature *138,* 723 (1936).

Hibdon, Pool and Kurbatov, *Transmutation of titanium.* Phys. Rev. *65,* 351 (1944).

Hill and Haxby, *A note on the disintegration of boron by protons.* Phys. Rev. *55,* 147 (1939).

Hill, *The p-n reactions in lithium and beryllium.* Phys. Rev. *57,* 567 (1940).

Hincks, *Evidence for the reaction O^{17} (n, α) C^{14}.* Phys. Rev. *70,* 770 (1946).

Hirzel and Wäffler, *Nuclear photoeffect under the emission of a proton in Cd: $Cd^{111} (\partial, p) Ag^{110}$* (in German). Helv. Phys. Acta *19,* 214 (1946).

Hirzel and Wäffler, *Nuclear photoeffect in emission of a proton* (in German). Helv. Phys. Acta *19,* 425 (1946).

Hole and Siegbahn, *The disintegration of Cl^{38}* (in English). Ark. Mat. Astr. Fys. *33A,* No. 9 (1946).

Holloway and Livingston, *Range and specific ionization of alpha particles.* Phys. Rev. *54*, 18 (1938).

Holloway and Moore, *Disintegration of N^{14} and N^{15} produced by deuteron bombardment.* Phys. Rev. *56*, 705 (1939).

Holloway, *The disintegration of N^{14} and N^{15} by deuterons.* Phys. Rev. *57*, 347 (1940).

Holloway and Bethe, *Cross section of the reaction N^{15} (p, α) C^{12}.* Phys. Rev. *57*, 747 (1940).

Holloway and Moore, *The disintegration of C^{12}, C^{13} and O^{16} by deuterons.* Phys. Rev. *57*, 1086 (1940).

Holloway and Moore, *Disintegration of N^{14} and N^{15} by deuterons.* Phys. Rev. *58*, 847 (1940).

Huber, Huber and Scherrer, *Determination of the mass of C_8^{14} from the nuclear reaction N (n, p) C* (in German). Helv. Phys. Acta *13*, 209 (1940).

Huber, *Investigation of nuclear reactions of nitrogen and sulphur with neutrons* (in German). Helv. Phys. Acta *14*, 163 (1941).

Huber, Rossel and Scherrer, *Nuclear reactions of fluorine with fast neutrons* (in German). Helv. Phys. Acta *14*, 314 (1941).

Huber, Lienhard, Scherrer and Wäffler, *Nuclear photoeffect on sulphur, aluminum and magnesium.* Phys. Rev. *60*, 910 (1941).

Huber Lienhard and Wäffler, *Nuclear photoeffect with titanium, nickel and copper, using lithium gamma rays* (in German). Helv. Phys. Acta *16*, 226 (1943).

Huber, Lienhard, Scherrer and Wäffler, *Excitation of isomer conditions by nuclear photoeffect* (in German). Helv. Phys. Acta *16*, 228 (1943).

Huber, Lienhard, Scherrer and Wäffler, *Nuclear photoeffect with the ejection of a proton. Mg^{26} (∂, p) Na^{25}* (in German). Helv. Phys. Acta *17*, 139 (1944).

Huber, Lienhard and Wäffler, *The nuclear photoeffect with lithium gamma radiation. II. The elements titanium to rubidium* (in German). Helv. Phys. Acta *17*, 195 (1944).

Hudson, Herb and Plain, *Excitation of the 455 Kev level of Li^7 by proton bombardment.* Phys. Rev. *57*, 587 (1941).

Hudspeth and Bonner, *Observation of H^1 and H^3 ranges from the disintegration of deuterium by deuterons.* Phys. Rev. *54*, 308 (1938).

Humphreys and Pollard, *Alpha-proton reactions of sodium and aluminum.* Phys. Rev. *59*, 942 (1941).

Humphreys and Watson, *Protons from deuteron bombardment of C^{13}.* Phys. Rev. *60*, 542 (1941).

Huntoon, Ellett, Bayley and van Allen, *Distribution in angle of protons from the deuteron-deuteron reaction.* Phys. Rev. *58*, 97 (1940).

Hurst, Latham and Lewis, *Production of radium and polonium by deuteron bombardment of bismuth.* Proc. Roy. Soc. *174*, 136 (1940).

Jacobs and McLean, *Resonance in B^{11}.* Proc. Iowa Acad. Sci. *48*, 308 (1941).

Jaeckel, *Disintegration of neon by neutrons.* Z. Phys. *96*, 151 (1935).

Jensen, *The relative efficiency of some (n, 2n) processes.* Z. Phys. *122*, 387 (1944).

Kahn, *Interaction of nuclear particles* (in English). Physica *4*, 403 (1937).

Kalbfell, *Internal conversion of gamma rays in element 43.* Phys. Rev. *54*, 543 (1938).

Kanne, *Disintegration of aluminum by polonium alpha particles.* Phys. Rev. *52*, 266 (1937).

Kapitza, *X-ray tracks in a strong magnetic field.* Proc. Roy. Soc. *106*, 602 (1924).

Kempton, Browne and Maasdorp, *Transmutation of Li^7 by deuterons.* Proc. Roy. Soc. *157*, 372 (1936).

Kempton, Browne and Maasdorp, *Angular distributions of the protons and neutrons emitted in transmutations of deuterium.* Proc. Roy. Soc. *157*, 386 (1936).

Kennedy, Seaborg and Segrè, *Nuclear isomerism in zinc.* Phys. Rev. *56*, 1095 (1939).

Kikuchi, Husimi and Aoki, *Energy liberated at the addition of neutrons.* Z. Phys. *105,* 265 (1937).

Kikuchi, Watase, Itoh, Takeda and Yamaguchi, *On the disintegration of Na* ^{24}. Proc. Phys.-Math. Soc. Jap. (3) *21,* 381 (1939).

Kinsey, *Attempts at disintegration using lithium ions.* Phys. Rev. *50,* 386 (1936).

Kirchner, *Artificial disintegration of atomic nuclei.* Phys. Z. *34,* 777 (1933).

Kirchner and Neuert, *On the extensive disintegration of boron upon bombardment with protons.* Phys. Z. *34,* 897 (1933).

Kirchner, *Transformation of elements by fast hydrogen nuclei.* Ergeb. d. exakt. Naturwiss. *13,* 57 (1934).

Kirchner and Neuert, *On the ranges of fragments of lithium and boron under radiation of fast protons.* Phys. Z. *35,* 292 (1934).

Kirchner and Neuert, *The transmutation of beryllium by fast protons and the mass of Be^9.* Phys. Z. *36,* 36 (1935).

Kirchner and Neuert, *Transformation of beryllium by slow protons and the mass of $_4Be^9$.* Phys. Z. *36,* 54 (1935)

Kirchner, Neuert and Laaff, *Ionizing power of atomic nuclei emitted in nuclear transformation processes.* Phys. Z. *38,* 969 (1937).

Klaiber, Luebke and Baldwin, *Range-momentum measurements of particles emitted in nuclear disintegrations induced by 100 Mev x-rays.* Phys. Rev. *70,* 789 (1946).

Klarmann, *Artificial transmutation of magnesium by polonium alpha rays.* Z. Phys. *87,* 411 (1933).

Klarmann, *Transformation of aluminum by neutron bombardment.* Z. Phys. *95,* 221 (1935).

Klema and Barschall, *Saturation characteristics for alpha particles in purified gases.* Phys. Rev. *63,* 18 (1943).

König, *Disintegration of sodium.* Z. Phys. *90,* 197 (1934).

Korff and Clarke, *The production of neutrons and protons by the cosmic radiation at 14,125 feet.* Phys. Rev. *61,* 422 (1942).

Korsunsky, Lange and Spinell, *Threshold value for nuclear excitation of In*^{*115} *by x-rays* (in English). C. R., URSS, *26,* 144 (1940).

Korsunsky, Walther, Ivanov, Zypkin and Ganenko, *An investigation of "Bremstrahlung" by means of excited In^{115} nuclei.* J. Phys., USSR 7, 129 (1943).

Kovarik and Adams, *The disintegration constant of Th and the branching ratio of Th C.* Phys. Rev. *54,* 413 (1938).

Kovarik and Cork, *The bombardment of palladium with deuterons.* Phys. Rev. *51,* 383 (1937).

Krishnan and Gant, *Deuteron bombardment of silver.* Nature *144,* 547 (1939).

Krishnan and Banks, *A new type of disintegration produced by deuterons.* Nature *145,* 777 (1940).

Krishnan and Nahum, *Deuteron bombardment of the heavy elements.* Proc. Camb. Phil. Soc. *36,* 490 (1940).

Krishnan, *Deuteron bombardment of silver.* Proc. Camb. Phil. Soc. *36,* 500 (1940).

Krishnan, *Deuteron-tritium reaction in fluorine.* Nature *148,* 407 (1941).

Krishnan, *Deuteron bombardment of gold.* Proc. Camb. Phil. Soc. *37,* 186 (1941).

Krishnan and Banks, *New type of disintegration produced deuterons.* Proc. Camb. Phil. Soc. *37,* 317 (1941).

Krishnan and Nahum, *Deuteron bombardment of the heavy elements. II. Platinum.* Proc. Camb. Phil. Soc. *37,* 422 (1941).

Krishnan and Nahum, *Cross section measurements for disintegrations produced by deuterons in the heavy elements.* Proc. Roy. Soc. *180,* 321 (1942).

Krishnan and Nahum, *Excitation function measurements for disintegrations produced by deuterons in the heavy elements.* Proc. Roy. Soc. *180,* 333 (1942).

Kronig, *Nucleus $_4^8Be$* (in *English).* Physica *4,* 171 (1937).

Kruger, Stallmann, Shoupp, *Nuclear energy levels in B^{10}.* Phys. Rev. *56,* 297 (1939).

Kruger and Ogle, *Pair electrons formed in the field of an electron.* Phys. Rev. *67*, 282 (1945).

Kundu and Pool, *Deuteron-H^3 reaction in columbium and silver.* Phys. Rev. *71*, 140 (1947).

Kundu and Pool, *Results of H^3 bombardments on silver and rhodium.* Phys. Rev. *71*, 467 (1947).

Kurbatov and Kurbatov, *Determination of stable species $_{39}Y^{89}$, $_{40}Zr^{90}$, and $_{56}Ba^{134}$ produced by nuclear transmutations.* Phys. Rev. *71*, 466 (1947).

Kurie, *The angular distributions of protons projected by neutrons.* Phys. Rev. *43*, 1056 (1933).

Kurie, *A nitrogen disintegration by a very fast neutron.* Phys. Rev. *43*, 771 (1933).

Kurie, *New mode of disintegration induced by neutrons.* Phys. Rev. *45*, 904 (1934).

Kurie, *The disintegration of nitrogen by neutrons.* Phys. Rev. *47*, 97 (1935).

Ladenburg and Kanner, *The yield of protons and neutrons from the deuteron-deuteron reaction.* Phys. Rev. *51*, 1022 (1937).

Ladenburg and Wheeler, *Mass of the meson by the method of momentum loss.* Phys. Rev. *61*, 105 (1942).

Langsdorf and Segrè, *Nuclear isomerism in selenium and krypton.* Phys. Rev. *57*, 105 (1940).

Lark-Horovitz, Risser and Smith, *Nuclear excitation of indium with alpha particles.* Phys. Rev. *55*, 878 (1939).

Lauritsen and Crane, *Disintegration of boron by deutons and by protons.* Phys. Rev. *45*, 493 (1934).

Lawrance, *Production of Li^8 by bombardment of boron with neutrons.* Proc. Camb. Phil. Soc. *35*, 304 (1939).

Lawrence, Livingston and White, *The disintegration of lithium by swiftly moving protons.* Phys. Rev. *42*, 150 (1932).

Lawrence, Livingston and Lewis, *The emission of protons from various targets bombarded by deutons of high speed.* Phys. Rev. *44*, 56 (1933).

Lawrence and Livingston, *Three types of nuclear disintegration of calcium fluoride by bombarding protons of high energy.* Phys. Rev. *44*, 317 (1933).

Lawrence and Livingston, *Emission of protons and neutrons from various targets bombarded by three million volt deutons.* Phys. Rev. *45*, 220 (1934).

Lawrence, *Transmutations of sodium by deutons.* Phys. Rev. *47*, 17 (1935).

Lawrence, MacMillan and Henderson, *Transmutations of nitrogen by deutons.* Phys. Rev. *47*, 254 (1935).

Lawrence, McMillan and Henderson, *Transmutations of nitrogen by deutons.* Phys. Rev. *47*, 273 (1935).

Lewis, Livingston and Lawrence, *The emission of alpha particles from various targets bombarded by deutons of high speed.* Phys. Rev. *44*, 55 (1933).

Lewis, Livingston, Henderson and Lawrence, *Disintegration of deutons by high speed protons and the instability of the deuton.* Phys. Rev. *45*, 242 (1934).

Lewis and Bowden, *An analysis of the fine structure of the alpha particle groups from thorium C and of the long range groups from thorium C'.* Proc. Roy. Soc. *145*, 235 (1934).

Lewis, Burcham and Chang, *Alpha particles from the radioactive disintegration of a light element.* Nature *139*, 24 (1937).

Livingston, Henderson and Lawrence, *The disintegration of the elements by swiftly moving protons.* Phys. Rev. *44*, 316 (1933).

Livingston, Lewis and Lawrence, *The disintegration of nuclei by swiftly moving ions of the heavy isotope of hydrogen.* Phys. Rev. *44*, 317 (1933).

Livingston, Genevese and Konopinski, *The excitation of characteristic x-rays by protons.* Phys. Rev. *51*, 835 (1937).

Livingston and Bethe, *Nuclear Physics. Part III. Nuclear dynamics, experimental.* Rev. Mod. Phys. *9*, 245 (1937).

Livingston and Hoffman, *Slow neutron disintegration of B^{10} and Li^6.* Phys. Rev. *53*, 227 (1938).

Livingston and Wright, *(d, 2n) reaction in copper.* Phys. Rev. *58*, 656 (1940).

Magnan, *Energies of electrons and positrons emitted in certain nuclear reactions.* C. R., Paris *205*, 1147 (1937).

Maier-Leibnitz and Maurer, *Longest range proton groups in the disintegration of boron by alpha particles.* Z. Phys. *107*, 509 (1937).

Manley, Coon and Graves, *Cross section of D (d, n) He³ reaction.* Phys. Rev. *70*, 101 (1946).

Mann, *Nuclear transformations produced in copper by alpha particle bombardment.* Phys. Rev. *52*, 405 (1937).

Mann, *Nuclear transformations produced in zinc by alpha particle bombardment.* Phys. Rev. *53*, 212 (1938).

Mann, *Nuclear transformations in zinc by alpha particle bombardment.* Phys. Rev. *54*, 649 (1938).

Manning, Crenshaw and Young, *Distribution with angle of protons from d-d reactions.* Phys. Rev. *59*, 941 (1941).

Manning, Huntoon, Myers and Young, *Distribution in angle of protons from the d-d reaction.* Phys. Rev. *61*, 371 (1942).

Mano, *Retardation of the alpha rays of The C' in air.* C. R., Paris *194*, 1235 (1932).

Martin, *Energy levels in the nucleus Mn⁵⁶.* Phys. Rev. *71*, 127 (1947).

Martin, *Energy levels in the nucleus Mn⁵⁶.* Phys. Rev. *71*, 466 (1947).

Marvin, *Width of resonance process in boron.* Phys. Rev. *68*, 228 (1945).

Maurer and Fisk, *Transformation of boron by slow neutrons.* Z. Phys. *112*, 436 (1939).

May and Vaidyanathan, *Energy levels of light nuclei.* Proc. Roy. Soc. *155*, 519 (1936).

May and Hincks, *Evidence for an (n, α) reaction induced in O¹⁷ by thermal neutrons.* Canad. J. Res. 25, 77 (1947).

McCreary, Keurti and Van Voorhis, *Positron energy distribution of Si²⁷.* Phys. Rev. *57*, 351 (1940).

McLean, Becker, Fowler and Lauritsen, *Short range alpha particles from F¹⁹ + H¹.* Phys. Rev. *55*, 796 (1939).

McLean, Ellett and Jacobs, *Distribution in angle of the long range alpha particles from fluorine bombarded with protons.* Phys. Rev. *58*, 500 (1940).

McMillan and Lawrence, *Transmutation of aluminum by deutons.* Phys. Rev. *47*, 343 (1935).

Meerhaut, *Nuclear spectra of light elements on the photographic plate.* Phys. Z. *41*, 528 (1940).

Meerhaut, *Nuclear spectra of light elements on the photographic plate.* Z. techn. Phys. *21*, 280 (1940).

Meitner and Philipp, *On the interaction between neutrons and atomic nuclei.* Naturwiss. *20*, 929 (1932).

Meitner, *Energy spectrum of positrons from aluminum.* Naturwiss. *22*, 388 (1934).

Meitner, *Behavior of rare earths under neutron bombardment* (in German). Ark. Mat. Astr. Fys. *27A*, No. 17 (1941).

Meitner, *The resonance energy of the thorium capture process.* Phys. Rev. *60*, 58 (1941).

Miller, Duncanson and May, *Disintegration of boron by alpha particles.* Proc. Camb. Phil. Soc. *30*, 549 (1934).

Mitchell and Langer, *The nuclear isomers of In¹¹⁶.* Phys. Rev. *53*, 505 (1938).

Mitchell, Jurney and Ramsey, *Coincidence experiments on Ga⁷²*. Phys. Rev. *71*, 324 (1947).

Miwa, *Photodisintegration of deuteron by radium gamma rays.* Proc. Phys.-Math. Soc. Jap (3) *22*, 560 (1940).

Monod-Herzen, *Distribution of protons produced by neutron bombardment.* J. Phys. Rad. *5*, 95 (1934).

Murrell and Smith, *Transmutation of sodium by deuterons.* Proc. Roy. Soc. *173*, 410 (1939).

Myers and Langer, *Search for an excited state of the H³ nucleus.* Phys. Rev. *54*, 90 (1935).

Myers, *The angular distribution of resonance disintegration products.* Phys. Rev. *34*, 361 (1938).

Myers, Huntoon, Shull and Crenshaw, *Search for a short range group of protons in the d-d reaction.* Phys. Rev. *56*, 1104 (1939).

Myers and Van Atta, *On the photodisintegration of beryllium and deuterium*. Phys. Rev. *61*, 19 (1942).

Nahamias, *Attempt to detect the neutrino*. Proc. Camb. Phil. Soc. *31*, 99 (1935).

Nedelsky and Oppenheimer, *Production of positives by nuclear gamma rays*. Phys. Rev. *45*, 136 (1934). Errata, *45*, 283 (1934).

Neuert, *Ranges of particles emitted by light elements under proton bombardment*. Phys. Z. *36*, 629 (1935).

Neuert, *Angular distribution of the products of disintegration of light elements by hydrogen ions. Part I*. Phys. Z. *38*, 122 (1937).

Neuert, *Angular distribution of transmuted particles in transformation processes of light nuclei*. Ann. Phys., Leipzig, *36*, 437 (1939).

Newson, *Transmutation functions at high bombarding energies*. Phys. Rev. *51*, 620 (1937).

Newson and Borst, *The determination of nuclear reaction energies for a mixture of copper isotopes*. Phys. Rev *59*, 941 (1941).

Nishina, Yasaki, Kimura and Ikawa, *Artificial production of uranium Y from thorium*. Nature *142*, 874 (1938).

Norling, *On the coupling of the beta and gamma rays of radio-manganese and the energy levels of the stable iron nucleus*. Naturwiss. *27*, 432 (1939).

Norling, *The coincidence method and its applications to disintegration problems* (in English). Ark. Mat. Astr. Fys. *27A*, No. 27 (1941).

Northrup, Van Atta, Van de Graaff and Van Atta, *Some experiments on the irradiation of deuterium, beryllium and indium by x-rays*. Phys. Rev. *58*, 199 (1940).

Occhialini and Powell, *Nuclear disintegrations produced by slow charged particles of small mass*. Nature *159*, 186 (1947).

Ogle, Brown and Conklin, *Evidence for a gamma-proton reaction in Be*[9]. Phys. Rev. *71*, 378 (1947).

Oliphant and Rutherford, *Transmutation of elements by protons*. Proc. Roy. Soc. *141*, 259 (1933).

Oliphant, Kinsey and Rutherford, *Transmutation of lithium*. Proc. Roy. Soc. *141*, 722 (1933).

Oliphant, *Transformations by hydrogen ions*. Int. Conf. Phys., London *1*, 144 (1934).

Oliphant, Shire and Crowther, *Disintegration of the separated isotopes of lithium by protons and by heavy hydrogen*. Nature *133*, 377 (1934).

Oliphant, Harteck and Rutherford, *Transmutation effects observed with heavy hydrogen*. Nature *133*, 413 (1934).

Oliphant, Harteck and Rutherford, *Transmutation effects observed with heavy hydrogen*. Proc. Roy. Soc. *144*, 692 (1934).

Oliphant, Shire and Crowther, *Separation and transmutation of the isotopes of lithium*. Proc. Roy. Soc. *146*, 922 (1934).

Oliphant, Kempton and Rutherford, *Energy released in nuclear transformations*. Proc. Soc. Roy. *149*, 406 (1935).

Oliphant, Kempton and Rutherford, *Nuclear transformations of beryllium and boron, and the masses of light elements*. Proc. Roy. Soc. *150*, 241 (1935).

O'Neal and Goldhaber, *Decay constant of H*[3]. Phys. Rev. *58*, 574 (1940).

Ortner and Stetter, *H-rays from aluminum with Ra(B + C) as radiation source*. Z. Phys. *89*, 708 (1934).

Ortner and Protiwinsky, *The reactions of fast neutrons with nitrogen and neon nuclei*. Anz. Akad. Wien *76A*, 116 (1939).

Ortner and Protiwinsky, *The reaction of fast neutrons with nitrogen and neon nuclei*. S. B. Akad. Wiss. Wien, *148*, 349 (1939).

Pais, *Lifetime of the neutral meson*. Nature *156*, 715 (1945).

Paneth and Loleit, *Chemical detection of artificial transmutation of elements*. Nature *136*, 950 (1935).

Park and Mouzon, *Evidence against the existence of an excited state of He*[3]. Phys. Rev. *58*, 43 (1940).

Paton, *Protons emitted from boron and phosphorus bombarded by high energy alpha particles.* Phys. Rev. *46*, 229 (1934).

Paton, *Proton emission from boron and phosphorus under bombardment by high energy alpha rays.* Z. Phys. *90*, 586 (1934).

Paton, *Proton emission resulting from alpha ray bombardment of boron and phosphorus.* Phys. Rev. *47*, 197 (1935).

Paxton, *Positrons from deuteron activated phosphorus.* Phys. Rev. *49*, 206 (1936).

Perkins, *Nuclear disintegration by meson capture.* Nature *159*, 126 (1947).

Perlow, *Accurate measurement of the energy released in the disintegration of Li^6 by protons.* Phys. Rev. *58*, 218 (1940).

Pettersson, *Artificial disintegration of atoms and their packing fractions* (in English). Ark. Mat. Astr. Fys. *21A*, No. 1 (1928).

Pollard, *Experiments on the protons produced in the artificial disintegration of the nitrogen nucleus.* Proc. Roy. Soc. *141*, 375 (1933).

Pollard and Eaton, *Disintegration of nitrogen and boron and possible emission of deuterons.* Phys. Rev. *47*, 597 (1935).

Pollard and Brasefield, *The transmutation of potassium, chlorine and phosphorus by Th C' alpha particles.* Phys. Rev. *50*, 890 (1936).

Pollard and Brubaker, *Maximum energy of the protons from the bombardment of boron by alpha particles.* Phys. Rev. *52*, 762 (1937).

Pollard, *The transmutation of scandium by Th C' alpha particles.* Phys. Rev. *54*, 411 (1938).

Pollard and Watson, *Deuteron bombardment of the separated isotopes of neon.* Phys. Rev. *57*, 567 (1940).

Pollard, Davidson and Schultz, *Protons from the transmutation of boron by deuterons.* Phys. Rev. *57*, 1117 (1940).

Pollard and Humphreys, *Energy levels of Mg^{26} and Si^{30} formed by deuteron bombardment.* Phys. Rev. *59*, 466 (1941).

Pollard, *Yields of protons in transmutations.* Phys. Rev. *71*, 467 (1947).

Pontecorvo and Lazard, *Nuclear isomerism produced by x-rays of the continuous spectrum.* C. R., Paris *208*, 99 (1939).

Pool and Cork, *Deuteron bombardment of barium, lanthanum and cerium.* Phys. Rev. *51*, 1010 (1937).

Pose, *Detection of products of the atomic disintegration of aluminum.* Phys. Z. *30*, 780 (1929).

Pose, *Range groups of H-particles from aluminum.* Z. Phys. *64*, 1 (1930).

Pose, *Number and range of the nuclear protons liberated from aluminum and nitrogen by alpha rays.* Z. Phys. *95*, 84 (1935).

Pose, *Measurements of nuclear transformation processes in the presence of strong gamma rays.* Phys. Z. *37*, 154 (1936).

Powell, May, Chadwick and Pickavance, *Excited states of stable nuclei.* Nature *145*, 893 (1940).

Reddemann, *Nuclear isomerism in Sr^{87}.* Naturwiss. *28*, 110 (1940).

Reddemann, *On nuclear isomerism of Sr.* Z. Phys. *116*, 137 (1940).

Reid, *Synthesis by nuclear recoil.* Phys. Rev. *69*, 530 (1946).

Richardson and Emo., *The photodisintegration of the deuteron by the gamma radiation from Na^{24}.* Phys. Rev. *53*, 234 (1938).

Richardson and Emo, *The photodisintegration of the deuteron.* Phys. Rev. *51*, 1014 (1937).

Richardson and Wright, *The nuclear reaction (p, pn).* Phys. Rev. *70*, 445 (1946).

Ringo, *Velocity spectrum of alpha particles.* Phys. Rev. *58*, 942 (1940).

Roberts, Zandstra, Cortell and Myers, *Variation of range with angle of the disintegration alpha particles of Li^7.* Phys. Rev. *49*, 783 (1936).

Roberts, *Investigation of the deuteron-deuteron reaction.* Phys. Rev. *51*, 810 (1937).

Roberts and Heydenburg, *Further observations on the production of N^{13}.* Phys. Rev. *53*, 375 (1938).

Roberts, Downing and Deutsch, *A study of the radiations from the disintegration of* Br^{82}. Phys. Rev. *60*, 544 (1941).

Rogers, Bennett, Bonner and Hudspeth, *Resonances in the emission of protons from the reaction* $C^{12} + H^2$. Phys. Rev. *58*, 186 (1940).

Rosenblum, *Passage of alpha rays through matter.* Ann. Phys., Leipzig *10*, 408 (1928).

Rosenblum, *Recent progress in the study of the magnetic spectrum of alpha rays.* J. Phys. Rad. *1*, 438 (1930).

Rosenblum, Guillot and Perey, *Magnetic alpha spectra of actinium family.* C. R., Paris *204*, 175 (1937).

Rossi and Swartz, *The angular distribution of alphas from* $F^{19}(p, \alpha)O^{16}$. Phys. Rev. *65*, 83 (1944).

Rotblat, *Range of particles emitted in the disintegration of boron and lithium by slow neutrons.* Nature *138*, 202 (1936).

Roussinow and Yusephovich, *Energy of conversion electrons arising in the transformation of bromine isomers* (in English). C. R., URSS. *24*, 129 (1939).

Rousinow and Yusephovich, *X-ray emission from the isomers of radioactive bromine.* Phys. Rev. *55*, 979 (1939).

Roussinow and Igelnitzki, *On nuclear isomers with long life times* (in English). C. R., URSS *47*, 333 (1945).

Rubin, *Excitation energy of excited* Li^7. Phys. Rev. *59*, 216 (1941).

Rubin, *Excitation energy of excited* Li^7. Phys. Rev. *69*, 134 (1946).

Rubin, Fowler and Lauritsen, *Angular distribution of the* Li^7 *(p, α) α reaction.* Phys. Rev. *71*, 212 (1947).

Rubin, *Angular distribution of long range alpha particles from excited* Ne^{20}. Phys. Rev. *71*, 275 (1947).

Rumbaugh and Hafstad, *Disintegration experiments on the separated isotopes of lithium.* Phys. Rev. *50*, 681 (1936).

Rumbaugh, Roberts and Hafstad, *Transmutation processes of* Li^6 *and* Li^7. Phys. Rev. *51*, 1013 (1937).

Rumbaugh, Roberts and Hafstad, *Conservation of energy in the disintegration of* Li^8. Phys. Rev. *51*, 1106 (1937).

Rumbaugh, Roberts and Hafstad, *Nuclear transmutations of the lithium isotopes.* Phys. Rev. *54*, 657 (1938).

Rutherford and Chadwick, *Artificial disintegration of light elements.* Phil. Mag. (6) *42*, 809 (1921).

Rutherford and Chadwick, *The bombardment of elements by alpha particles.* Nature *113*, 457 (1924).

Saha, *Disintegration of fluorine by alpha particle bombardment.* Z. Phys. *110*, 473 (1938).

Savel, *Artificial disintegration.* Ann. Phys. Leipzig *4*, 88 (1935).

Scherrer, Huber and Rossel, *Nuclear reactions of fluorine with fast neutrons* (in German). Helv. Phys. Acta *14*, 618 (1941).

Schiff, *Thresholds for slow neutron induced reactions.* Phys. Rev. *70*, 106 (1946).

Schiff, *Thresholds for slow neutron induced reactions.* Phys. Rev. *70*, 562 (1946).

Schintlmeister, *Range of Th alpha rays.* S. B. Akad. Wiss. Wien *146*, 371 (1937).

Schnetzler, *Nuclear processes in lithium upon irradiation with alpha rays.* Z. Phys. *95*, 84 (1935).

Schnetzler, *Nuclear processes in lithium on bombardment with alpha rays.* Z. Phys. *95*, 302 (1935).

Schrader and Pollard, *Protons from the deuteron bombardment of the separated isotopes of chlorine.* Phys. Rev. *59*, 277 (1941).

Schultz, *Disintegration of the deuteron under fast alpha particle bombardment.* Phys. Rev. *51*, 1023 (1937).

Schultz, *Disintegration of the deuteron by alpha particles.* Phys. Rev. *53*, 622 (1938).

Schultz, Davidson and Ott, *Protons from carbon and aluminum bombarded by deuterons.* Phys. Rev. *58*, 1043 (1940).

Schultz and Watson, *Protons from the separated isotopes of carbon and neon under deuteron bombardment.* Phys. Rev. *58*, 1047 (1940).

Seaborg and Segrè, *Nuclear isomerism in element 43.* Phys. Rev. *55*, 808 (1939).

Segrè and Seaborg, *Nuclear isomerism in element 43.* Phys. Rev. *54*, 772 (1938).

Seren, Freidlander and Turkel, *The reaction* $_{38}Sr^{86}$ (n, ∂) $_{38}Sr^{87}$ *produced by thermal neutrons*. Phys. Rev. *71*, 454 (1947).

Shepherd, Haxby and Hill, *A search for protons from lithium bombarded by Ra C' alpha particles*. Phys. Rev. *52*, 674 (1937).

Sherr, *Evidence for a new type of nuclear reaction*. Phys. Rev. *57*, 937 (1940).

Sherr and Bainbridge, *Transmutation of mercury by fast neutrons*. Phys. Rev. *59*, 937 (1941).

Sherr, Bainbridge and Anderson, *Transmutation of mercury by fast neutrons*. Phys. Rev. *60*, 473 (1941).

Shinohara and Hatoyama, *Pair production by gamma rays from fluorine bombarded with protons*. Tokyo Inst. Phys. Chem. Res. *38*, 253, 326 (1941).

Shrader and Pollard, *Protons from deuteron bombardment of separated chlorine isotopes*. Phys. Rev. *58*, 199 (1940).

Shrader and Pollard, *Protons from the deuteron bombardment of the separated isotopes of chlorine*. Phys. Rev. *59*, 277 (1941).

Siegbahn and Petersson, *Coincidence on N^{13} and C^{11}* (in English). Ark. Mat. Astr. Fys. *32B*, No. 5 (1945).

Siegbahn, *The disintegration of Mn^{56}* (in English). Ark. Mat. Astr. Fys. *33A*, No. 10 (1946).

Siegbahn, *The disintegration of Na^{24} and P^{32}*. Phys. Rev. *70*, 127 (1946).

Siegbahn and Hole, *The disintegration of I^{128}*. Phys. Rev. *70*, 133 (1946).

Sirkar and Bhattacharyva, *On the evidence for the existence of neutral mesotrons*. Phys. Rev. *67*, 365 (1945).

Sizoo and Friele, *On the nuclear isomerism of $^{104}_{45}Rh$* (in English). Physica *10*, 57 (1943).

Skaggs, *A precise determination of the energy released in the production of deuterium from beryllium by proton bombardment*. Phys. Rev. *56*, 24 (1939).

Skaggs and Graves, *Excited state of Li^7 from the reaction Be^9 (d, α) Li^7*. Phys. Rev. *57*, 1087 (1940).

Smith, *The energies released in the reactions Li^7 (p, α) He and Li^6 (d, α)*

He^4, *and the masses of light atoms*. Phys. Rev. *56*, 548 (1939).

Smith and Murrell, *Disintegration of separated isotopes of boron under proton and deuteron bombardment*. Proc. Camb. Phil. Soc. *35*, 298 (1939).

Smith and Pollard, *Energy levels of S^{33} by deuteron bombardment of S^{32}*. Phys. Rev. *59*, 942 (1941).

Smith, *Isomeric states in indium 112*. Phys. Rev. *61*, 389 (1942).

Smith and Groetzinger, *On the positive particles appearing near beta ray emitters*. Phys. Rev. *70*, 96 (1946).

Smith and Groetzinger, *On the positive particles emitted by radioactive electron source*. Phys. Rev. *70*, 102 (1946).

Soonawala, *Some nuclear disintegrations*. Indian J. Phys. *19*, 185 (1945).

Stegmann, *Excitation of the nitrogen nucleus to H ray emission by alpha rays from polonium*. Z. Phys. *95*, 72 (1935).

Stephens and Bonner, *Disintegration of boron by deuterons*. Phys. Rev. *52*, 527 (1937).

Stetter, *The disintegration of the elements*. Phys. Z. *37*, 88 (1936).

Stetter, *Excitation of the $^{17}_{8}O$ nucleus*. Z. Phys. *100*, 652 (1936).

Steudel, *Artificial disintegration of aluminum and nitrogen*. Z. Phys. *77*, 139 (1932).

Streib, Fowler and Lauritsen, *Transmutation of fluorine by protons*. Phys. Rev. *59*, 253 (1941).

Swartz, Rossi, Jennings and Inglis, *The angular distribution of alphas from Li^7 (p, α) α*. Phys. Rev. *65*, 80 (1944).

Szalay, *Neutron excitation functions for $^{27}_{13}Al$ and $^{10}_{5}B$*. Z. Phys. *112*, 29 (1939).

Tatel and Cork, *Formation of radium E and Po by deuteron bombardment of bismuth*. Phys. Rev. *71*, 159 (1947).

Taylor and Goldhaber, *Detection of nuclear disintegration in a photographic emulsion*. Nature *135*, 341 (1935).

Taylor, *Disintegration of boron by neutrons*. Proc. Phys. Soc., Lond. *47*, 873 (1935).

Taylor and Dabholkar, *Ranges of alpha particles in photographic emulsions*. Proc. Phys. Soc., Lond. *48*, 285 (1936).

Tomlinson, *Nuclear electron pairs from O^{16}*. Phys. Rev. *60*, 159 (1941).

Traubenberg, Eckhardt and Gebauer, *Atom disintegration at low potentials*. Naturwiss. *21*, 26 (1933).

Tsien, Bachelet and Bouissières, *Paths and fine structures of the alpha rays of protactinium*. J. Phys. Rad. 7, 167 (1946).

Tsien, Bachelet and Bouissières, *Fine structure of the alpha rays from protactinium*. Phys. Rev. *69*, 39 (1946).

Tuve and Hafstad, *Emission of disintegration particles from targets bombarded by protons and by deuterium ions at 1200KV*. Phys. Rev. *45*, 651 (1934).

Valente and Zagor, *Fast neutron resonance with nitrogen*. Phys. Rev. *69*, 55 (1946).

Valley and McCreary, *Internal conversion electrons from Br^{80}*. Phys. Rev. *55*, 666 (1939).

Valley, *Internal conversion in rhenium*. Phys. Rev. *59*, 686 (1941).

Valley, *Internal conversion in mercury*. Phys. Rev. *60*, 167 (1941).

Van Allen, Ellett and Bayley, *Cross section for the reaction $H^2 + H^2 —> H^1 + H^3$*. Phys. Rev. *56*, 383 (1939).

Van Allen and Smith, *The absolute number of quanta from the bombardment of fluorine with protons*. Phys. Rev. *59*, 108 (1941).

Van Allen and Smith, *The absolute cross section for the photodisintegration of deuterium by 6.2 Mev quanta*. Phys. Rev. *59*, 618 (1941).

Van Heerden and Nimck-Blok, *The radiation excited by slow neutrons in cadmium* (in English). Physica *10*, 13 (1943).

Waldman, Waddel, Callihan and Schneider, *The resonance processes in the disintegration of boron by protons*. Phys. Rev. *54*, 1017 (1938).

Waldman and Collins, *Nuclear excitation of lead by x-rays*. Phys. Rev. *57*, 338 (1940).

Waldman and Wiedenbeck, *The nuclear excitation of indium by x-rays and electrons*. Phys. Rev. *63*, 60 (1943)

Walke, Thompson and Holt, *K-electron capture and internal conversion in Cr^{51}*. Phys. Rev. *57*, 171 (1940).

Wambacher, *Nuclear disintegration in a photographic emulsion by cosmic rays*. S. B. Akad. Wiss. Wien *149*, 157 (1940).

Watase, *A study of the disintegrations of Cl^{38}, Al^{28} and Na^{24} with the use of the method of coincidence counting*. Proc. Phys.-Math. Soc. Jap. (3) *23*, 618 (1941).

Welles, *Deuteron bombardment of oxygen*. Phys. Rev. *59*, 679 (1941).

Wertenstein, *Remarks on the article by Crane and Halpern, "New experimental evidence for the neutrino."* Phys. Rev. *54*, 306 (1938).

Whitmer and Pool, *Low voltage disintegration of lithium with lithium*. Phys. Rev. *47*, 795 (1935).

Wiedenbeck, *Nuclear excitation of cadmium*. Phys. Rev. *66*, 36 (1944).

Wiedenbeck, *The nuclear isomerism of gold*. Phys. Rev. *67*, 53 (1945).

Wiedenbeck and Marhoefer, *A new method for determining thresholds in ∂-n processes*. Phys. Rev. *67*, 54 (1945).

Wiedenbeck, *The nuclear excitation of silver and cadmium*. Phys. Rev. *67*, 92 (1945).

Wiedenbeck, *The nuclear excitation of rhodium*. Phys. Rev. *67*, 267 (1945).

Wiedenbeck, *The excitation of heavy nuclei*. Phys. Rev. *68*, 1 (1945).

Wiedenbeck, *The nuclear excitation of krypton and rhodium*. Phys. Rev. *68*, 237 (1945).

Wiedenbeck, *The lifetimes of metastable states*. Phys. Rev. *69*, 47 (1946).

Wiedenbeck, *Note on the lifetime of metastable states*. Phys. Rev. *69*, 567 (1946).

Wilkins and Crawford, *A cloud chamber of the alpha rays of actino-uranium*. Phys. Rev. *54*, 316 (1938).

Williams and Wells, *Evidence from the efficiency curves for the nature of the disintegration process for boron*. Phys. Rev. *50*, 186 (1936).

Williams, Wells, Tate and Hill, *A resonance process in the disintegration of boron by protons*. Phys. Rev. *51*, 434 (1937).

Williams, Shepherd and Haxby, *Evidence for the instability of He5*. Phys. Rev. *51*, 888 (1937).

Williams, Shepherd and Haxby, *The disintegration of lithium by deuterons*. Phys. Rev. *51*, 1011 (1937).

Williams, Shepherd and Haxby, *The disintegration of lithium by deuterons*. Phys. Rev. *52*, 390 (1937).

Wilson, *Investigation of the disintegration of boron by slow neutrons*. Proc. Roy. Soc. *177* 382 (1941).

Winand, *Energy liberated by ionium*. J. Phys. Rad. *8*, 429 (1937).

Yalow and Goldhaber, *A "fast" gamma ray transition*. Phys. Rev. *67*, 59 (1945).

Young, Plain, McLean and Ellett, *Distribution in angle of alpha particles from Li7 + H1*. Phys. Rev. *57*, 1083 (1940).

Young, Ellett and Plain, *Distribution in angle of alpha particles from lithium bombarded with protons*. Phys. Rev. *58*, 498 (1940).

Yu, *Disintegration by consecutive orbital electron captures:* $_{56}Ba^{131} -> {}_{55}Cs^{131} -> {}_{54}Xe^{131}$. Phys. Rev. *71*, 382 (1947).

Zagor and Valente, *Heavy particle groups from the neutron disintegrations of nitrogen and neon*. Phys. Rev. *71*, 133 (1945).

Zah-Wei, *Disintegration of* $_{17}Cl^{34}$. Phys. Rev. *70*, 782 (1946).

Zeleny, Brasefield, Bock and Pollard, *Alpha particles from lithium ions striking hydrogen compounds*. Phys. Rev. *46*, 318 (1934).

Zhdanov, Perfilov and Deisenrod, *Anomalous rate of nuclear disintegration effected by cosmic rays*. Phys. Rev. *65*, 202 (1944).

Zlotowski, *A nuclear disintegration induced by the cosmic radiation*. Phys. Rev. *56*, 484 (1939).

CHAPTER FIVE

Radioactivity

Abelson, *The identification of x-rays emitted during the decay of radioactive elements*. Phys. Rev. *55*, 424 (1939).

Adams, *Non K-electron capture by nuclei of relatively low isotopic number*. Phys. Rev. *66*, 358 (1944).

Alexeeva, *Artificial radioactivity produced by slow neutrons in antimony*. C. R., URSS *17*, 13 (1937).

Alexeeva, *Long-period radioactivity in Ag, Cs and In activated by slow neutrons*. C. R., URSS *18*, 553 (1938).

Alichanian, Alichanow and Dzelepow, *Artificial radioactivity* (in English). Phys. Z. Sowjet. *10*, 78 (1936).

Alichanow Alichanian and Dzelepow, *A new type of artificial beta radioactivity*. Nature *133*, 871 (1934).

Alichanow, Alichanian and Dzelepow, *Energy spectrum of positive electrons ejected by radioactive nitrogen*. Nature *133*, 950 (1934).

Alichanow, Alichanian and Dzelepow, *Limits of the energy spectra of positrons and electrons from artificial radioactive elements*. Nature *134*, 254 (1934).

Allen, Pool, Kurbatov and Quill, *Artificial radioactivity of Ti45*. Phys. Rev. *60*, 155 (1941).

Allen, Pool, Kurbatov and Quill, *Artificial radioactivity of Ti45*. Phys. Rev. *60*, 425 (1941).

Allison, *Efficiencies of production and half lives of radio-carbon and radionitrogen*. Proc. Camb. Phil. Soc. *32*, 179 (1936).

Alvarez and Cornog, *Radioactive hydrogen*. Phys. Rev. *57*, 248 (1940).

Alvarez and Cornog, *Radioactive hydrogen—a correction*. Phys. Rev. *58*, 197 (1940).

Amaki, Iiomori and Sugimoto, *Artificial radioactivity of chromium*. Tokyo Inst. Phys. Chem. Res. *37*, 395 (1940).

Amaki, Iiomori and Sugimoto, *Radio isotopes of chromium*. Phys. Rev. *57*, 751 (1940).

Amaldi, D'Agostino, Fermi, Pontecorvo, Rasetti and Segrè, *Artificial radioactivity produced by neutron bombardment. Part II*. Proc. Roy. Soc. *149*, 522 (1935).

Andersen, *Induced radioactivity of mercury*. Nature *137*, 457 (1936).

Andersen, *A radioactive isotope of iron.* Nature *138*, 76 (1936).

Andersen, *Radioactive sulphur isotope.* Z. phys. Chem. *32*, 237 (1936).

Antonoff, *On the position of uranium Y in radioactive series.* Phys. Rev. *68*, 288 (1945).

Atterling, Bohr and Sigurgeirsson, *Neutron induced radioactivity in lutecium and ytterbium* (in English). Ark. Mat. Astr. Fys. *32A*, No. 2 (1945).

Banks, Chalmers and Hopwood, *Induced radioactivity produced by neutrons liberated from heavy water by radium gamma rays.* Nature *135*, 99 (1935).

Barendregt, Griffioen and Sizoo, *The positron-spectrum emitted by radio-phosphorus $_{15}P^{32}$* (in English). Physica *7*, 860 (1940).

Barkas, Carlson, Henderson and Moore, *Concentration of radioactive bromine produced by gamma radiation.* Phys. Rev. *58*, 577 (1940).

Barnes, Du Bridge, Wiig, Buch and Strain, *Proton-induced radioactivity in heavy nuclei.* Phys. Rev. *51*, 775 (1937).

Barnes and Valley, *A long period activity induced in copper.* Phys. Rev. *53*, 946 (1938).

Barnes and Aradine, *Radioactivity by nuclear excitation. II. Excitation by protons.* Phys. Rev. *55*, 50 (1939).

Barnes, *Proton activation of indium and cadmium.* Phys. Rev. *56*, 414 (1939).

Becker and Gaerttner, *The half life of B^{12}.* Phys. Rev. *56*, 854 (1939).

Bertl, Oboril and Sitte, *Artificial radioactivity at low potentials.* Z. Phys. *106*, 463 (1937).

Bethe and Henderson, *Evidence for incorrect assignment of the supposed Si^{27} radioactivity of 6.7 minute half life.* Phys. Rev. *56*, 1060 (1939).

Bjerge and Westcott, *Radioactivity induced by bombardment with neutrons of different energies.* Nature *134*, 177 (1934).

Bjerge and Westcott, *Radioactivity induced by neutron bombardment.* Nature *134*, 286 (1934).

Bjerge, *Induced radioactivity of short period.* Nature *137*, 865 (1936).

Bjerge, *Radio-helium.* Nature *138*, 400 (1936).

Bjerge, *Radioactive neon.* Nature *139*, 757 (1937).

Bohr and Hole, *Radioactivity induced by neutrons and deuterons in ruthenium* (in English). Ark. Mat. Astr. Fys. *32A*, No. 15 (1946).

Bothe and Gentner, *Artificial radioactivity induced by gamma rays.* Naturwiss. *25*, 90 (1937).

Bothe and Gentner, *Wavelength dependence of the nuclear photoeffect with regard to the radioactive isotope of selenium.* Z. Phys. *112*, 45 (1939).

Bothe, *Activation of rare earths by thermal neutrons.* Z. Naturforsch. *1*, 173 (1946).

Bradt and Scherrer, *Disintegration constant and half-life period of The C'* (in English). Helv. Phys. Acta *16*, 229, 259 (1943).

Bradt, Gugelot, Huber, Medicus, Preiswerk and Scherrer, *Metastable states of the silver nuclei Ag^{107} and Ag^{109}* (in German). Helv. Phys. Acta *18*, 256 (1945).

Bradt, Gugelot, Huber, Medicus, Preiswerk and Scherrer, *Radioactive transformation of Cr^{51}* (in German). Helv. Phys. Acta *18*, 259 (1945).

Bramley and Brewer, *The radioactivity of potassium.* Phys. Rev. *53*, 502 (1938).

Brandt, *Artificial activity of radio-aluminum (Al^{26}) and the radio-chlorine (Cl^{34}).* Z. Phys. *108*, 726 (1938).

Bretscher, *Radioactivity of Be^{10}.* Nature *146*, 94 (1940).

Brown, *Laboratory experiments on radioactive recoil.* Amer. J. Phys. *9*, 373 (1941).

Buck, *Proton induced radioactivities. III. Zinc and selenium targets.* Phys. Rev. *54*, 1025 (1938).

Burcham, Goldhaber and Hill, *Radioactivity produced in scandium by fast neutrons.* Nature *141*, 510 (1938).

Cacciapuoti and Segrè, *Radioactive isotopes of element 43.* Phys. Rev. *52*, 1252 (1937).

Cacciapuoti, *Decay constant of P^{32}.* Nuovo Cim. *15*, 213 (1938).

Cacciapuoti, *Radioactivity induced in Mo by deuterons.* Nuovo Cim. *15*, 425 (1938).

Cacciapuoti, *Radioactive isotopes of element 43.* Phys. Rev. *55*, 110 (1939).

Chamberlain, Williams and Yuster, *Half-life of uranium (234).* Phys. Rev. *70*, 580 (1946).

Chamberlain, Gofman, Segrè and Wahl, *Range measurements of alpha particles from 94²³⁹ and 94²³⁸.* Phys. Rev. *71*, 529 (1947).

Chang, Goldhaber and Sagane, *Radioactivity produced by gamma rays and neutrons of high energy.* Nature *139*, 962 (1937).

Chang and Szalay, *Formation of radio-aluminum (Al²⁸) and the resonance effect of Mg²⁵.* Proc. Roy. Soc. *159*, 72 (1937).

Cichocki and Soltan, *Radio-silicon produced by bombardment of sulphur with rapid neutrons.* C. R., Paris *207*, 423 (1938).

Clancy, *Induced radioactivity of krypton.* Phys. Rev. *58*, 88 (1940)).

Clancy, *The induced radioactivity of xenon.* Phys. Rev. *59*, 686 (1941).

Clancy, *The induced radioacivity of krypton and xenon.* Phys. Rev. *60*, 87 (1941).

Clark, *Radiations from radioactive gold, tungsten and dysprosium.* Phys. Rev. *61*, 242 (1942).

Cockcroft, Gilbert and Walton, *Production of induced radioactivity by high velocity protons.* Nature *133*, 328 (1934).

Cockcroft, Gilbert and Walton, *High velocity positive ions. Part IV. Production of induced radioactivity by protons and diplons.* Proc. Roy. Soc. *148*, 225 (1935).

Collar, Cork and Smith, *Radioactive barium from cesium.* Phys. Rev. *59*, 937 (1941).

Conn, Brosi, Swartout, Cameron, Carter and Hill, *Confirmation of assignment of 2.6 hr Ni to a mass number of 65.* Phys. Rev. *70*, 768 (1946).

Cook and McDaniel, *Radiations from radioactive Co⁵⁶.* Phys. Rev. *62*, 412 (1942).

Cook, Jurney and Langer, *On the disintegration scheme of Na²⁴.* Phys. Rev. *70*, 985 (1946).

Cork, Richardson and Kurie, *Radiations emitted by radio-aluminum.* Phys. Rev. *49*, 208 (1936).

Cork and Lawson, *Induced radioactivity in cadmium and indium.* Phys. Rev. *56*, 291 (1939).

Cork and Halpern, *Radioactive isotopes of gold.* Phys. Rev. *58*, 201 (1940).

Cork and Smith, *Radioactive isotopes of barium from cesium.* Phys. Rev. *60*, 480 (1941).

Cornog and Libby, *Production of radioactive hydrogen by neutron bombardment of boron and nitrogen.* Phys. Rev. *59*, 1046 (1941).

Corson, MacKenzie and Segrè, *Possible production of radioactive isotopes of element 85.* Phys. Rev. *57*, 459 (1940).

Corson, MacKenzie and Segrè, *Artificially radioactive element 85.* Phys. Rev. *58*, 672 (1940).

Coven, *Evidence of increased radioactivity of the atmosphere after the atomic bomb test in New Mexico.* Phys. Rev. *68*, 279 (1945).

Crane and Lauritsen, *Radioactivity from carbon and boron oxide bombarded with deutons and the conversion of positrons into radiation.* Phys. Rev. *45*, 430 (1934).

Crane and Lauritsen, *Further experiments with artificially produced radioactive substances.* Phys. Rev. *45*, 497 (1934).

Crane and Lauritsen, *Radioactivity produced by artificially accelerated particles.* Phys. Rev. *45*, 746 (1934).

Creutz, Fox and Sutton, *Radioactivity induced in neon and aluminum by protons.* Phys. Rev. *57*, 567 (1940).

Creutz, Delsasso, Sutton, White and Barkas, *Radioactivity produced by proton bombardment of bromine and iodine.* Phys. Rev. *58*, 481 (1940).

Cuer and Lattes, *Radioactivity of samarium.* Nature *158*, 197 (1946).

Curie and Joliot, *Penetration of atoms by very penetrating rays excited in the heavy nuclei.* C. R., Paris, *194*, 876 (1932).

Curie and Joliot, *New type of radioactivity.* C. R., Paris *198,* 254 (1934). Erratum, *198,* 408 (1934).

Curie and Joliot, *Chemical separation of new radioactive elements.* C. R., Paris *198,* 559 (1934).

Curie and Joliot, *Induced radioactivity.* J. Phys. Rad. *5,* 153 (1934).

Curie and Preiswerk, *Activation of thulium by slow neutrons.* C. R., Paris *203,* 787 (1936).

Curie and Tsien, *Comparison of radiation of radioactive isotopes formed from uranium and thorium.* J. Phys. Rad. *10,* 495 (1939).

Curran and Strothers, *Radioactivity of F^{20}.* Proc. Camb. Phil. Soc. *36,* 252 (1940).

Curtis and Richardson, *Radiations from radioactive indium.* Phys. Rev. *57,* 1121 (1940).

Danysz and Zyw, *Radiofluorine* (in French). Acta Phys. Polonica *3,* 485 (1934).

Darling, Curtis and Cork, *Radioactivity in iron by deuteron bombardment,* Phys. Rev. *51,* 1010 (1937).

Davidson, *Energy levels, mass and radioactivity of A^{41}.* Phys. Rev. *57,* 244 (1940).

Davidson, *Production of F^{18} by deuterons.* Phys. Rev. *57,* 1086 (1940).

DeBenedetti and McGowan, *A metastable state of 22 microseconds in Ta^{181}.* Phys. Rev. *70,* 569 (1946).

DeBenedetti and McGowan, *A metastable state of half-life about 10^{-6} second in Re^{187}.* Phys. Rev. *71,* 380 (1947).

Delsasso, Ridenour, Sheer and White, *Artificial radioactivity produced by protons.* Phys. Rev. *55,* 113 (1939).

Deutsch, Downing, Elliott, Irvine and Roberts. *Disintegration schemes of radioactive substances. IV. Fe^{59}.* Phys. Rev. *62,* 3 (1942).

Deutsch and Elliott, *Disintegration schemes of radioactive substances. VII. Mn^{54} and Co^{58}.* Phys. Rev. *65,* 211 (1944).

Deutsch, Elliott and Roberts, *Disintegration schemes of radioactive sub-stances. VIII. Co^{60}.* Phys. Rev. *66,* 193 (1945).

De Wire, Pool and Kurbatov, *Radioactive isotopes of praseodymium.* Phys. Rev. *61,* 564 (1942).

Dickson, McDaniel and Konopinski, *Another criterion in the identification and prediction of half lives.* Phys. Rev. *57,* 351 (1940).

Diemer and Groendijk, *Some measurements on the 66 hour period of gold* (in English). Physica *11,* 396 (1946).

Dode and Pontecorvo, *Radioactive cadmium.* C. R., Paris *207,* 287 (1938).

Downing and Roberts, *Coincidence study of the decay of Br^{82}.* Phys. Rev. *59,* 940 (1941).

Downing, Deutsch and Roberts, *Disintegration scheme of the yttrium activity of 100 day half-period.* Phys. Rev. *60,* 470 (1941).

Downing, Deutsch and Roberts, *Disintegration schemes of radioactive substances. III. I^{131}.* Phys. Rev. *61,* 686 (1942).

Du Bridge, Barnes and Buck, *Proton induced radioactivity in oxygen.* Phys. Rev. *51,* 995 (1937).

Du Bridge, Barnes, Buck and Strain, *Proton induced radioactivies.* Phys. Rev. *53,* 447 (1938).

Du Bridge and Marshall, *A radioactive isomer of Sr^{87}.* Phys. *56,* 706 (1939).

Du Bridge and Seaborg, *Artificial radioactivity.* Chem. Rev. *27,* 199 (1940).

Du Bridge and Marshall, *Radioactive isotopes of Sr, Y and Zr.* Phys. Rev. *57,* 348 (1940).

Du Bridge and Marshall, *Radioactive isotopes of Sr, Y and Zr.* Phys. Rev. *58,* 7 (1940).

Duffield and Calvin, *Preparation of high specific induced radioactivity by neutron bombardment of metal chelate compounds.* J. Amer. Chem. Soc. *68,* 1129 (1946).

Dunworth, *Radiations from radiomanganese.* Nature *143,* 1065 (1939).

Dunworth, *A determination of the half value periods of radium C' and thorium C', with a note on time lags in a Geiger counter.* Nature *144,* 152 (1939).

Eckhardt, *Generation of radioactive elements by bombardment of lithium and magnesium with thorium C' alpha radiation.* Ann. Phys., Leipzig *29*, 497 (1937).

Edwards and Pool, *Radiactive tellurium from antimony bombardment.* Phys. Rev. *69*, 140 (1946).

Edwards and Pool, *Radioactive isotopes of Mo and Ma.* Phys. Rev. *69*, 253 (1946).

Elliott and King, *Extension of the radioactive series, $Z = N \pm 1$.* Phys. Rev. *60*, 489 (1941).

Elliott and Deutsch, *Disintegration scheme of Co^{58}.* Phys. Rev. *63*, 219 (1943).

Elliott, Deutsch and Roberts, *The disintegration scheme of Na^{24}.* Phys. Rev. *63*, 386 (1943).

Elliott and Deutsch, *Radiations from the 100 day arsenic activity.* Phys. Rev. *63*, 457 (1943).

Elliott and Deutsch, *Disintegration schemes of radioactive substance. VI. Mn^{56} and Co^{56}.* Phys. Rev. *64*, 321 (1943).

Ellis and Henderson, *Artificial radioactivity produced in Mg by alpha particles.* Proc. Roy. Soc. *156*, 358 (1936).

Ellis and Henderson, *The period of radionitrogen.* Nature *135*, 429 (1935).

Ellis and Henderson, *Induced radioactivity by bombarding magnesium with alpha particles.* Nature *136*, 755 (1935).

Enns, *Radioactivities produced by proton bombardment of palladium.* Phys. Rev. *56*, 872 (1939).

Erbacher and Philipp, *The identification of the artificial radioactive elements produced by neutrons and their use in chemistry as indicators.* Z. angew. Chem. *48*, 409 (1935).

Erbacher and Philip, *Production of artificially radioactive halogens in minute amounts from the stable isotopes.* Ber. Deutsch. Chem. Ges. *69*, 893 (1936).

Erbacher, *Extraction of artificially radioactive P^{32} in unweighable amounts from CS_2.* Z. phys. Chem. *42*, 173 (1939).

Ewing, Perry and McCreary, *Radioactivities induced in Mo by fast protons.* Phys. Rev. *55*, 1136 (1939).

Fahlenbrach, *Artificial radioactivity.* Z. Phys. *94*, 607 (1935).

Fahlenbrach, *Discussion of artificial radioactivity (Curie-Joliot process). II.* Z. Phys. *96*, 503 (1935).

Fajans and Stewart, *Induced radioactivity in europium.* Phys. Rev. *56*, 625 (1939).

Fajans and Voigt, *Radioactivity produced by bombarding thallium.* Phys. Rev. *58*, 177 (1940).

Fajans and Sullivan, *Induced radioactivity of rhenium and tungsten.* Phy. Rev. *58*, 276 (1940).

Fajans and Voigt, *A note on the radiochemisty of europium.* Phys. Rev. *60*, 533 (1941).

Fajans and Voigt, *Artificial radioactive isotopes of thallium, lead and bismuth.* Phys. Rev. *60*, 619 (1941).

Fajans and Voigt, *The use of uranium lead in the assignment of artificial radioactive isotopes.* Phys. Rev. *60*, 626 (1941).

Feather and Dainty, *Absorption and coincidence experiments with Au.[198]* Proc. Camb. Phil. Soc. *40*, 57 (1944).

Fermi, *Artificial radioactivity by neutron bombardment.* Int. Conf. Phys. London *1*, 75 (1934).

Fermi, *Radioactivity induced by neutron bombardment.* Nature *133*, 757 (1934).

Fermi, Amaldi, D'Agostino, Rasetti and Segrè, *Artificial radioactivity produced by neutron bombardment.* Proc. Roy. Soc. *146*, 483 (1934).

Fleischmann, *Nuclei, radioactivity.* Phys. in regelmäss. Ber. *8*, 17 (1940).

Flügge, *The artificial production of naturally radioactive elements.* Naturwiss. *29*, 462 (1941).

Fomin and Houtermans, *Radioactivity produced in tantalum by neutron bombardment* (in German). Phys. Z. Sowjet. *9*, 273 (1936).

Fowler, Delsasso and Lauritsen, *Radioactive elements of low atomic number.* Phys. Rev. *49*, 561 (1936).

Fowler and Lauritsen, *Artificial radio-activity produced by alpha particles.* Phys. Rev. *51*, 1103 (1937).

Friedlander and Wu, *Radioactive isotopes of mercury.* Phys. Rev. *63*, 227 (1943).

Frisch, *Induced radioactivity of Na and P.* Nature *133*, 721 (1934).

Frisch, *Induced radioactivity of fluorine and calcium.* Nature *136*, 220 (1935).

Geiger and Nuttall, *The ranges of the α particles from various radioactive substances and a relation between range and period of transformation.* Phil. Mag. (6) *22*, 613 (1911).

Goldhaber, Hill and Szilard, *Radioactivity induced by nuclear excitation. I. Excitation by neutrons.* Phys. Rev. *55*, 47 (1939).

Goldhaber, *Activation of Ag (225 d) by resonance neutrons.* Phys. Rev. *70*, 89 (1946).

Goldhaber and Sturm, *The decay of Cb94 (6.6min).* Phys. Rev. *70*, 111 (1946).

Goodman and Pool, *Radioactive isotopes in the osmium region.* Phys. Rev. *70*, 112 (1946).

Goodman and Pool, *Radioactive isotopes of Re, Os and Ir.* Phys. Rev. *71*, 288 (1947).

Groendijk and deVries, *On the activities caused by n-∂ processes and the absorption coefficient for thermal neutrons* (in English). Physica *10*, 381 (1943).

Grahame, *Search for short-lived radiochlorine.* Phys. Rev. *54*, 972 (1938).

Grahame and Walke, *Preparation and properties of long-lived radiochlorine.* Phys. Rev. *60*, 909 (1941).

Grosse, Booth and Dunning, *The fourth (4n + 1) radioactive series.* Phys. Rev. *59*, 322 (1941).

Guében, *Activation of silver by neutrons.* Nature *138*, 1095 (1936).

Gugelot, Huber, Medicus, Preiswerk, Scherrer and Steffen, *On the radioactive decay of element 43* (in German). Helv. Phys. Acta *19*, 418 (1946).

Hafstad and Tuve, *Some observations on the induced radioactivity which follows bombardment by deuterium ions.* Phys. Rev. *45*, 767 (1934).

Hafstad and Tuve, *Artificial radioactivity using carbon targets.* Phys. Rev. *45*, 902 (1934).

Hafstad and Tuve, *Induced radioactivity using carbon targets.* Phys. Rev. *47*, 506 (1935).

Hafstad and Tuve, *Carbon radioactivity and the resonance transmutations by protons.* Phys. Rev. *48*, 306 (1935).

Hahn, *On a radioactive substance in uranium.* Ber. Deutsch. Chem. Ges. *54*, 1131 (1921).

Hahn, *On uranium Z and its mother substance.* Z. phys. Chem. *103*, 461 (1923).

Hahn and Meitner, *The artificial transmutation of Th by neutrons. Formation of the previously undiscovered radioactive series.* Naturwiss. *23*, 320 (1935).

Hahn, Strassmann and Walling, *Production of weighable quantities of the Sr isotope 87 as transmutation products of rubidium from a Canadian mica.* Naturwiss. *25*, 189 (1937).

Hancock and Butler, *Radioactive isotopes of rubidium.* Phys. Rev. *57*, 1088 (1940).

Harteck, Knauer and Schaeffer, *Radio-arsenic.* Z. Phys. *109*, 153 (1938).

Hayden and Inghram, *Assignment of mass to 46 hr samarium and 9.2 hr europium by a mass spectrograph.* Phys. Rev. *70*, 89 (1946).

Helmholz, Pecher and Stout, *Radioactive Rb from deuteron bombardment of Sr.* Phys. Rev. *59*, 902 (1941).

Helmholz, *Long-lived radioactive Cd from deuteron bombardment of Ag.* Phys. Rev. *60*, 160 (1941).

Helmholz, *Possible assignment of the long-lived yttrium.* Phys. Rev. *62*, 301 (1942).

Helmholz, *Isotopic assignment of Cd and Ag activities.* Phys. Rev. *70*, 982 (1946).

Hemmendinger and Smythe, *The radioactive isotope of rubidium.* Phys. Rev. *51*, 1052 (1937).

Hemmendinger, *Proton induced activity of manganese.* Phys. Rev. *55*, 604 (1939).

Hemmendinger, *Proton induced radioactivity of manganese.* Phys. Rev. *58*, 929 (1940).

Henderson, Livingston and Lawrence, *Artificial radioactivity produced by deuton bombardment.* Phys. Rev. *45*, 428 (1934).

Henderson, *Two radioactivities produced in Mg by deuteron bombardment and their excitation functions.* Phys. Rev. *48*, 480 (1935).

Henderson, *Two radioactive substances from magnesium after deuteron bombardment.* Phys. Rev. *48*, 855 (1935).

Henderson, Ridenour, White and Henderson, *The radioactivity of K^{38}.* Phys. Rev. *51*, 1107 (1937).

Henderson and Ridenour, *Artificial radioactivity produced by alpha particles.* Phys. Rev. *52*, 40 (1937).

Henderson and Doran, *The existence of radioactive Al^{29}.* Phys. Rev. *56*, 123 (1939).

Hevesy, *Artificial radioactivity of scandium* (in English). Kgl. Danske Vid. Sels. *13*, No. 3 (1925).

Hevesy and Levi, *Radiopotassium and other artificial radioelements.* Nature *135*, 580 (1935).

Hevesy and Levi, *Artificial radioactivity of dysprosium and other rare earth elements.* Nature *136*, 103 (1935).

Hevesy, *Natural and artificial radioactivity of potassium,* Nature *135*, 96 (1935).

Hevesy, *The radioactivity of calcium.* Naturwiss, *23*, 583 (1935).

Hevesy and Levi, *Artificial activity of hafnium and some other elements* (in English). Kgl. Danske Vid. Sels. *15*, No. 11 (1938).

Heyden and Wefelmeier, *A natural radioactivity of cassiopeium.* Naturwiss, *26*, 612 (1938).

Heyn, *Radioactivity induced by fast neutrons according to the (n, 2n) reaction.* Nature *139*, 842 (1937).

Heyn, *Radioactivity of Co, Ni, Cu and Zn induced by neutrons* (in English). Physica *4*, 160 (1937).

Heyn, *Radioactivity of Ni, Cu and Zn* (in English). Physica *4*, 1224 (1937).

Hibdon, Pool and Kurbatov, *The relative intensities and characteristic radiations of radioactive scandium.* Phys. Rev. *63*, 462 (1943).

Hibdon, Pool and Kurbatov. *Radioactive scandium. I.* Phys. Rev. *67*, 289 (1945).

Hibdon and Pool, *Radioactive scandium. II.* Phys. Rev. *67*, 313 (1945).

Hill and Valley, *A study of radioactive Be^7.* Phys. Rev. *55*, 678 (1939).

Hoag, *Production and half life of Cl^{33}.* Phys. Rev. *57*, 937 (1940).

Hönigschmid, and Hirschbold-Wittner, *The atomic weight of phosphorus.* Z. anorg. allg. Chem. *243*, 355 (1940).

Hole, *On neutron induced radioactivities in Rh* (in English). Ark. Mat. Astr. Fys. *32A*, No. 3 (1945).

Hosemann, *Radioactivity of samarium.* Z. Phys. *99*, 405 (1936).

Houtermans, *Half life of radiotantulum.* Naturwiss. *28*, 578 (1940).

Howland, Templeton and Perlman, *Artificial radioactive isotopes of Po, Bi and Pb.* Phys. Rev. *71*, 552 (1947).

Huber, Lienhard and Wäffler, *A new activity induced in cerium and neodymium by the nuclear photoeffect* (in German). Helv. Phys. Acta *17*, 251 (1944).

Hughes, Eggler and Huddleston, *The half life of Be^{10}.* Phys. Rev. *71*, 269 (1947).

Hull and Seelig, *The half life of iodine (128),* Phys. Rev. *60*, 553 (1941).

Hurst and Walke, *The induced radioactivity of K.* Phys. Rev. *51*, 1033 (1937).

Hurst and Pool, *K electron capture in radioactive silver.* Phys. Rev. *65*, 60 (1944).

Hurst and Pool, *Radioactive silver.* Phys. Rev. *65*, 351 (1944).

Inghram, *Mass spectrographic observation of radioactive C^{14}.* Phys. Rev. *70*, 111 (1946).

Inghram and Hayden, *Artificial activities produced in europium and holmium by slow neutron bombardment.* Phys. Rev. *71*, 130 (1947).

Inghram and Hayden, *Some neutron induced activities in the rare earths.* Phys. Rev. *71*, 144 (1947).

Inghram, Hayden, and Hess, *Neutron induced activities in lutecium, ytterbium and dysprosium*, Phys. Rev. *71*, 270 (1947).

Jensen, *Radiations from radioactive cobalt*. Phys. Rev. *59*, 936 (1941).

Jensen, *Radiations from radioactive cobalt*. Phys. Rev. *60*, 430 (1941).

Jnanananda, *Radioactive isotope of gold, $_{79}Au^{198}$, and low energy range of its spectrum*. Phys. Rev. 70, 812 (1946).

Johnson and Hamblin, *Radioactive isotopes of bromine*. Nature *138*, 504 (1936).

Kalbfell and Coolley, *Radio isotopes of Ba and Cs*. Phys. Rev. *58*, 91 (1940).

Kamen, *Production and isotopic assignment of long-lived radioactive sulphur*. Phys. Rev. *60*, 537 (1941).

Kamen, *Radioactive sulphur from neutron activation of chlorine isotope*. Phys. Rev. *62*, 303 (1942).

Kennedy and Seaborg, *Isotopic identification of induced radioactivity by bombardment of separated isotopes: 37 min. Cl^{38}*. Phys. Rev. 57, 843 (1940).

Kent, Cork and Wadey, *Radioactive selenium from arsenic*. Phys. Rev. *61*, 389 (1942).

Kent and Cork, *Radioactive tellurium from antimony*. Phys. Rev. *62*, 297 (1942).

King, Henderson and Risser, *The production of artificial radioactivity by alpha particles*. Phys. Rev. *55*, 1118 (1939).

King and Henderson, *Radioactive isotopes of indium from alpha bombardment of silver*. Phys. Rev. *56*, 1169 (1939).

King and Henderson, *Radioactive indium isotopes produced by alpha particles on silver*. Phys. Rev. *57*, 71 (1940).

King and Elliott, *Short-lived radioactivities of $_{14}Si^{27}$, $_{16}S^{31}$ and $_{18}A^{35}$*. Phys. Rev. *58*, 846 (1940).

King and Elliott, *Short-lived radioactivities of $_{14}Si^{27}$, $_{16}S^{31}$, $_{18}A^{35}$ and $_{21}Sc^{41}$*. Phys. Rev. *59*, 108 (1941).

Kittel, *Radioactivity of Li^8*. Phys. Rev. *55*, 515 (1939).

Klemperer, *Radioactivity of potassium and rubidium*. Proc. Roy. Soc. *148*, 638 (1935).

Knol and Veldkamp, *Artificially radioactive elements with very short periods* (in English). Physica *3*, 145 (1936).

Kraus and Cork, *Radioactive isotopes of palladium and silver from palladium*. Phys. Rev. *52*, 763 (1937).

Kuerti and Van Voorhis, *Induced radioactivity produced by bombarding aluminum with protons*. Phys. Rev. *56*, 614 (1939).

Kundu and Pool, *Mo^{93} and Cb^{94} from columbium*. Phys. Rev. 70, 111 (1946).

Kurbatov and Kurbatov, *Isolation of radioactive yttrium and some of its properties in minute concentrations*. J. Phys. Chem. *46*, 441 (1942).

Kurbatov, MacDonald, Pool and Quill, *Further progress on the study of the radioactive isotopes of the Nd-Il-Sm region*. Phys. Rev. *61*, 106 (1942).

Kurbatov and Pool, *Progress report on the radioactivities in the illinium region*. Phys. Rev. *63*, 463 (1943).

Kurbatov and Pool, *Isolation of pure radioactive cerium*. Phys. Rev. *65*, 61 (1944).

Kurbatov and Kurbatov, *Progress in the determination of the number of artificially produced radioactive atoms*. Phys. Rev. *67*, 60 (1945).

Kurtschatov, Kurtschatov, Myssowsky and Roussinow, *Artificial radioactivity produced by neutron bombardment without capture of the neutron*. C. R., Paris *200*, 1201 (1935).

Kurtschatow, Nemenow and Selinow, *Artificial radioactivity of ruthenium bombarded by neutrons*. C. R., Paris *200*, 2162 (1935).

Kurtschatow, Latyschew, Nemenow and Selinow, *Artificial radioactivity by neutron bombardment*. Phys. Z. Sowjet. *8*, 589 (1935).

Langer, Mitchell and McDaniel, *Coincidence measurement in As^{76}*. Phys. Rev. *57*, 347 (1940).

Laslett, *A long period positron activity*. Phys. Rev. *50*, 388 (1936).

Laslett, *A long period positron activity. Na^{22}*. Phys. Rev. *52*, 529 (1937).

Latimer, Hull and Libby, *Radiocesium activated by neutrons*. J. Amer. Chem. Soc. *57*, 781 (1935).

Law, Pool, Kurbatov and Quill, *Radioactive isotopes of Nd, Il and Sm.* Phys. Rev. *59,* 936 (1941).

Lawrence, *Radioactive sodium produced by deuton bombardment.* Phys. Rev. *46,* 746 (1934).

Lawrence, McMillan and Thornton, *The transmutation functions for some cases of deuteron induced radioactivity.* Phys. Rev. *48,* 493 (1935).

Lawson and Cork, *The radioactive isotopes of indium.* Phys. Rev. *52,* 531 (1937).

Lawson and Cork, *Energy spectra of radioactive indium.* Phys. Rev. *57,* 356 (1940).

Lawson and Cork, *Radioactive isotopes of indium.* Phys. Rev. *57,* 982 (1940).

Lecoin, *The penetrating radiation of radium* E. C. R., Paris *224,* 912 (1947).

Levi, *Radiosulphur.* Nature *145,* 588 (1940).

Lewis and Wynn-Williams, *The ranges of alpha particles from radioemanations and "A" products and from polonium.* Proc. Roy. Soc. *136,* 349 (1932).

Libby, *Radioactivity of neodymium and samarium.* Phys. Rev. *46,* 196 (1934).

Libby, *Natural radioactivity of lutecium.* Phys. Rev. *56,* 21 (1939).

Libby, *Atmospheric helium three and radiocarbon from cosmic radiation.* Phys. Rev. *69,* 671 (1946).

Livingood and Snell, *Search for radioactivity induced by 800kv electrons.* Phys. Rev. *48,* 851 (1935).

Livingood, *Deuteron induced radioactivities.* Phys. Rev. *50,* 425 (1936).

Livingood and Seaborg, *Deuteron induced radioactivity in Tin.* Phys. Rev. *50,* 435 (1936).

Livingood and Seaborg, *Radioactive antimony isotopes.* Phys. Rev. *52,* 135 (1937).

Livingood, Seaborg and Fairbrother, *Radioactive isotopes of Mn, Fe and Co.* Phys. Rev. *52,* 135 (1937).

Livingood and Seaborg, *Radio isotopes of nickel.* Phys. Rev. *53,* 765 (1938).

Livingood and Seaborg, *Long-lived radiocobalt isotopes.* Phys. Rev. *43,* 847 (1938).

Livingood and Seaborg, *Radioactive iodine isotopes.* Phys. Rev. *53,* 1015 (1938).

Livingood and Seaborg, *Radioactive isotopes of iron.* Phys. Rev. *54,* 51 (1938).

Livingood and Seaborg, *Long-lived radioactive silver.* Phys. Rev. *54,* 88 (1938).

Livingood and Seaborg, *Long period radioactive zinc.* Phys. Rev. *54,* 239 (1938).

Livingood and Seaborg, *Radioactive manganese isotopes.* Phys. Rev. *54,* 391 (1938).

Livingood and Seaborg, *Radioactive isotopes of iodine.* Phys. Rev. *54,* 775 (1938).

Livingood and Seaborg, *Radioactive antimony from $I+n$ and $Sn+D$. Phys. Rev. *55,* 414 (1939).

Livingood and Seaborg, *Radioactive isotopes of zinc.* Phys. Rev. *55,* 457 (1939).

Livingood and Seaborg, *New periods of radioactive tin.* Phys. Rev. *55,* 667 (1939).

Livingood and Seaborg, *Long-lived radioactive Fe^{55}.* Phys. Rev. *55,* 1268 (1939).

Livingood and Seaborg, *Induced radioactivities.* Rev. Mod. Phys. *12,* 30 (1940).

Livingood and Seaborg *Radioactive isotopes of cobalt.* Phys. Rev. *60,* 913 (1941).

Livingston, Henderson and Lawrence, *Radioactivity induced by neutron bombardment.* Phys. Rev. *46,* 325 (1934).

Livingston and McMillan, *Production of radioactive oxygen.* Phys. Rev. *46,* 437 (1934).

Lougher and Rowlands, *Radioactivity in osmium.* Nature *153,* 374 (1944).

Lyford and Bearden, *Radioactivity of samarium and "columnar ionization."* Phys. Rev. *45,* 743 (1934).

Madsen, *Radioactive isotopes of nickel and copper.* Nature *138,* 722 (1936).

Maier-Leibnitz, *Coincidence measurements with radioactive sodium isotopes.* Z. Phys. *122,* 233 (1944).

Marsh and Sugden, *Artificial radioactivity of the rare earth elements.* Nature *136,* 102 (1935).

Mattauch and Hauk, *The isotopic constitution and the atomic weight of neodymium.* Naturwiss. *25,* 780 (1937).

Maurer and Ramm, *Investigation of the 19 min. isotope of Mo and the isotope of element 43 derived from it.* Naturwiss. *29,* 368 (1941).

Maurer and Ramm, *Investigation of the 19 minute isotope of Mo and the resulting isotope of element 43.* Z. Phys. *119,* 334 (1942).

Maurer and Ramm, *Artificial radioactive isotopes of lead and its neighboring elements, using U and Th lead.* Z. Phys. *119,* 602 (1942).

McLennan, Grimmett and Read, *Production of radioactivity by neutrons.* Nature *135,* 505 (1935).

McLennan and Rann, *Radioactivity of some rare earths induced by neutron bombardment.* Nature *136,* 831 (1935).

McMillan and Livingston, *Artificial radioactivity produced by deuteron bombardment of nitrogen.* Phys. Rev. *47,* 452 (1935).

McMillan, *Artificial radioactivity of very long life.* Phys. Rev. *49,* 875 (1936).

McMillan, Kamen and Ruben, *Neutron induced radioactivity of the noble metals.* Phys. Rev. *52,* 375 (1937).

McMillan and Abelson, *Radioactive element 93.* Phys. Rev. *57,* 1185 (1940).

McMillan, *Seven day uranium activity.* Phys. Rev. *58,* 178 (1940).

McMillan and Ruben, *Radioactivity of Be[10].* Phys. Rev. *70,* 123 (1946).

Meitner, *Radiations emitted by Sc[46]* (in German). Ark. Mat. Astr. Fys. *28B,* No. 14 (1942).

Meitner, *The disintegration scheme of Sc[46]* (in German). Ark. Mat. Astr. Fys. *32A,* No. 6 (1945).

Meitner, *Radioactive nuclei produced by irradiation of copper with deuterons* (in German). Ark. Mat. Astr. Fys. *33A,* No. 3 (1946).

Meye, *Induced radioactivity of light nuclei.* Z. Phys. *105,* 232 (1937).

Meyer, Hess and Paneth, *New range determinations in polonium, ionium and actinium preparations.* A. B. Akad. Wiss. Wien. *123,* 1459 (1914).

Migdal, *Ionization of atoms accompanying alpha and beta decay.* J. Phys., USSR *4,* 449 (1941).

Milatz and ten Kate, *The quantitativity of the Geiger Muller counter. The spectrum of Ra E* (in English). Physica *7,* 779 (1940).

Miller, *Auger electrons resulting from K capture.* Phys. Rev. *67,* 309 (1945).

Minakawa, *Neutron induced radioactivity of W.* Phys. Rev. *57,* 1189 (1940).

Minakawa, *Long-lived activity of rhodium.* Phys. Rev. *60,* 689 (1941).

Mitchell, *Long period activity in Cd irradiated with neutrons.* Phys. Rev. *51,* 995 (1937).

Mitchell, *Long period activity of indium produced by slow neutrons.* Phys. Rev. *53,* 269 (1938).

Mitchell, Langer and Brown, *Radiations from radioactive lanthanum (140).* Phys. Rev. *71,* 140 (1947).

Moquin and Pool, *Artificial radioactivity of columbium and zirconium.* Phys. Rev. *65,* 60 (1944).

Motta, Boyd and Brosi, *Production and isotopic assignment of a 90-day activity in element 43.* Phys. Rev. *71,* 210 (1947).

Mounce, Pool and Kurbatov, *Radioactive isotopes of lanthanum.* Phys. Rev. *61,* 389 (1942).

Mulder, Hoeksema and Sizoo, *Measurements on the period of radioactive phosphorus* (in English). Physica 7, 849 (1940).

Nahmias and Walen, *Artificial radioactive elements.* C. R., Paris *203,* 71 (1936).

Nahmias and Walen, *Artificial radioactive products of short period.* C. R., Paris *203,* 176 (1936).

Nahmias and Walen, *Induced radioactivity.* J. Phys. Rad. *8,* 153 (1937).

Naidu, *Induced radioactivity of nickel and tin.* Nature *137,* 578 (1936).

Neddermeyer and Anderson, *Energy spectra of positrons ejected by artificially stimulated radioactive substances.* Phys. Rev. *45,* 498 (1934).

Nelson, Pool and Kurbatov, *The radioactive isotopes of nickel and their assignments.* Phys. Rev. *61,* 428 (1942).

Nelson, Pool and Kurbatov, *The characteristic radiations of Co^{60}*. Phys. Rev. *62*, 1 (1942).

Nelting, *Molybdenum-thorium*. Z. Phys. *115*, 469 (1940).

Neuninger and Rona, *On the artificial activity of thulium*. Wien Akad. Anz. *72A*, 275 (1935).

Neuninger and Rona, *Artificial radioactivity of thorium*. S. B. Akad. Wiss. Wien *145*, 479 (1936).

Newson, *The radioactivity of oxygen, silicon and phosphorus under deuteron bombardment*. Phys. Rev. *48*, 482 (1935).

Newson, *The radioactivity induced in oxygen by deutron bombardment*. Phys. Rev. *48*, 790 (1935).

Newson, *The radioactivity induced in Si and P by deuteron bombardment*. Phys. Rev. *51*, 624 (1937).

Newson and Borst, *Experiments with radioactive recoil atoms*. Phys. Rev. *57*, 1083 (1940).

Norling, *On the radioactivity of gold* (in English). Ark. Mat. Astr. Fys. *27B*, No. 9 (1941).

Norris and Inghram, *Half life determination of carbon (14) with a mass spectrometer and low absorption counter*. Phys. Rev. *70*, 772 (1946).

O'Conner, Pool and Kurbatov, *Artificial radioactivity of Cr^{49}*. Phys. Rev. *62*, 413 (1942).

Oeser and Tuck, *Radioactive isotopes of copper*. Nature *139*, 1110 (1937).

Oldenberg, *Artificial radioactivity of tantalum*. Phys. Rev. *53*, 35 (1938).

O'Neal and Goldhaber, *Radioactive hydrogen from the transmutation of Be by deuterons*. Phys. Rev. *57*, 1086 (1940).

O'Neal, *On the activation of Cl by slow neutrons*. Phys. Rev. *59*, 109 (1941).

Oppenheimer and Tomlinson, *The positron spectra of N^{13} and Na^{22}*. Phys. Rev. *56*, 858 (1939).

Osborne and Deutsch, *Disintegration scheme of the 21-minute isomer of Mn^{52}*. Phys. Rev. *71*, 467 (1947).

Peacock and Deutsch, *Disintegration schemes of radioactive substances. IX.* Mn^{52} *and* V^{48}. Phys. Rev. *69*, 306 (1946).

Perey and Lecoin, *Radiation of Ac K*. J. Phys. Rad. *10*, 439 (1939).

Perrier, Santangelo and Segrè, *Radioactive isotopes of Zn and Co*. Phys. Rev. *53*, 104 (1938).

Philipp and Riedhammer, *The energy relations for 44 hr lanthanum and element 93*. Z. Naturforsch. *1*, 372 (1946).

Plessett, *Nuclear excitations from radioactive decay*. Phys. Rev. *62*, 181 (1942).

Polessitsky, *Chemical investigation of short-lived artificial radio-elements*. Phys. Z. Sowjet. *12*, 339 (1937).

Pollard, *Energy and half life of Be^{10}*. Phys. Rev. *57*, 241 (1940).

Pool and Cork, *Radioactive isotopes from aluminum*. Phys. Rev. *51*, 383 (1937).

Pool, Cork and Thornton, *Radioactivity due to neutron ejection produced by fast neutrons*. Phys. Rev. *51*, 890 (1937).

Pool, Cork and Thornton, *A survey of the radioactivity produced by high energy neutron bombardment*. Phys. Rev. *52*, 239 (1937).

Pool, *Radioactivity in silver induced by fast neutrons*. Phys. Rev. *53*, 116 (1938).

Pool and Quill, *Radioactivity induced in the rare earth elements by fast neutrons*. Phys. Rev. *53*, 437 (1938).

Pool and Kurbatov, *The artificial radioactivities of cerium*. Phys. Rev. *63*, 463 (1943).

Pool and Edwards, *Radioactive zirconium and columbium*. Phys. Rev. *67*, 60 (1945).

Poole and Paul, *Short-lived radioactivity from Li bombarded with neutrons*. Nature *158*, 482 (1946).

Preiswerk and Halban, *Radioactive elements produced by neutrons*. C. R., Paris *201*, 722 (1935).

Rajam, Capron and de Hemptenne, *Periods of radiorhodiums and radiosilvers obtained by bombardment with slow neutrons*. Ann. Soc. sci. Brux. *59*, 403 (1939).

Ramsey, Montgomery and Montgomery, *Artificial radioactivity produced by cosmic rays*. Phys. Rev. *53*, 196 (1938).

Reddemann and Strassmann, *On the production of the silver isotope with a half life of 24 min. by Be neutrons.* Naturwiss. *25*, 458 (1938).

Reid, Dunning, Weinhouse and Grosse, *Half-life of C^{14}.* Phys. Rev. *70*, 431 (1946).

Reid and Keston, *Long life radio-iodine.* Phys. Rev. *70*, 987 (1946).

Richardson, *Radiations from radioactive substances: Au^{198}, Eu^{152}, Ag^{106}, Cu^{64} and N^{13}.* Phys. Rev. *55*, 609 (1939).

Ridenour, Henderson, Henderson and White, *Radioactivity induced by alpha particles.* Phys. Rev. *51*, 1013 (1937).

Ridenour and Henderson, *Artificial radioactivity produced by alpha particles.* Phys. Rev. *51*, 1102 (1937).

Ridenour and Henderson, *Artificial radioactivity produced by alpha particles.* Phys. Rev. *52*, 139, 889 (1937).

Ridenour, Delsasso, White and Sherr, *Artificial radioactivity produced by protons.* Phys. Rev. *53*, 770 (1938).

Riezler, *Radioactive sodium with mass 25.* Phys. Z. *45*, 91 (1944).

Risser, *Long period radioactivity in Co induced by slow neutrons.* Phys. Rev. *51*, 1013 (1937).

Risser, *Neutron induced radioactivity of long life in Co.* Phys. Rev. *52*, 768 (1937).

Risser, Lark-Horovitz and Smith, *Activation of indium by alpha particles.* Phys. Rev. *57*, 355 (1940).

Roberts, Heydenburg and Locher, *Radioactivity of Be^{7}.* Phys. Rev. *53*, 1016 (1938).

Roberts, Elliott, Downing, Peacock and Deutsch, *Disintegration schemes of radioactive substances. V. I^{130}.* Phys. Rev. *64*, 268 (1947).

Rona, Scheichenberger and Stangle, *Artificial activity of thorium.* S. B. Akad. Wiss. Wien *147*, 209 (1938).

Rosenblum and Chamie, *Fine structure of alpha radiation of radiothorium.* C. R., Paris *194*, 1154 (1932).

Rosenblum, Guillot and Perey, *On the intensity of the groups of fine structure of magnetic alpha spectra of radioactinium and its descendants.* C. R., Paris *202*, 1274 (1936).

Rotblat, *Induced radioactivity of nickel and cobalt.* Nature *136*, 515 (1935).

Rotblat, *Artificial radioactivity produced by fast neutrons and their inelastic collisions.* Nature *139*, 1110 (1937).

Rotblat, *Application of the coincidence method to testing the lifetime level scheme of Ra C.* Nature *144*, 248 (1939).

Rotblat, *Radiations from bromine (82).* Nature *148*, 371 (1941).

Ruark, *Alpha, beta and gamma rays of the actinium family.* Phys. Rev. *45*, 564 (1934).

Ruark and Fussler, *On the half lives of K, Ru, Nd and Sa.* Phys. Rev. *48*, 151 (1935).

Ruben and Kamen, *Radioactive carbon of long half·life.* Phys. Rev. *57*, 549 (1940).

Ruben and Kamen, *Long-lived radioactive carbon: C^{14}.* Phys. Rev. *59*, 349 (1941).

Rutherford, Lewis, Bowden, *Analysis of long range alpha particles from radium C′ by the magnetic focusing method.* Proc. Roy. Soc. *142*, 347 (1933).

Sagane, *Radioactivity induced in S.* Phys. Rev. *50*, 1141 (1936).

Sagane, Kojima and Ikawa, *Radioactive arsenic isotopes.* Phys. Rev. *54*, 149 (1938).

Sagane, *Radioactive isotopes of Cu, Zn, Ga and Ge.* Phys. Rev. *53*, 212 (1938).

Sagane, Kojima, Miyamoto and Ikawa, *Preliminary report on the radioactivity produced in Y, Zr and Mo.* Phys. Rev. *54*, 542 (1938).

Sagane, Kojima, Miyamoto and Ikawa, *Neutron induced radioactivity in columbium.* Phys. Rev. *54*, 970 (1938).

Sagane, *Radioactive isotopes of Cu, Zn, Ga and Ge.* Phys. Rev. *55*, 31 (1939).

Sagane, Kojima, Miyamoto and Ikawa, *Radioactive As isotopes.* Proc. Phys.-Math. Soc. Jap. (3) *21*, 660 (1939).

Sagane, Kojima, Miyamoto and Ikawa, *New radioactive isotope of masurium.* Phys. Rev. *57*, 750 (1940).

Sagane, Kojima, Miyamoto and Ikawa, *Neutron induced radioactivity in Cb (Nb).* Proc. Phys.-Math. Soc. Jap. (3) *22*, 174 (1940).

Sagane, Miyamoto and Ikawa, *Radioactivity produced in germanium*. Phys. Rev. *59*, 904 (1941).

Sagane, Kojima, Miyamoto and Ikawa, *Radioactive isotopes of molybdenum and their daughter products*. Proc. Phys.-Math. Soc. Jap. (3) *24*, 499 (1942).

Saha, *The β-∂ coupling in radioactive $_{33}As^{76}$*. Naturwiss. *27*, 786 (1939).

Sampson, Ridenour and Bleakney, *The isotopes of Co and their radioactivity*. Phys. Rev. *50*, 382 (1936).

Sargent, *Half periods of Ac B, Ac C'' and UX*. Canad. J. Res. (A) *17*, 103 (1939).

Schaeffer and Harteck, *Electron and positron spectra of radioactive As*. Z. Phys. *113*, 287 (1939).

Scheichenberger, *Artificial activity upon irradiation of europium, rubidium and cesium*. Wien. Akad. Anz. *75*, 18 (1938).

Seaborg, Livingood and Kennedy, *Radioactive tellurium further production and separation of isomers*. Phys. Rev. *55*, 794 (1939).

Seaborg, *Artificial radioactivity*. Chem. Rev. 27, 199 (1940).

Seaborg, Livingood and Kennedy, *Radioactive isotopes of tellurium*. Phys. Rev. *57*, 363 (1940).

Seaborg, Livingood and Friedlander, *Radioactive isotopes of Ge*. Phys. Rev. *59*, 320 (1941).

Seaborg, Gofman and Kennedy, *Radioactive isotope of protactinium*. Phys. Rev. *59*, 321 (1941).

Seaborg and Friedlander, *Radioactive isotopes of osmium*. Phys. Rev. *59*, 400 (1941).

Seaborg, McMillan, Kennedy and Wahl, *Radioactive element 94 from deuterons on uranium*. Phys. Rev. *69*, 366 (1946).

Seaborg, Wahl and Kennedy, *Radioactive element 94 from deuterons on uranium*. Phys. Rev. *69*, 367 (1946).

Segrè, *Artificial radioactivity*. Nuovo Cim. *12*, 232 (1935).

Seren, Engelkemeir, Sturm, Freidlander and Turkel, *Forty-three day $_{48}Cd^{115}$*. Phys. Rev. *71*, 409 (1947).

Shrader, Saxon and Snell, *An attempt to observe the radioactivity of the neutron*. Phys. Rev. *70*, 791 (1946).

Siegbahn and Slätis, *Radiation from active N* (in English). Ark. Mat. Astr. Fys. *32A*, No. 9 (1945).

Siegbahn and Slätis, *Radioactivity of active N*. Nature *156*, 568 (1945).

Siegbahn and Deutsch, *The decay scheme of Cs^{134}*. Phys. Rev. *71*, 483 (1947).

Silveira, *Natural radioactivity by neutron emission* (in French). Portugaliae Physica *1*, 167 (1945).

Sizoo and Koene, *Period of radioactive phosphorus $_{15}P^{32}$* (in English). Physica *3*, 1053 (1936).

Slätis, *On neutron induced activities in cesium* (in English). Ark. Mat. Astr. Fys. *32A*, No. 16 (1946).

Smythe and Hemmendinger, *The radioactive isotope of K*. Phys. Rev. *51*, 178 (1937).

Snell, *Radioactive argon*, Phys. Rev. *49*, 555 (1936).

Snell, *A new radioactive isotope of fluorine*. Phys. Rev. *51*, 143 (1937).

Snell, *The radioactivities of Br*. Phys. Rev. *51*, 1011 (1937).

Snell, *The radioactive isotopes of bromine: isomeric forms of Br^{80}*. Phys. Rev. *52*, 1007 (1937).

Solomon, *Half Life of C^{11}*. Phys. Rev. *60*, 279 (1941).

Soltan and Wertenstein, *Isomeric radioisotopes of bromine*. Nature *144*, 76 (1938).

Sommers and Sherr, *Activity of N^{16} and He^6*. Phys. Rev. *69*, 21 (1946).

Sosnowski, *Artificial radioactivity excited in gold and the complexity of the radiation*. C. R., Paris *200*, 391 (1935).

Sosnowski, *Artificial radioactivity of iridium*. C. R., Paris *200*, 922 (1935).

Sosnowski, *Artificial radioactivity of bismuth*. C. R., Paris *200*, 1027 (1935).

Starke, *Concentration of the artificial radioactive arsenic isotope $_{33}^{76}As$*. Naturwiss. *28*, 631 (1940).

Stewart, Lawson and Cork, *Induced radioactivity in strontium and yttrium*. Phys. Rev. *52*, 901 (1937).

Stewart, *Induced radioactivity in strontium and yttrium; nuclear isomers in strontium*. Phys. Rev. *56*, 629 (1939).

Strain, *Proton induced radioactivities. II. Nickel and copper targets.* Phys. Rev. *54*, 1021 (1938).

Strassmann and Walling, *The separation of a pure strontium isotope (87) from an old rubidium bearing lepidolith and the half life of rubidium.* Ber. Deutsch. Chem. Ges. *71*, 1 (1938).

Strassmann and Hahn, *On the isolation and some of the properties of element 93.* Naturwiss. *30*, 256 (1942).

Sugden, *Radioactivity of some rarer elements produced by neutron bombardment.* Nature *135*, 469 (1935).

Sullivan, *Decay schemes for isotopes W^{187} and W^{185}.* Phys. Rev. *68*, 277 (1945).

Sullivan, Sleight and Gladrow, *Discovery, identification and characterization of 2.8 d Ru^{97}.* Phys. Rev. *70*, 778 (1946).

Surugue, *The rays and excitation levels of some radioactive bodies.* J. Phys. Rad. 7, 145 (1946).

Swartout, Boyd, Cameron, Keim and Larson, *Mass assignment of 2.6 hr nickel.* Phys. Rev. *70*, 232 (1946).

Szilard and Chalmers, *Detection of neutrons liberated from beryllium by gamma rays: a new technique for inducing radioactivity.* Nature *134*, 494 (1934).

Szilard and Chalmers, *Radioactivity induced by neutrons.* Nature *135*, 98 (1935).

Tape and Cork, *Induced radioactivity in Te.* Phys. Rev. *53*, 676 (1938).

Thibaud, *The radiation emitted by radioactive substances in beta disintegration.* C. R., Paris *223*, 984 (1946).

Thibaud, *The light particles accompanying beta disintegration.* C. R., Paris *224*, 739 (1947).

Thomson and Saxton, *Attempt to detect radioactivity produced by positrons.* Phil. Mag. *23*, 241 (1937).

Thompson and Rowlands, *Dual decay of K.* Nature *152*, 103 (1943).

Thornton, *Artificial radioactivity induced in arsenic, nickel and cobalt under deuteron bombardment.* Phys. Rev. *49*, 207 (1936).

Thornton and Cork, *Induced radioactivity in lead.* Phys. Rev. *51*, 383 (1937).

Thornton, *The radioactivity produced in Ni by deuteron bombardment.* Phys. Rev. *51*, 893 (1937).

Thornton, *The radioactive isotopes of zinc.* Phys. Rev. *53*, 326 (1938).

Tuck, *Radioactive sulphur for biochemical experiments.* J. Chem. Soc. P. 1292 (1939).

Turner, *Radioactive isotopes of vanadium.* Phys. Rev. *58*, 679 (1940).

Urry, *Radioactive determination of small amounts.* Amer. J. Sci. *239*, 191 (1941).

Van Voorhis, *The artificial radioactivity of copper, a branch reaction.* Phys. Rev. *50*, 895 (1936).

Veldkamp and Knol, *Radioactivity induced in lithium by neutrons* (in English). Physica *4*, 166 (1937).

Veldkamp and de Vries, *Radioactivity induced in ruthenium by neutrons* (in English). Physica *4*, 1229 (1937).

Victorin, *Production of the long-lived radioactive cobalt isotope from nickel bombarded by fast neutrons.* Proc. Camb. Phil. Soc. *34*, 612 (1938).

Voge, *Exchange reactions with radiosulphur.* J. Amer. Chem Soc. *61*, 1032 (1939).

de Vries and Veldkamp, *Radioactivity induced by neutrons in ruthenium* (in English). Physica *5*, 249 (1938).

de Vries and Diemer, *Artificial radioactivity of Pb produced by neutrons* (in English). Physica *6*, 599 (1939).

Walen and Nahmias, *Artificial radioactivities and search for the negative proton.* C. R., Paris *203*, 1149 (1937).

Walke, *The induced radioactivity of Ca.* Phys. Rev. *51*, 439 (1937).

Walke, *The induced radioactivity of titanium.* Phys. Rev. *51*, 1011 (1937).

Walke, *Radioactive isotopes of Sc from Ca and K by alpha particle bombardment.* Phys. Rev. *52*, 400 (1937).

Walke, *A new radioactive isotope of K.* Phys. Rev. *52*, 663 (1937).

Walke, *The induced radioactivity of Sc.* Phys. Rev. *52*, 669 (1937).

Walke, *The induced radioactivity of Ti and Va.* Phys. Rev. *52*, 777 (1937).

Walke, Williams and Evans, *K electron capture, nuclear isomerism and long period activities of Ti and Sc.* Proc. Roy. Soc. *171*, 360 (1939).

Walke, *Radioactive isotopes of Sc and their properties.* Phys. Rev. *57*, 163 (1940).

Walke, Thompson and Holt, *Radioactive isotopes of calcium and their suitability as indicators in biological invesgations.* Phys. Rev. *57*, 177 (1940).

Wang, *Radioactivity of the neutron.* Nature *155*, 574 (1945).

Ward, Wynn-Williams and Cave, *The rate of emission of alpha particles from radium.* Proc. Roy. Soc. *125*, 713 (1929).

Ward, *Half value period of $^{13}_{7}N$.* Proc. Camb. Phil. Soc. *35*, 523 (1939).

Waring and Chang, *Formation of radiophosphorus (P³⁰).* Proc. Roy. Soc. *157*, 652 (1936).

Watase, Itoh and Takeda, *Radioactive Zn⁶⁵.* Proc. Phys-Math. Soc. Jap. (3) *22*, 90 (1940).

Weimer, Kurbatov and Pool, *Radioactive argon A³⁷.* Phys. Rev. *60*, 469 (1941).

Weimer, Pool and Kurbatov, *Transmutation of barium.* Phys. Rev. *63*, 59 (1943).

Weimer, Pool and Kurbatov, *Radioactive isotopes of lanthanum.* Phys. Rev. *63*, 67 (1943).

Weimer, Pool and Kurbatov, *Radioactivities in Ba obtained from cesium.* Phys. Rev. *64*, 43 (1943).

Weimer, Kurbatov and Pool, *K electron capture in radioactive argon A³⁷.* Phys. Rev. *66*, 209 (1944).

Wertenstein, *An artificial radio-element from nitrogen.* Nature *133*, 564 (1934).

White, Delsasso, Fox and Creutz, *Short-lived radioactivities induced in fluorine, sodium and magnesium by high energy protons.* Phys. Rev. *56*, 512 (1939).

Wiedenbeck, *Radioactive isotopes in the columbium region.* Phys. Rev. *70*, 435 (1946).

Wilkins and Dempster, *The radioactive isotope of samarium.* Phys. Rev. *54*, 315 (1938).

Wu and Friedlander, *Radioactive isotopes of Hg.* Phys. Rev. *60*, 747 (1941).

Wu and Segrè, *Artificial radioactivity of some rare earths.* Phys. Rev. *61*, 203 (1942).

Wu and Segrè, *Radioactive xenons.* Phys. Rev. *67*, 142 (1945).

Yalow and Goldhaber, *X-ray coincidences from Te¹²¹ (125 days).* Phys. Rev. *66*, 36 (1944).

Yasaki and Watanabe, *Deuteron induced radioactivity in oxygen.* Nature *141*, 787 (1938).

Yost, Ridenour and Shinohara, *Chemical identification of the radio-elements produced from carbon and bromine by deuton bombardment.* J. Chem. Phys. *3*, 133 (1935).

Zimens and Hedvall, *Radiometric analysis of U traces with unknown U:Ra ratio* (in German). Ark. Kemi Min. Geol. *22A*, No. 25 (1946).

Zirkler, *Experiments with radioactive Tl isotope ThC²⁻.* Z. Phys. Chem. *187*, 103 (1940).

Zünti and Bleuler, *Two activities S³⁷ and P³⁴ induced in chlorine by rapid neutrons* (in German). Helv. Phys. Acta *18*, 263 (1945).

Zumstein, Kurbatov and Pool, *Disintegration of the long life radioactive isotope of tantalum.* Phys. Rev. *63*, 59 (1943).

Zyw, *Induced radioactivity of K.* Nature *134*, 64 (1934).

CHAPTER SIX

Beta Radiation

Alichanian, Alichanow and Dzelepow, *On the formation of the beta ray spectrum of Ra E in he vicinity of the upper limit and the mass of the neutrino.* Phys. Rev. *53*, 766 (1938).

Alichanian and Nikitin, *The shape of the beta spectrum of Th C and the mass of the neutrino.* Phys. Rev. *53*, 767 (1938).

Alichanow, Alichanian and Dzelepow, *Beta spectra of some radioactive elements.* Nature *135*, 393 (1935).

Alichanow, Alichanian and Dzelepow, *Beta ray spectra of artificially produced radioactive elements.* Nature *136*, 257 (1935).

Amaki and Sugimoto, *Beta ray spectra of radioactive Sb and Na.* Tokyo Inst. Phys. Chem. Res. *34,* 1650 (1938).

Backus, *The beta ray spectra of Cu^{64} at low energies.* Phys. Rev. *68,* 59 (1945).

Bacon, Grisewood and van der Merwe, *The beta ray spectrum of Mn^{56}.* Phys. Rev. *52,* 668 (1937).

Bacon, Grisewood and van der Merwe, *The beta ray spectrum of I^{128}.* Phys. Rev. *54,* 315 (1938).

Bacon, Grisewood and van der Merwe, *The radioactivity of Mn^{56} and I^{129}.* Phys. Rev. *59,* 531 (1941).

Barindrecht and Sizoo, *Further measurements concerning the production of positive electrons by beta particles* (in English). Physica 7, 490 (1940).

Barkas, Creutz, Delsasso, Sutton and White, *Beta radiation from Si^{27} and P^{30}.* Phys. Rev. 58, 383 (1940).

Barschall, Harris, Kanner, Turner, *Penetrating beta particles from U activated by neutrons.* Phys. Rev. *55,* 989 (1939).

Bayley and Crane, *Experiments on Li^{8} and B^{12}.* Phys. Rev. *51,* 1012 (1937).

Bayley and Crane, *The beta ray spectra of Li^{8} and B^{12}.* Phys. Rev. *52,* 604 (1937).

Beers, *Direct determination of the charge of the beta particle.* Phys. Rev. *63,* 77 (1943).

Bjerge and Brostrøm, *Beta ray spectrum of radio-helium.* Nature *138,* 500 (1936).

Bleuler, Bollmann and Zünti, *Beta decay of A^{41}* (in German). Helv. Phys. Acta *19,* 419 (1946).

Bothe, *The beta radiation of Po.* Z Phys. *96,* 607 (1935).

Bradt, Heine and Scherrer, *Conversion lines in the beta spectrum of UX* (in German). Helv. Phys. Acta *16,* 455 (1943).

Bradt, Gugelot, Huber, Medicus, Preiswerk and Scherrer, *Test of the Fermi theory of beta disintegration by determination of the probability of taking up a K electron and the e^+ emission of 6.7 hr cadmium* (in German). Helv. Phys. Acta *18,* 351 (1945).

Bradt, Gugelot, Huber, Medicus, Preiswerk and Scherrer, *K capture and e^+ emission of 6.7 hr Cd: an experimental verification of the Fermi theory of beta decay.* Phys. Rev. *68,* 57 (1945).

Brown and Mitchell, *Beta ray spectra of radioactive manganese, arsenic and indium.* Phys. Rev. *50,* 593 (1936).

Brown, *Beta ray energy of H^3.* Phys. Rev. *59,* 954 (1941).

Buechner and Van de Graaff, *Calorimetric experiment on the radiation losses of 2 Mev electrons.* Phys. Rev. *70,* 174 (1946).

Crane, Delsasso, Fowler and Lauritsen, *Short-lived beta radioactivity.* Phys. Rev. *48,* 484 (1935).

Crane and Halpern, *Further experiments on the recoil of the nucleus in beta decay.* Phys. Rev. *56,* 232 (1939).

Crittenden and Bacher, *Nuclear isomerism in rhodium.* Phys. Rev. *54,* 862 (1938).

Crittenden, *The beta ray spectra of Mg^{27}, Cu^{62} and the nuclear isomers of Rh^{104}.* Phys. Rev. *56,* 709 (1939).

Curie, Savitch and da Silva, *Neutron induced beta radioactivities from U.* J. Phys. Rad. *9,* 440 (1938).

Curtis and Cork, *Beta radiation from activated isotopes of arsenic.* Phys. Rev. *53,* 681 (1938).

Deutsch, *Absence of high energy beta rays from Br^{82}.* Phys. Rev. *61,* 672 1942.

Dzelepow, Kopjova and Vorobjov, *Beta ray spectrum of K^{40}.* Phys. Rev. *69,* 538 (1946).

Edwards and Pool, *Characteristic x-rays excited by beta particles.* Phys. Rev. *69,* 49 (1946).

Ellis, *Beta ray disintegration.* Int. Conf. Phys., London *1,* 43 (1934).

Flammersfeld, *Lower limit of continuous beta spectrum of Ra E.* Z. Phys. *112,* 727 (1939).

Gaerttner, Turin and Crane, *Beta ray spectra of several slow neutron activated substances.* Phys. Rev. *49,* 793 (1936).

Gibert, *Analysis of beta ray spectra* (in French). Portugaliae Physica *1,* 15 (1943).

Haggstrom, *The beta ray spectrum of $_{91}$ekatantalum.* Phys. Rev. *59,* 322 (1941).

Haggstrom, *The beta ray spectra of rubidium (86), strontium (89), ekatantalum and protactinium.* Phys. Rev. *62,* 144 (1942).

Hales and Jordan, *The beta ray spectrum of antimony (124).* Phys. Rev. *62,* 553 (1942).

Hales and Jordan, *The beta ray spectrum of antimony (124).* Phys. Rev. *64,* 202 (1943).

Harper and Roberts, *Precise measurements of the energies of beta rays from radium (B + C).* Proc. Roy. Soc. *178,* 170 (1941).

Henderson, *The upper energy limit of the K^{40} beta ray spectrum.* Phys. Rev. *71,* 323 (1947).

Huber, Lienhard, Scherrer and Wäffler, *Simple absorption method for determining the energy of weak beta spectra* (in German). Helv. Phys. Acta *18,* 221 (1945).

Hull, Libby and Latimer, *Beta rays of actinium.* J. Amer. Chem. Soc. *57,* 1649 (1935).

Jacobsen, *On the recoil of the nucleus in beta decay.* Phys. Rev. *70,* 789 (1946).

Jnanananda, *The beta radiations of uranium X.* Phys. Rev. *69,* 570 (1946).

Kanner and Harris, *Search for beta and delayed gamma radiation from the deuteron-deuteron reaction.* Phys. Rev. *56,* 839 (1939).

Karlik and Bernert, *On a supposed beta radiation of radium A and the natural existence of element 85.* Naturwiss. *30,* 685 (1942).

Kikuchi, Aoki and Husimi, *Emission of beta rays from substances bombarded with neutrons.* Nature *138,* 841 (1936).

Kikuchi, Watase, Itoh, Takeda and Yamaguchi, *The beta ray spectrum of N^{13}.* Proc. Phys-Math. Soc. Jap. (3) *21,* 41, 52 (1939).

Kurie, Richardson and Paxton, *Further data on the energies of beta rays emitted from artificially produced radioactive bodies.* Phys. Rev. *49,* 203 (1936).

Kurie, Richardson and Paxton, *The radiations emitted from artificially produced radioactive substances. I. The upper limits and shapes of the beta spectra from several elements.* Phys. Rev. *49,* 368 (1936).

Labhart and Medicus, *The beta disintegration of Ra A and the formation of element 85* (in German). Helv. Phys. Acta *16,* 225 (1943).

Labhart and Medicus, *Beta decay of Ra A* (in German). Helv. Phys. Acta *16,* 392 (1943).

Langer, Mitchell and McDaniel, *Coincidences between beta and gamma rays in indium.* Phys. Rev. *56,* 380 (1939).

Langer, Mitchell and McDaniel, *Coincidences between beta and gamma rays in manganese.* Phys. Rev. *56,* 422 (1939).

Langer, Mitchell and McDaniel, *Coincidences between beta and gamma rays in Na^{24}.* Phys. Rev. *56,* 962 (1939).

Lawson, *The beta ray spectra of P, Na and Co.* Phys. Rev. *56,* 131 (1939).

Lee and Libby, *The beta rays of mesothorium 1 and radium D.* Phys. Rev. *55,* 252 (1939).

Lecoin, *Beta spectra.* J. Phys. Rad. (7) *9,* 81 (1938).

Lewis and Bohm, *The low energy beta-spectrum of Cu^{64},* Phys. Rev. *69,* 129 (1946).

Li, *Absolute intensities of the strong beta ray lines of Ra (B + C), Th (B + C) and Ac (B + C).* Proc. Roy. Soc. *158,* 571 (1937).

Libby and Lee, *Energies of the soft beta radiations of Ru and other bodies. Method for their determination.* Phys. Rev. *55,* 245 (1939).

Lutz, Pool and Kurbatov, *Beta ray spectra of the artificially radioactive isotopes, Pb^{205} and Bi^{207}.* Phys. Rev. *65,* 61 (1944).

Lyman, *The beta ray spectra of radium E and radioactive phosphorus.* Phys. Rev. *51,* 1 (1937).

Lyman, *The beta ray spectrum of N^{13} and the mass of the neutrino.* Phys. Rev. *55,* 234 (1939).

Lyman, *Evidence for the composite character of the N^{13} beta ray spectrum.* Phys. Rev. *55,* 1123 (1939).

Magnan, *Energy spectra of high energy beta and gamma rays studied by means* of the spectrometer with magnetic focusing. Ann. Phys., Paris 15, 5 (1941).

Marinelli, Brinckerhoff and Hine, Average energy of beta rays emitted by radioactive isotopes. Rev. Mod. Phys. 19, 25 (1947).

Marshak, Forbidden transitions in beta decay and orbital electron capture and spins of nuclei. Phys. Rev. 61, 431 (1942).

Marshall, Beta radiations of UX_1 and UX_2. Proc. Roy Soc. 173, 391 (1939).

Meitner and Orthmann, An absolute determination of the energy of the primary beta rays of radium E. Z. Phys. 60, 143 (1930).

Meitner, A simple method for the investigation of secondary electrons excited by gamma rays and the interference of these electrons with the measurements of primary beta ray spectra. Phys. Rev. 63, 73 (1943). Errata, Phys. Rev. 63, 384 (1943).

Miller and Curtiss, Beta and gamma ray energies of several radioactive isotopes. Phys. Rev. 70, 983 (1946).

Mitchell, Langer and McDaniel, Investigations of beta and gamma rays from Sb^{122} and Sb^{124}. Phys. Rev. 57, 347 (1940).

Moore, Absorption measurements of beta rays. Phys. Rev. 57, 355 (1940).

Nahmias, On the distribution of periods of the beta radioelements. J. Phys. Rad. 7, 266 (1945).

Naidu and Siday, Beta ray spectra of induced radioactive elements resulting from neutron bombardment. Proc. Phys. Soc., Lond. 48, 330 (1936).

Neary, Beta ray spectrum of Ra E. Proc. Roy. Soc. 175, 71 (1940).

Neary, The absorption of the primary beta radiation from radium in lead and platinum and the specific gamma ray. dose rate at a filtration of 0.5 mm of platinum. Brit. J. Radiol. 15, 104 (1942).

Nielsen, The energy spectrum of H^3 beta rays. Phys. Rev. 60, 160 (1941).

Nishina, Yasuki, Ezoe, Kimura and Ikawa, Induced beta activity of uranium by fast neutrons. Phys. Rev. 57, 1182 (1940).

Norling, Beta and gamma radiation of radioactive arsenic. Z. Phys. 111, 158 (1938).

O'Conor, The beta ray spectrum of Ra E. Phys. Rev. 52, 303 (1937).

O'Neal, Note on the beta ray energy of H^3. Phys. Rev. 60, 359 (1941).

Osborne and Peacock, Beta and gamma ray spectra of La^{140}. Phys. Rev. 69, 679 (1946).

Perey and Lecoin, Beta spectrum of actinium K. Nature 144, 326 (1939).

Perey, Relation between the maximum energy of beta rays of artificial radioelements and the superficial mass absorbing half of the beam. C. R., Paris 218, 74 (1944).

Perey, Contribution to the study of beta rays of artificial radioelements. J. Phys. Rad. 6, 28 (1945).

Pontecorvo, Nuclear isomerism and internal conversion. Phys. Rev. 54, 542 (1938).

Pool, Potential nuclear monokinetic electron sources. J. Appl. Phys. 15, 716 (1944).

Rajam, Capron and de Hemptinne, Evaluation of the beta upper energy limits with simple absorption data. Nature 144, 202 (1939).

Rall and Wilkinson, Gamma and beta ray energies of some radioactive isotopes as measured by a thin magnetic lens beta ray spectrometer. Phys. Rev. 71, 321 (1947).

Rasetti, Evidence for the radioactivity of slow mesotrons. Phys. Rev. 59, 706 (1941).

Rathgeber, Influence of intercrystalline forces on beta ray absorption. Phys. Rev. 69, 239 (1946).

Renard, A measurement of the number of quanta emitted by beta decay of vanadium 52. C. R., Paris 223, 945 (1946).

Roaf, Energies of beta particles from UX_2. Nature 145, 223 (1940).

Rogers, McReynolds and Rogers, Determination of the masses and velocities of three radium B beta particles. The relativistic mass of the electron. Phys. Rev. 57, 379 (1940).

Sagane, Kojima and Miyamoto, *Beta rays from radioactive isotopes* Ni^{63}, Zn^{63}, Zn^{65}, Zn^{69}, Zn^{69*}, Ga^{68}, Ga^{70}, Ga^{72}, Ge^{71}, Ge^{75}, *and* Ge^{77}. Proc. Phys.-Math. Soc. Jap. (3) *21*, 728 (1939).

Saha and Saha, *Nuclear energetics and beta activity.* Nature *158*, 6 (1946).

Sargent, *Maximum energy of the beta rays from UX and other bodies.* Proc. Roy. Soc. *139*, 659 (1933).

Sargent, *The beta rays of actinium B and actinium C" partial spectra.* Phys. Rev. *54*, 232 (1938).

Sargent, *Beta and gamma rays of actinium B and actinium C".* Canad. J. Res. *17A*, 82 (1939).

Seren, Moyer and Sturm, *Absorption curve of 31 sec.* $^8O^{19}$ *beta rays and cross section for production by thermal neutrons.* Phys. Rev. *70*, 561 (1946).

Siegbahn and Bohr, *The* $\beta+$ *spectrum of* C^{11} (in English). Ark. Mat. Astr. Fys. *30B*, No. 3 (1944).

Siegbahn, *Beta radiation from active phosphorus and sodium.* Nature *153*, 221 (1944).

Sinma and Yamasaki, *Beta ray spectra of* Cu^{62}, Cu^{64} *and* Cu^{66}. Tokyo Inst. Phys. Chem. Res. *35*, 16 (1938).

Sinma and Yamasaki, *Beta radioactivities of rhenium.* Phys. Rev. *55*, 320 (1939).

Smith, *Beta ray spectra of scandium.* Phys. Rev. *59*, 937 (1941).

Smith, *Beta ray spectra of scandium.* Phys. Rev. *61*, 578 (1942).

Tape, *Beta spectra associated with iodine.* Phys. Rev. *55*, 1135 (1939).

Tape, *Beta spectra of iodine.* Phys. Rev. *56*, 965 (1939).

Tcheng and Yang, *Absorption coefficient of beta rays.* Phys. Rev. *60*, 616 (1941).

Townsend, *Beta ray spectra of light elements.* Proc. Roy. Soc. *177*, 357 (1941).

Tyler, *The beta and gamma radiations from* Cu^{64} *and* Eu^{152}. Phys. Rev. *56*, 125 (1939).

Watase and Itoh, *Beta ray spectrum of radiochlorine.* Proc. Phys.-Math. Soc. Jap. (3) *21*, 626 (1939).

Watson and Pollard, *Mass and beta ray energies of* Na^{23}. Phys. Rev. *57*, 1082 (1940).

Watts and Williams, *Beta rays from* H^3. Phys. Rev. *70*, 640 (1946).

Weil and Barkas, *The beta radiation of* As^{76}. Phys. Rev. *56*, 485 (1939).

Weil, *Beta ray spectra of arsenic, rubidium and krypton.* Phys. Rev. *62*, 229 (1942).

Widdowson and Champion. *Upper limits of continuous beta ray spectra.* Proc. Phys. Soc., Lond. *50*, 185 (1938).

Winchester, *Influence of intercrystalline forces on beta ray absorption.* Phys. Rev. *70*, 437 (1946).

Wright and McReynolds, *The beta rays from radium E.* Phys. Rev. *54*, 231 (1938).

Wu, *The continuous X-rays excited by the beta particles of* $_{15}P^{32}$. Phys. Rev. *59*, 481 (1941).

Yamasaki and Sinma, *Beta radioactivity of Re.* Tokyo Inst. Phys. Chem. Res. *37*, 10 (1940).

Yuasa, *On the continuous spectrum of beta rays emitted by radiovanadium.* C. R., Paris *215*, 414 (1942).

Zlotowski, *A microcalorimetric determination of the mean energy of beta rays from radium E.* Phys. Rev. *60*, 483 (1941).

CHAPTER SEVEN

Gamma Radiation

Aoki, *Gamma ray excitation by fast neutrons.* Nature *139*, 372 (1937).

Bak and Nikolaevskaya, *Gamma rays from radioactive iodine* (I^{128}). C. R., URSS *22*, 312 (1939).

Bennett and Bonner, *Emission of gamma rays from the disintegration of C by deuterons.* Phys. Rev. *57*, 1086 (1940).

Bennett and Bonner, *Gamma ray resonances from the bombardment of carbon by deuterons.* Phys. Rev. *58*, 183 (1940).

Bennett, Bonner and Watt, *The energy of the gamma rays from the disintegration of fluorine by protons and deuterons.* Phys. Rev. *59*, 793 (1941).

Bennett, Bonner, Richards and Watt,

High energy gamma rays from Li + D. Phys. Rev. *59,* 904 (1941).

Beringer, *The angular correlation of successive gamma rays.* Phys. Rev. *63,* 23 (1943).

Bernet, Herb and Parkinson, *Gamma rays from fluorine due to proton bombardment.* Phys. Rev. *54,* 398 (1938).

Bernstein, Preston, Wolfe and Slattery, *High energy gamma rays from U(235) fission products.* Phys. Rev. *71,* 463 (1947).

Bleuler, Scherrer, Walter and Zünti, *Gamma radiation of N^{16}* (in German). Helv. Phys. Acta *19,* 421 (1946).

Bonner, Becker, Rubin and Streib, *gamma rays from $C + H^2$.* Phys. Rev. *59,* 215 (1941).

Bonner and Richards, *4.9 Mev gamma rays from Li + H^2.* Phys. Rev. *60,* 167 (1941).

Bothe and Becker, *Artificial excitation of nucleus gamma rays.* Z. Phys. *66,* 289 (1930).

Bradt and Scherrer, *The 93Kev gamma line of UX_1* (in German). Helv. Phys. Acta *19,* 307 (1946).

Bradt and Scherrer, *The 93Kev gamma line of UX_1.* Phys. Rev. *71,* 141 (1947).

Burling, *Gamma rays from sodium bombarded by protons.* Phys. Rev. *60,* 340 (1941).

Clay and Kivieser, *Ionization by gamma rays in gases at high pressures* (in English). Physica 7, 721 (1940).

Constantinov and Latyshev, *Internal conversion of gamma rays from Ra C.* J. Phys., USSR *5,* 239 (1941).

Cork and Pidd, *The absorption of gamma radiation in copper and lead.* Phys. Rev. *66,* 227 (1944).

Cork, *Gamma ray absorption.* Phys. Rev. *67,* 53 (1945).

Cork and Shull, *Internally converted gamma radiation from tantalum (182).* Phys. Rev. *71,* 467 (1947).

Crane and Lauritsen, *Gamma rays from nuclear transformations.* Int. Conf. Phys., Lond. *1,* 130 (1934).

Crane, Delsasso, Fowler and Lauritsen, *High energy gamma rays from lithium and fluorine bombarded with protons.* Phys. Rev. *46,* 531 (1934).

Crane, Delsasso, Fowler and Lauritsen, *Gamma rays from boron bombarded with deutons.* Phys. Rev. *46,* 1109 (1934).

Crane, Delsasso, Fowler and Lauritsen, *Gamma rays from the disintegration of beryllium by deuterons and protons.* Phys. Rev. *47,* 782 (1935).

Crane, Delsasso, Fowler, and Lauritsen, *gamma rays from nitrogen bombarded with deuterons.* Phys. Rev. *48,* 100 (1935).

Crane, Delsasso, Fowler and Lauritsen, *Gamma rays from boron bombarded with protons.* Phys. Rev. *48,* 102 (1935).

Crane, Delsasso, Fowler and Lauritsen, *Cloud chamber studies of the gamma radiation from lithium bombarded with protons.* Phys. Rev. *48,* 125 (1935).

Crane, Halpern and Oleson, *Gamma rays from Be and N bombarded with neutrons.* Phys. Rev. *57,* 13 (1940).

Creutz, Barkas and Furman, *Gamma radiation from rhenium.* Phys. Rev. *58,* 1008 (1940).

Curran, Dee and Petržilka, *Excitation of gamma radiation in processes of proton capture by light elements.* Proc. Roy. Soc. *169,* 269 (1938).

Currand and Strothers, *Excitation of gamma rays in process of proton capture.* Proc. Roy. Soc. *172,* 72 (1939).

Curran, Dee and Strothers, *Measurement of gamma ray energies.* Proc. Roy. Soc. *174,* 546 (1940).

Curtis and Richardson, *Gamma radiation associated with indium.* Phys. Rev. *53,* 942 (1938).

Curtis, *The gamma radiation from radioactive cobalt.* Phys. Rev. *55,* 1136 (1939).

Davidson and Latyshev, *The photo effect from hard gamma rays.* J. Phys., USSR *6,* 15 (1942).

De Benedetti and Kerner, *The gamma rays of Po^{210}.* Phys. Rev. *71,* 122 (1947).

Dee, Curran and Strothers, *Emission of gamma rays from fluorine under proton bombardment.* Nature *143,* 759 (1939).

Delsasso, Fowler and Lauritsen, *Energy and absorption of the gamma radiation from $Li^7 + H^1$.* Phys. Rev. *51,* 391 (1937).

Delsasso, Fowler and Lauritsen, *Gamma radiation from fluorine bombarded with protons.* Phys. Rev. *51,* 527 (1937).

Deutsch and Roberts, *Energies of gamma rays from Br^{82}, I^{131}, I^{130}, Mn^{56}, Mn^{54}, As^{74}.* Phys. Rev. *60,* 362 (1941).

Deutsch, Roberts and Elliott, *Gamma rays from long period activities.* Phys. Rev. *61,* 389 (1942).

De Vault and Libby, *Evidence for gamma radioactivity of 4.5 hour Br^{80} from radiobromate.* Phys. Rev. *55,* 322 (1939).

Elliott, Deutsch and Roberts, *Measurement of the energy of the gamma radiation from Na^{24}.* Phys. Rev. *61,* 99 (1924).

Ellis and Aston, *Absolute intensities and conversion coefficients of the gamma rays of Ra B and Ra C.* Proc. Roy. Soc. *129,* 180 (1930).

Ellis, *Association of gamma rays with the alpha particle group of Th C.* Proc. Roy. Soc. *136,* 396 (1932).

Ellis, *Gamma rays of Th C and of the Th C bodies.* Proc. Roy. Soc. *138,* 318 (1932).

Ellis, *Gamma rays of Ra (B + C) and Th (C + C').* Proc. Roy. Soc. *143,* 350 (1934).

Feather, *Gamma radiations emitted in nuclear processes.* Rep. Phys. Soc. Progr. Phys. *7,* 66 (1941).

Fleischman, *Output of gamma radiation excited by slow neutrons.* Z. Phys. *100,* 307 (1936).

Fowler, Gaerttner and Lauritsen, *The gamma radiation from boron bombarded by protons.* Phys. Rev. *53,* 628 (1938).

Fowler, Lauritsen, *Low energy gamma radiations from lithium bombarded with protons.* Phys. Rev. *56,* 841 (1939).

Fowler and Lauritsen, *Gamma radiation from $N^{15} H^{1}$.* Phys. Rev. *58,* 192 (1940).

Gaerttner and Crane, *The gamma radiation from lithium bombarded with protons.* Phys. Rev. *51,* 49 (1937).

Gaerttner and Crane, *Experiments on the gamma radiation from lithium and fluorine bombarded with protons.* Phys. Rev. *52,* 582 (1937).

Gaerttner, Fowler and Lauritsen, *The gamma radiation with boron bombarded by deuterons.* Phys. Rev. *55,* 27 (1939).

Gaerttner and Pardue, *Gamma radiation from nitrogen bombarded with neutrons.* Phys. Rev. *57,* 386 (1940).

Gamertsfelder, *A high energy gamma ray from radio-yttrium (100 days).* Phys. Rev. *63,* 60 (1943).

Gamertsfelder, *High energy gamma ray from radio-yttrium (100 days).* Phys Rev. *66,* 288 (1944).

Gentner, *The absorption of gamma rays in heavy elements as a function of the wavelength.* J. Phys. Rad. *6,* 274 (1935).

Gentner, *On a gamma radiation in the bombardment of boron with fast protons.* Naturwiss. *25,* 12 (1937).

Goldhaber, Klaiber and Scharff-Goldhaber, *Investigation of high energy gamma rays of Na^{24} by the photoneutron method.* Phys. Rev. *65,* 61 (1944).

Good, *The angular distribution of gamma rays in Na^{24}, Co^{60} and Y^{88}.* Phys. Rev. *70,* 978 (1946).

Gray, *Photo-electric absorption of gamma rays.* Proc. Camb. Phil. Soc. *27,* 103 (1931).

Griffiths and Szilard, *Gamma rays excited by capture of neutrons.* Nature *139,* 323 (1937).

Groetzinger and Smith, *Absorption of 2.8 Mev gamma rays in lead.* Phys. Rev. *67,* 53 (1945).

Groshev, *Formation of pairs in gases by gamma rays from Th C".* J. Phys., USSR *5,* 115 (1941).

Gugelot, Huber, Medicus, Preiswerk, Scherrer and Steffen, *Hard gamma radiation in positron annihilation* (in German). Helv. Phys. Acta *19,* 418 (1946).

Guthrie, *Energies of gamma rays from Ni^{63}, Mn^{56}, Zn^{69}, Ga^{67}.* Phys. Rev. *60,* 746 (1941).

Hafstad, Tuve and Brown, *Gamma ray emission of various targets under bombardment by deuterium ions.* Phys. Rev. *45,* 746 (1934).

Halpern and Crane, *The pair internal conversion coefficient in the $F + H^{1}$ reaction and measurements on the* gamma ray spectrum. Phys. Rev. *55*, 260 (1939).

Halpern and Crane, *Gamma rays from B + D.* Phys. Rev. *55*, 415 (1939).

Helmholz, *Energy and multipole order of nuclear gamma rays.* Phys. Rev. *60*, 415 (1941).

Helmholz, *Gamma rays from radioactive Hg.* Phys. Rev. *61*, 204 (1942).

Herb, Kerst and McKibben, *Gamma rays from light elements due to proton bombardment.* Phys. Rev. *51*, 691 (1937).

Hirzel and Wäffler, *The gamma radiation emitted in decay of K^{40}* (in German). Helv. Phys. Acta *19*, 216 (1946).

Hole, Holtsmark and Tangen, *Gamma rays in the bombardment of magnesium with protons.* Naturwiss, *28*, 399 (1940).

Huber, Lienhard, Scherrer and Wäffler, *Nuclear photo-effect with lithium gamma rays* (in German). Helv. Phys. Acta *15*, 312 (1942).

Huber, Lienhard, Scherrer and Wäffler, *Nuclear photo-effect with gamma rays. I. Light elements up to calcium* (in German). Helv. Phys. Acta *16*, 33 (1943).

Hudson, Herb and Plain, *Excitation of the 455 Kev level of Li^7 by proton bombardment.* Phys. Rev. *57*, 587 (1940).

Hushley, *Gamma rays from Be caused by proton bombardment.* Phys. Rev. *67*, 34 (1945).

Itoh and Watase, *Gamma rays emitted from Rn and MsTh-I and their daughter elements.* Proc. Phys.-Math. Soc. Jap. (3) *23*, 142 (1941).

Itoh, *Gamma rays emitted from Na^{24}, Mg^{27}, Al^{28}, and Cl^{38}.* Proc. Phys.-Math. Soc. Jap. (3) *23*, 605 (1941).

Kanne and Ragan, *Gamma rays from $Li + H^1$.* Phys. Rev. *54*, 480 (1938).

Kikuchi, Aoki and Husimi, *Excitation of gamma rays by slow neutrons.* Proc. Phys.-Math. Soc. Jap. (3) *17*, 369 (1935).

Kikuchi, Aoki and Husimi, *Excitation of gamma rays in boron.* Nature *137*, 745 (1936).

Kikuchi, Aoki and Husimi, *Energy of gamma rays excited by slow neutrons.* Nature *137*, 186 (1936).

Kikuchi, Aoki and Husimi, *Excitation of gamma rays by fast neutrons.* Nature *137*, 398 (1936).

Kikuchi, Husimi and Aoki, *Quantum energy of gamma rays excited by slow neutrons.* Nature *137*, 992 (1936).

Kikuchi, Husimi and Aoki, *Excitation of gamma rays by neutrons, II.* Proc. Phys.-Math. Soc. Jap. (3) *18*, 35 (1936).

Kikuchi, Aoki and Husimi, *Excitation of gamma rays by neutrons.* Proc. Phys.-Math. Soc. Jap. (3) *18*, 115 (1936).

Kikuchi, Husimi and Aoki, *Excitation of gamma rays by neutrons.* Proc. Phys.-Math. Soc. Jap. (3) *18*, 188 (1936).

Kikuchi, Aoki and Kusimi, *Excitation of gamma rays by neutrons from Ra + Be.* Proc. Phys.-Math. Soc. Jap. (3) *18*, 297 (1936).

Kikuchi and Aoki, *On the "neutron electron interaction" proposed by us, and the gamma rays in the D-D reaction.* Proc. Phys.-Math. Soc. Jap. (3) *21*, 20 (1939).

Kikuchi, Watase, Itoh, Takeda and Yamaguchi, *Gamma ray accompanying the disintegration of Na^{24}.* Proc. Phys.-Math. Soc. Jap. (3) *21*, 260 (1939).

Kikuchi, Watase and Itch, *On the angular relation of the two gamma quanta radiated in cascades from an atomic nucleus.* Z. Phys. *119*, 185 (1942).

Kovacs, *The excitation function of the gamma radiation generated by Po-∂ bombardment of sodium.* Phys. Rev. *70*, 895 (1946).

Krebs, *The gamma ray transition of the Br isomer of Br^{80} (4.5 hr).* Naturwiss. *30*, 121 (1942).

Kruger and Green, *The gamma ray spectrum of B^{10}.* Phys. Rev. *52*, 773 (1937).

Kruger and Ogle, *Gamma rays emitted during the radioactive transitions $Sb^{124} - > Tl^{124}$, $Na^{24} - > Mg^{24}$.* Phys. Rev. *67*, 273 (1945).

Latyschev and Kulchitsky, *The recoil electron spectrum of gamma rays from Th active deposit.* J. Phys., USSR *4*, 515 (1941).

Lauritsen and Crane, *Gamma rays from*

lithium bombarded with protons. Phys. Rev. *45*, 63 (1934).

Lauritsen and Crane, *Gamma rays from carbon bombarded with deutons.* Phys. Rev. *45*, 345 (1934).

Lauritsen and Fowler, *Gamma radiation from C^{13} + H^1.* Phys. Rev. *58*, 193 (1940).

Lauritsen, Lauritsen and Fowler, *Gamma radiation from Be^9 + H^1.* Phys. Rev. *71*, 279 (1947).

Lawson and Cork, *Internally converted gamma rays from radioactive gold.* Phys. Rev. *58*, 580 (1940).

Lea, *Secondary gamma rays excited by the passage of neutrons through matter.* Proc. Roy. Soc. *150*, 637 (1935).

Maier-Leibnitz, *The gamma spectrum of Be^7.* Naturwiss. *26*, 614 (1938).

Maier-Leibnitz, *Absolute counter measurements on gamma rays.* Z. Naturforsch. *1*, 243 (1946). .

Mandeveille, *Gamma rays from Sc^{48}.* Phys. Rev. *62*, 555 (1942).

Mandeville, *Gamma rays from As^{76}.* Phys. Rev. *63*, 91 (1943).

Mandeville, *Gamma rays from Na^{24}.* Phys. Rev. *62*, 309 (1942).

Mandeville, *Gamma rays from Na^{24} and La^{140}.* Phys. Rev. *63*, 387 (1943).

Mandeville, *The energies of the gamma rays from radioactive scandium, gallium, tungsten and lanthanum.* Phys. Rev. *64*, 147 (1943).

Mandeville and Fulbright, *The energies of the gamma rays from Sb^{122}, Cd^{115}, Ir^{192}, Mn^{54}, Zn^{65}, and Co^{60}.* Phys. Rev. *64*, 265 (1943).

McMillan, *Some gamma rays accompanying artificial nuclear disintegrations.* Phys. Rev. *46*, 868 (1934).

Meitner and Hupfield, *On the absorption law for hard gamma rays.* Z. Phys. *67*, 147 (1931).

Members of the Nuclear Physics Laboratory of the Osaka Imperial University, *Gamma ray spectra of V^{52} and Mn^{56}.* Proc. Phys.-Math. Soc. Jap. (3) *24*, 818 (1942).

Mitchell and Langer, *Energy of the gamma rays of radio-indium and radio-manganese.* Phys. Rev. *52*, 137 (1937).

Mitchell, Langer and McDaniel, *Study of the nuclear radiations from antimony and arsenic.* Phys. Rev. *57*, 1107 (1940).

Mouzon, Park and Richards, *Gamma rays from uranium activated by neutrons.* Phys. Rev. *55*, 668 (1939).

Nonaka, *Resonance capture of slow neutrons and emission of gamma rays.* Nature *144*, 831 (1939).

Nonaka, *On the resonance capture of slow neutrons and the emission of gamma rays.* Proc. Phys.-Math. Soc. Jap. (3) *21*, 594 (1939).

Nonaka, *Resonance capture of slow neutrons and emission of gamma rays. Part II.* Proc. Phys.-Math. Soc. Jap. (3) *22*, 551 (1940).

Nonaka, *Excitation of gamma rays by fast neutrons of different energy.* Phys. Rev. *59*, 681 (1941).

Nonaka, *On the resonance capture of slow neutrons and emission of gamma rays. III. Resonance groups of silver.* Proc. Phys.-Math. Soc. Jap. (3) *24*, 33 (1942).

Ogle and Kruger, *Energy of gamma rays from $Na^{24}->Mg^{24}$.* Phys. Rev. *65*, 61 (1944).

Plain, Herb, Hudson and Warren, *Gamma rays from Al due to proton bombardment.* Phys. Rev. *57*, 187 (1940).

Pontecorvo, *On a soft radiation emitted upon the capture of neutrons by nuclei.* C. R., Paris *207*, 423 (1938).

Rasetti, *Gamma rays produced by neutron absorption.* Z. Phys. *97*, 64 (1935).

Richardson, *Gamma rays emitted by several artificially produced radioactive elements.* Phys. Rev. *49*, 203 (1936).

Richardson and Kurie, *The radiations emitted from the artificially produced radioactive substances. II. Gamma rays from several elements.* Phys. Rev. *50*, 999 (1936).

Richardson, *The radiations produced from artificially produced radioactive sources. IV. Further studies on the gamma rays from several elements.* Phys. Rev. *53*, 124 (1938).

Richardson, *Gamma radiation from N^{13}.* Phys. Rev. *53*, 610 (1938).

Richardson, *The gamma radiation from long-lived yttrium.* Phys. Rev. *60*, 188 (1941).

Roberts, *Spectrum of radium (B + C) gamma rays.* Phil. Mag. (7) *36*, 264 (1945).

Roberts, *Absorption of radium (B + C) gamma rays.* Proc. Roy. Soc. *183*, 338 (1945).

Rose, *Gamma radiation from the proton bombardment of C.* Phys. Rev. *53*, 844 (1938).

Ruhlig, *Search for gamma rays from the deuteron-deuteron reaction.* Phys. *54*, 308 (1938).

Salgueiro, *Spectrography of gamma rays emitted by the active deposit in the slow evolution of radon* (in French). Portugaliae Physica *1*, 67 (1944).

Savel, *Complex radiation excited in Al by alpha particles.* C. R., Paris *198*, 368 (1934).

Savel, *Complex radiation excited in light elements by alpha particles.* C. R., Paris *198*, 1404 (1934).

Scharff-Goldhaber, *Energy of the hard gamma rays of radio-yttrium (100 days).* Phys. Rev. *59*, 937 (1941).

Scharff-Goldhaber, Goldhaber and Yalow, *Attempt to detect nuclear resonance absorption of gamma rays.* Phys. Rev. *67*, 59 (1945).

Schwarz and Pool, *Gamma rays from tungsten.* Phys. Rev. *70*, 102 (1946).

Schwarz and Pool, *Gamma rays from tungsten and molybdenum.* Phys. Rev. *71*, 122 (1947).

Shepherd, Haxby and Williams, *The gamma rays from B and Be under proton bombardment and from Li under deuteron bombardment.* Phys. Rev. *52*, 247 (1937).

Sheppard, *Absorption of gamma radiation in terrestrial materials.* Phys. Rev. *64*, 379 (1943).

Siday, *The gamma ray transition of radio-bromine.* Proc. Roy. Soc. *178*, 189 (1941).

Siegbahn, *Intensities of gamma rays, studied by means of their Compton secondaries.* Proc. Roy. Soc. *188*, 541 (1947). ⊙

Silveira, *Absorption of gamma rays emitted by U I and its immediate descendants* (in French). Portugaliae Physica *1*, 151 (1944).

Silveira, *On the absorption of the gamma radiation emitted by the UX complex* (in French). Portugaliae Physica *1*, 175 (1945).

Sizoo and Willemsen, *Absorption of radium gamma rays* (in English). Physica *5*, 100 (1938).

Sizoo and Eijkman, *Gamma radiation from $^{198}_{79}Au$* (in English). Physica *6*, 332 (1939).

Speh, *Gamma rays of Li and F under alpha particle bombardment.* Phys. Rev. *50*, 689 (1936).

Stahel and Walton, *Experiments to determine a new high energy gamma ray component of Ra C* (in German). Helv. Phys. Acta *14*, 326 (1941).

Szalay and Zimonyi, *The excitation function of the gamma radiation which is excited by bombarding Be^9, B^{10}, B^{11}, and Al^{27} with Po-particles.* Z. Phys. *115*, 639 (1940).

Tandberg, *Some convenient expressions for calculating the absorption of gamma rays and X-rays* (in English). Ark. Mat. Astr. Fys. *27B*, No. 3 (1940).

Te-Tchao and Surugue, *On the gamma ray of low energy of radio-actinium.* C. R., Paris *218*, 591 (1944).

Traubenberg, Eckardt and Gebauer, *Gamma rays produced during the disintegration of Li.* Z. Phys. *80*, 557 (1933).

Tsien, *Intensity of gamma rays of radium D.* C. R., Paris *218*, 503 (1944).

Tsien and Marty, *On photoelectrons of gamma rays of Ra D.* C. R., Paris *220*, 688 (1945).

Tsien and Marty, *On the existence of gamma rays of weak energy of radium D.* C. R., Paris, *221*, 177 (1945).

Tsien, *The gamma rays of radium D.* Phys. Rev. *69*, 38 (1946).

Tyler, *The beta and gamma ray spectra of Cu^{64} and Eu^{152}.* Phys. Rev. *55*, 1136 (1939).

Valley, *Magnetic spectrograph investigation of N^{13} gamma radiation.* Phys. Rev. *56*, 838 (1939).

Valley and McCreary, *The internally converted gamma rays of several radio-*

active elements. Phys. Rev. *56*, 863 (1939).

Watase and Itoh, *Gamma rays of N^{13}.* Proc. Phys.-Math. Soc. Jap (3) *21*, 389 (1939).

Webster, *Artificial production of nuclear gamma radiation.* Proc. Roy. Soc. *136*, 428 (1932).

Wilson, *Dependence of the secondary electronic emission by gamma radiation upon the direction of the radiation.* Proc. Phys. Soc., Lond. *53*, 613 (1941).

Zlotowski and Williams, *The energy of gamma rays accompanying the decay of Be^7.* Phys. Rev. *62*, 29 (1942).

Zuber, *Effective cross section for the materialization of 2.62 Mev gamma rays in argon* (in German). Helv. Phys. Acta *15*, 38 (1942).

Zuber, *The problem of resonance absorption of 2.62 Mev gamma rays in lead* (in German). Helv. Phys. Acta *16*, 407 (1943).

Zuber, *Resonance absorption of gamma rays* (in German). Helv. Phys. Acta *16*, 429 (1943).

CHAPTER EIGHT

Neutrons

Aebersold and Anslow, *Fast neutron absorption in gases, walls and tissue.* Phys. Rev. *69*, 1 (1946).

Agnew, Bright and Froman, *Distribution of neutrons in the atmosphere.* Phys. Rev. *70*, 102 (1946).

Amaldi and Fermi, *On the absorption and the diffusion of slow neutrons.* Phys. Rev. *50*, 899 (1936).

Amaldi and Fermi, *On the absorption of slow neutrons.* Ric. Sci. (6) *2*, 443 (1935).

Amaldi and Fermi, *Absorption of slow neutrons.* Ric. Sci. (7) *1*, 56 (1936).

Amaldi and Fermi, *Mean free path of slow neutrons in paraffin wax.* Ric. Sci. (7) *1*, 223 (1936).

Amaldi and Fermi, *Groups of slow neutrons.* Ric. Sci. (7) *1*, 310 (1936).

Amaldi and Fermi, *Diffusion of slow neutrons.* Ric. Sci. (7) *1*, 393 (1936).

Amaldi and Fermi, *Absorption and diffusion of slow neutrons.* Ric. Sci. (7) *1*, 454 (1936).

Amaldi and Fermi, *On the absorption and the diffusion of slow neutrons.* Phys. Rev. *50*, 899 (1936).

Amaldi, Hafstad and Tuve, *Neutron yields from artificial sources.* Phys. Rev. *51*, 896 (1937).

Anderson, Fermi and Marshall, *Production of low energy neutrons by filtering through graphite.* Phys. Rev. *70*, 815 (1946).

Arsenjewa-Heil, Heil and Westcott, *Influence of temperature on the 'groups' of slow neutrons.* Nature *138*, 462 (1936).

Bayley, Curtis, Gaerrtner and Goudsmit, *Diffusion of slow neutrons.* Phys. Rev. *50*, 461 (1936).

Bennett and Watt, *Yield of neutrons and beta rays from $Li + H^2$.* Phys. Rev. *60*, 166 (1941).

Bennett, Mandeville and Richards, *The yield function and angular distribution of the $D + D$ neutrons.* Phys. Rev. *69*, 418 (1946).

Bennett, Mandeville and Richards, *The yield function and angular distribution of the $D + D$ neutrons.* Phys. Rev. *70*, 101 (1946).

Bennett and Richards, *Neutrons from $C^{12} + D$.* Phys. Rev. *70*, 118 (1946).

Bernardini, *Excitation of neutrons on Be.* Z. Phys. *85*, 555 (1933).

Bernadini and Bocciarelli, *Energy and intensity of groups of neutrons emitted by Po-Be.* Atti Accad. Lincei *24*, 59 (1936).

Bernardini, *Neutrons of high energy from Po + Be and the nuclear levels of C^{12}.* Ric. Sci. *8*, 33 (1937).

Bernstein, *On the existence of a resonance absorption of neutrons in graphite.* Phys. Rev. *70*, 107 (1946).

Bloch, Hamermesh and Staub, *Neutron polarization and ferromagnetic saturation.* Phys. Rev. *62*, 303 (1942).

Bloch, Hamermesh and Staub, *Neutron polarization and ferromagnetic saturation.* Phys. Rev. *64*, 47 (1943).

Bloch, Condit and Staub, *Neutron polarization and ferromagnetic saturation.* Phys. Rev. *70,* 972 (1946).

Bonner, *Ionization of gases by neutrons.* Phys. Rev. *43,* 871 (1933).

Bonner and Mott-Smith, *Energy distributions of the neutrons from fluorine.* Phys. Rev. *45,* 552 (1934).

Bonner, *Collisions of neutrons with atomic nuclei.* Phys. Rev. *45,* 601 (1934).

Bonner and Mott-Smith, *Energy spectra of the neutrons from the disintegration of fluorine, boron and beryllium by alpha particles.* Phys. Rev. *46,* 258 (1934).

Bonner and Brubaker, *The energy spectrum of the neutrons from the disintegration of Be by deutrons.* Phys. Rev. *47,* 910 (1935).

Bonner and Brubaker, *Neutrons from the disintegration of deuterium by deuterons.* Phys. Rev. *49,* 19 (1936).

Bonner, *The energy of the neutrons from the disintegration of C by deuterons.* Phys. Rev. *53,* 496 (1938).

Bonner, Hudspeth and Bennett, *Resonances in the emission of neutrons from the reaction $C^{12} + H^2$.* Phys. Rev. *58,* 185 (1940).

Bonner, *A precise determination of the energy of the neutron from the deuteron-deuteron reaction.* Phys. Rev. *59,* 237 (1941).

Booth and Hurst, *Iso-energetic neutrons.* Proc. Roy. Soc. *161,* 248 (1937).

Borst, Ulrich, Osborne and Hasbrouck, *Diffraction of neutrons and neutron absorption spectra.* Phys. Rev. *70,* 108 (1946).

Borst, Ulrich, Osborne and Hasbrouck, *Neutron diffraction and nuclear resonance structure.* Phys. Rev. *70,* 557 (1946).

Bothe and Jensen, *Absorption of thermal neutrons in carbon.* Z. Phys. *122,* 749 (1944).

Bradt, *Neutrons from U* (in German). Helv. Phys. Acta *12,* 553 (1939).

Brasch, Lange, Waly, Banks, Chalmers, Szilard and Hopwood, *Liberation of neutrons from beryllium by X-rays.*

Radioactivity induced by means of electron tubes. Nature *134,* 880 (1934).

Breit and Wigner, *Capture of slow neutrons.* Phys. Rev. *49,* 519 (1936).

Browne and Dantzig, *On the theory of slowing down neutrons in substances with periodic absorption lines.* Phys. Rev. *71,* 463 (1947).

Carroll and Dunning, *The interaction of slow neutrons with gases.* Phys. Rev. *54,* 541 (1938).

Chadwick, *Possible existence of a neutron.* Nature *129,* 312 (1932).

Chadwick, *Existence of a neutron.* Proc. Roy. Soc. *136,* 692 (1932).

Chadwick, *The neutron.* Proc. Roy. Soc. *142,* 1 (1933).

Collie, *Absorption of slow neutrons.* Nature *137,* 614 (1936).

Coltman and Goldhaber, *Capture cross sections for slow neutrons.* Phys. Rev. *69,* 411 (1946).

Condon and Breit, *The energy distribution of neutrons slowed by elastic impacts.* Phys. Rev. *49,* 229 (1936).

Coon and Barschall, *Angular distribution of 2.5 Mev neutrons scattered by deuterium.* Phys. Rev. *70,* 592 (1946).

Corinaldesi, *On the measurement of the density of thermal neutrons.* Nuovo. Cim (9) *3,* 131 (1946).

Cork and Thornton, *The loss of neutrons by neutron bombardment and the radioactive isotopes of scandium.* Phys. Rev. *53,* 866 (1938).

Coster, de Vries and Diemer, *The resonance levels for neutron capture. III* (in English). Physica *10,* 281 (1943).

Coster, de Vries and Diemer, *The resonance levels for neutron capture. IV. Overlapping of levels* (in English). Physica *10,* 312 (1943).

Crane, Lauritsen and Soltan, *Artificial production of neutrons.* Phys. Rev. *44,* 514 (1933).

Crane, Lauritsen and Soltan, *Production of neutrons by high speed deutons.* Phys. Rev. *44,* 692 (1933).

Crane, Lauritsen and Soltan, *Artificial production of neutrons.* Phys. Rev. *45,* 507 (1934).

Curie and Joliot, *Mass of the neutron.* C. R., Paris, *197,* 237 (1933).

Curie and Joliot, *Experimental proofs of the existence of the neutron.* J. Phys. Rad. *4*, 21 (1933).

Curie and Joliot, *Emission of neutrons.* J. Phys. Rad. *4*, 278 (1933).

Danysz, Rotblat, Wertenstein and Zyw, *Experiments on the Fermi effect.* Nature *134*, 970 (1934).

Denison, *Attempt to find neutron-like particles accompanying beta ray emission.* Phys. Rev. *45*, 557 (1934).

Diemer, *Overlapping measurements on Cd-resonance neutrons* (in English). Physica *11*, 391 (1946).

Döpel, *Measurement of neutron intensities by rhodium Fermi electrons.* Phys. Z. *37*, 96 (1936).

Döpel, *Neutron emission from Be when bombarded by H, D and He.* Z. Phys. *104*, 666 (1937).

Duchon, *Ion optics of two neutron generators.* J. Phys. Rad. *6*, 290 (1945); *7*, 64 (1946).

Dunning and Pegram, *Neutron emission.* Phys. Rev. *45*, 295 (1934).

Dunning and Pegram, *Scattering of neutrons by H^1 $_2$O, H^2 $_2$O, paraffin, Li, B, and C, and the production of radioactive nuclei by neutrons found by Fermi.* Phys. Rev. *45*, 768 (1934).

Dunning, Pegram, Fink and Mitchell, *Slow neutrons.* Phys. Rev. *47*, 970 (1935).

Dunning, Pegram, Fink and Mitchell. *Interaction of neutrons with matter.* Phys. Rev. *48*, 265 (1935).

Dunning, Pegram, Fink, Mitchell, and Segrè, *Velocity of slow neutrons by mechanical velocity selector.* Phys. Rev. *48*, 704 (1935).

Dunning, Powers and H. G. Beyer, *Experiments on the magnetic properties of the neutron.* Phys. Rev. *51*, 54 (1937).

Dunning, Manley, Hoge and Brickwedde, *The interaction of neutrons with normal and parahydrogen.* Phys. Rev. *52*, 1076 (1937).

Eriksson, *On the distribution of the neutron energies in a moderator of infinite size* (in English). Ark. Mat. Astr. Fys. *33B*, No. 5, (1946).

Feenberg, *Interaction between neutrons and the mass of the neutron.* Phys. Rev. *45*, 649 (1934).

Feeny, Lapointe and Rasetti, *Resonance absorption of neutrons in rhodium, antimony and gold.* Phys. Rev. *61*, 469 (1942).

Feeny and Rasetti, *Resonance absorption neutrons in manganese, gallium, and palladium.* Canad. J. Res. *23A*, 12 (1945).

Feeny, *Absorption of thermal neutrons in indium.* Canad. J. Res. *23A*, 73 (1945).

Fenning, Graham and Seligman, *The ratio of the capture cross sections of Li and B for thermal neutrons.* Canad. J. Res. *25A*, 73 (1947).

Fermi and Rasetti, *Slow neutrons.* Nuovo Cim. *12*, 201 (1935).

Fermi, *Motion of neutrons in hydrogeneous substances.* Ric. Sci. 7, 13 (1935).

Fermi, Amaldi and Wick, *On the albedo of slow neutrons.* Phys. Rev. *53*, 493 (1938).

Fermi and Zinn, *Reflection of neutrons on mirrors.* Phys. Rev. *70*, 103 (1946).

Fields, Russell, Sachs and Wattenberg, *Total cross sections measured with photo-neutrons.* Phys. Rev. *71*, 508 (1947).

Fink, Dunning, Pegram and Mitchell, *The velocities of slow neutrons.* Phys. Rev. *49*, 103 (1936).

Fink, Dunning, Pegram and Segrè, *Production and absorption of slow neutrons in hydrogenic materials.* Phys. Rev. *49*, 199 (1936).

Fink, *The production and absorption of thermal energy neutrons.* Phys. Rev. *50*, 738 (1936).

Fleischmann and Gentner, *Wavelength dependence of nuclear photo-effect of beryllium.* Z. Phys. *100*, 440 (1936).

Fleischmann, *Energy liberated at neutron capture.* Z. Phys. *103*, 113 (1936).

Frisch and Sörensen, *Velocity of "slow neutrons."* Nature *136*, 258 (1935).

Frisch and Placzek, *Capture of slow neutrons.* Nature *137*, 357 (1936).

Frisch, von Halban and Koch, *Temperature equilibrium of C neutrons.* Nature *139*, 922 (1937).

Frisch, von Halban and Koch, *The magnetic field acting upon neutrons inside magnetized iron.* Nature *140*, 360 (1937).

Frisch, von Halban and Koch, *On the slowing down and capture of neutrons in hydrogeneous substances* (in English). Kgl. Danske Akad. *15*, No. 10 (1938).

Frisch, von Halban and Koch, *Some experiments on the magnetic properties of free neutrons.* Phys. Rev. *53*, 719 (1938).

Frisch, *The total cross sections of C and H for neutrons of energies from 35 to 490 Kev.* Phys. Rev. *70*, 589 (1946).

Frisch, *Total cross section of hydrogen for neutrons of energies from 35 to 490 Kev.* Phys. Rev. *70*, 792 (1946).

Fryer, *A study of velocity selected neutrons.* Phys. Rev. *62*, 303 (1942).

Fryer, *Transmission of velocity-selected neutrons through magnetized iron.* Phys. Rev. *70*, 235 (1946).

Fünfer and Bothe, *Interaction of neutrons and gamma rays with Be.* Z. Phys. *122*, 769 (1944).

Furry, *An application of the theory of neutron absorption in aqueous solutions.* Phys. Rev. *51*, 592 (1937).

Gamertsfelder and Goldhaber, *The diffusion length of C neutrons in water.* Phys. Rev. *62*, 556 (1942).

Gamertsfelder and Goldhaber, *A reproducible neutron standard.* Phys. Rev. *69*, 368 (1946).

Gamow, *Possibility of selective phenomena for fast neutrons.* Phys. Rev. *49*, 946 (1936).

Gehlen, *Determination of the effective cross section of commercial aluminum for the capture of slow neutrons.* Z. Phys. *121*, 268 (1943).

Gibson, Seaborg and Grahame, *On the interaction of fast neutrons with lead.* Phys. Rev. *51*, 370 (1937).

Goldberger and Seitz, *Refraction and diffraction of neutrons by crystals.* Phys. Rev. *70*, 116 (1946).

Goldhaber and O'Neil, *Slowing down of low energy neutrons in water.* Phys. Rev. *69*, 834 (1941).

Goldsmith and Rasetti, *On the resonance capture of slow neutrons.* Phys. Rev. *50*, 328 (1936).

Goldsmith and Manley, *Absorption of resonance neutrons.* Phys. Rev. *51*, 382 (1937).

Goldsmith and Cohen, *Mass of the neutron from the nuclear reaction* $H^2 + H^2 -> He^3 + n^1$. Phys. Rev. *45*, 850 (1934).

Goldsmith and Manley, *Absorption of resonance neutrons.* Phys. Rev. *61*, 1022 (1937).

Goloborodko and Rosenkewitsch, *Angular distribution of photo-neutrons from Be.* Phys. Z. Sowjet. *11*, 78 (1937).

Goloborodko, *The spectra of photo-neutrons (Ra Th + Be) and (Ra + Be).* J. Phys. USSR *5*, 15 (1941).

Goloborodko, *Spectra and energy of photoneutrons (Ra Th, Be) and (Ra, Be).* J. Phys. USSR *5*, 19 (1941).

Goudsmit, *On the slowing down of neutrons.* Phys. Rev. *49*, 406 (1936).

Graham and von Halban, *On the angular distribution of neutrons in the photodisintegration of the deuteron.* Rev. Mod. Phys. *17*, 297 (1945).

von Halban and Preiswerk, *Resonance levels for neutron capture.* C. R., Paris *202*, 133 (1936).

von Halban and Preiswerk, *Diffraction of neutrons.* C. R., Paris *203*, 73 (1936).

von Halban and Preiswerk, *Resonance levels for neutron absorption* (in German). Helv. Phys. Acta *9*, 318 (1936).

von Halban and Preiswerk, *Cross section measurements with slow neutrons of different velocities.* Nature *137*, 905 (1936).

von Halban and Preiswerk, *Slow neutrons.* J. Phys. Rad. *8*, 29 (1937).

von Halban, Kowarski and Savitch, *Simple capture of thermal neutrons and of resonance neutrons by uranium.* C. R., Paris *208*, 1396 (1939).

Halpern and Holstein, *On the passage of neutrons through ferromagnets.* Phys. Rev. *59*, 960 (1941).

Halpern, Hamermesh and Johnson, *The passage of neutrons through crystals and polycrystals.* Phys. Rev. *59*, 981 (1941).

Halpern and Holstein, *On the dipolarization of neutron beams by magnetic fields.* Proc. Nat. Acad. Sci. *28*, 112 (1942).

Hanstein, *Neutron-proton interaction.* Phys. Rev. *57*, 1045 (1940).

Hanstein, *Interaction experiments with resonance neutrons.* Phys. Rev. *59*, 489 (1941).

Harington and Stewart, *Capture cross section for thermal neutrons of Cd, Li, B, Ba, Hg and H. Canad.* J. Res. *19A*, 33 (1941).

Havens and Rainwater, *The slow neutron cross sections of In, Au, Ag, Sb, Li and Hg as measured with a neutron beam spectrometer.* Phys. Rev. *70*, 154, (1946).

Havens, Wu, Rainwater and Meaker, *Slow neutron velocity spectrometer studies. II. Au, In, Ta, W, Pt, Zr.* Phys. Rev. *71*, 165 (1947).

Haxel and Volz, *On the absorption of neutrons in aqueous solutions.* Z. Phys. *120*, 493 (1943).

Hereward, Lawrence, Paneth and Sargent, *Measurement of the diffusion length of thermal neutrons in graphite.* Canad. J. Res. *25A*, 15 (1947).

Hereward, Lawrence, Munn, Paneth and Sargent, *The diffusion length of thermal neutrons in heavy water containing $Li_2\,CO_3$.* Canad. J. Res. *25A*, 26 (1947).

Hill, *Studies with the neutrons from p-n reactions in Li and Be.* Phys. Rev. *55*, 1117 (1939).

Hill, *The relative distribution in energy of the neutrons from the (p, n) reactions in Li and Be.* Phys. Rev. *57*, 1076 (1940).

Hoffmann and Bethe, *Neutron absorption limit of cadmium.* Phys. Rev. *51*, 1021 (1937).

Hoffmann and Livingston, *The neutron absorption limit in cadmium.* Phys. Rev. *52*, 1228 (1937).

Hoffmann and Livingston, *Angular distributions of slow neutrons from a paraffin source.* Phys. Rev. *53*, 1020 (1938).

Hoffman and Bacher, *Photographic effects produced by cadmium and other elements under neutron bombardment.* Phys. Rev. *54*, 644 (1938).

Hopwood and Chalmers, *Directed diffusion or canalization of slow neutrons.* Nature *135*, 341 (1935).

Hornbostel and Valente, *On the resonance levels for neutron capture of iodine and indium.* Phys. Rev. *55*, 108 (1939).

Horvath and Salant, *The absorption of resonance neutrons by Bo, Cl, Co and Mn.* Phys. Rev. *59*, 154 (1941).

Horvay, *On the slowing down of neutrons in water.* Phys. Rev. *50*, 897 (1936).

Houtermans, *On a semi-empirical relation between the productivity of a neutron source and the maximum obtainable density of slow neutrons in a hydrogeneous medium.* Phys. Z. *45*, 258 (1945).

Hudspeth and Dunlap, *Low energy neutrons from the deuteron-deuteron reaction.* Phys. Rev. *55*, 587 (1939).

Hughes, *Radiative capture cross sections for fast neutrons.* Phys. Rev. *70*, 106 (1946).

Inghram, Hess and Hayden, *Neutron cross sections for mercury isotopes.* Phys. Rev. *71*, 561 (1947).

Jaeckel, *Resonance capture of slow neutrons by tungsten nuclei.* Z. Phys. *104*, 762 (1937).

Jaeckel, *Neutron resonance levels of Ir and Rh and comparison of yields.* Z. Phys. *107*, 669 (1937).

Jaeckel, *Neutron resonance levels for Ir and Rh.* Z. Phys. *110*, 330 (1938).

Jensen, *Retardation of neutrons in carbon, water and heavy water.* Z. Phys. *122*, 756 (1944).

Kimura, *Energy of photoneutrons liberated from deuterons by Ra C gamma rays.* Mem. Coll. Sci. Kyoto Imp. Univ. *22*, 237 (1939).

Korff and Hamermesh, *The energy distribution and number of cosmic ray neutrons in the free atmosphere.* Phys. Rev. *69*, 155 (1946).

Korff and Hamermesh, *The multiple production of neutrons by cosmic radiation.* Phys. Rev. *70*, 429 (1946).

Koyenuma, *An approximate formula for the effective reaction cross section σ of slow neutrons for a 1/v absorber.* Ann. Phys., Leipzig *43*, 279 (1943).

Kupferberg and Korff, *A determination of the rate of production in paraffin of neutrons by the cosmic radiation at sea level.* Phys. Rev. *65*, 253 (1944).

Ladenburg, Roberts and Sampson, *Some investigations of neutrons from different sources.* Phys. Rev. *48,* 467 (1935).

Landenburg and Kanner, *On the neutrons from the deuteron-deuteron reactions.* Phys. Rev. *52,* 911 (1937).

Lamb, *A note on the capture of slow neutrons in hydrogeneous substances.* Phys. Rev. *51,* 187 (1937).

Lamb, *Capture of neutrons by atoms in a crystal.* Phys. Rev. *55,* 190 (1939).

Laporte, *Absorption coefficients for thermal neutrons.* Phys. Rev. *52,* 72 (1937).

Lapp, Van Horn and Dempster, *The neutron absorbing isotopes in samarium and gadolinium.* Phys. Rev. *70,* 104 (1946).

Laughlin and Kruger, *Neutron spectra of Li, Al, and Be bombarded with 10 Mev deuterons.* Phys. Rev. *71,* 484 (1947).

Lauterjung, *On the effective cross section of some of the elements to slow neutrons.* Ann. Phys., Leipzig *41,* 177 (1942).

Lewis, *A theory of orbital neutrons.* Phys. Rev. *50,* 857 (1936).

Libby and Long, *The production and property of low temperature neutrons.* Phys. Rev. *52,* 593 (1937).

Libby, *Reactions of high energy atoms produced by slow neutron capture.* J. Amer. Chem. Soc. *62,* 1930 (1940).

Livingston and Hoffmann, *Neutron absorption limit in cadmium.* Phys. Rev. *51,* 1021 (1937).

Lukirsky and Coreva, *Slowing down of neutrons by nuclei of heavy elements* (in English). C. R., URSS *3,* 411 (1936).

Manley, Goldsmith and Schwinger, *Neutron energy levels.* Phys. Rev. *51,* 1022 (1937).

Manley, Goldsmith and Schwinger, *The resonance absorption of slow neutrons in indium.* Phys. Rev. *55,* 107 (1939).

Manley, Haworth and Luebke, *The mean life of neutrons in water and the hydrogen capture cross section.* Phys. Rev. *61,* 152 (1942).

Manley, Haworth and Luebke, *The velocity dependence of the absorption of boron for slow neutrons.* Phys. Rev. *69,* 405 (1946).

Marshall, *Total cross sections of various materials for indium resonance neutrons.* Phys. Rev. *70,* 107 (1946).

Maurer, *Excitation function and energy distribution of neutrons emitted from boron by bombardment with alpha particles from polonium.* Z. Phys. *107,* 721 (1937).

McDaniel, *Slow neutron resonances in indium.* Phys. Rev. *70,* 832 (1946).

Meitner and Philipp, *The electron orbits resulting from neutron excitation.* Naturwiss *21,* 286 (1933).

Meitner and Philipp, *Further experiments with neutrons.* Z. Phys. *87,* 484 (1934).

Meitner, *Capture cross sections for thermal neutrons in thorium, lead and U^{238}.* Nature *145,* 422 (1940).

Mescheryakov, *On the absorption of fast neutrons by heavy nuclei* (in English). C. R., URSS *48,* 555 (1945).

Mitchell, Rasetti, Fink and Pegram, *Some experiments with photoneutrons.* Phys. Rev. *50,* 189 (1936).

Mitchell, *The absorption of neutrons detected by boron and lithium.* Phys. Rev. *49,* 453 (1936).

Mitchell and Powers, *Bragg reflections of slow neutrons.* Phys. Rev. *50,* 486 (1936).

Moon and Tillman, *Neutrons of thermal energies.* Proc. Roy. Soc. *153,* 476 (1936).

Mott-Smith and Bonner, *Energy distribution of neutrons from boron.* Phys. Rev. *45,* 554 (1934).

Moyer, Peters and Schmidt, *Absorption of slow neutrons by Cd^{113}.* Phys. Rev. *69,* 666 (1946).

Moyer, Peters and Schmidt, *Isotopic absorption of slow neutrons in cadmium.* Phys. Rev. *70,* 446 (1946).

Muehlhause, *A method for measuring total cross sections for low energy neutrons.* Phys. Rev. *63,* 60 (1943).

Muehlhause and Goldhaber, *Slowing down of R neutrons into C neutrons in lead.* Phys. Rev. *66,* 36 (1944).

Murphy, Bright, Whitaker, Korff and Clarke, *Relative efficiencies of radioactive neutron sources.* J. Franklin Inst. *231,* 357 (1941).

O'Neal and Scharff-Goldhaber, *Determination of absolute neutron intensities.* Phys. Rev. *69,* 368 (1946).

O'Neal, *The slowing down of low energy neutrons in water. II. Determination of photo-neutron energies.* Phys. Rev. *70,* 1 (1946).

Oppenheimer and Volkoff, *On massive neutron cores.* Phys. Rev. *55,* 374 (1939).

Ornstein and Uhlenbeck, *Motion of neutrons through paraffin* (in English). Physica *4,* 478 (1937).

Perrin and Elsasser, *Theory of the selective capture of slow neutrons by certain nuclei.* J. phys. Rad. *6,* 194 (1935).

Placzek, *Diffusion of thermal neutrons.* Phys. Rev. *60,* 166 (1941).

Placzek, *On the theory of the slowing down of neutrons in heavy substances.* Phys. Rev. *69,* 423 (1946).

Pollard, Schultz and Brubaker. *Emission of neutrons from Cl and A under alpha particle bombardment.* Phys. Rev. *51,* 140 (1937).

Pollard, Schultz and Brubaker, *Emission of neutrons from argon, chlorine, aluminum and some heavier elements under alpha particle bombardment.* Phys. Rev. *53,* 351 (1938).

Pontecorvo, *Diffusion of neutrons of uniform speed through protons.* C. R., Paris *206,* 1003 (1938).

Pontecorvo and Wick, *Neutron diffusion. Parts I and II.* Ric. Sci. *1,* 134 (1946).

Pool, Cork and Thornton, *Evidence for the simultaneous ejection of three neutrons from elements bombarded with fast neutrons.* Phys. Rev. *52,* 41 (1937).

Pool, *Fast neutrons.* Phys. Rev. *53,* 707 (1938).

Pose, *Spontaneous neutron emission of U and Th.* Z. Phys. *121,* 293 (1943).

Powers, Fink and Pegram, *The absorption of neutrons slowed down by paraffin at different temperatures.* Phys. Rev. *49,* 650 (1936).

Rabi, *The effective neutron collision radius.* Phys. Rev. *43,* 838 (1933).

Raether, *On the application of neutron diffusion through boundary surfaces to the measurement of collision cross sections.* Z. Naturforsch. *1,* 367 (1946).

Rainwater and Havens, *Neutron beam spectrometer studies of boron, cadmium, and the energy distribution from paraffin.* Phys. Rev. *70,* 107 (1946).

Rainwater and Havens, *Neutron beam spectrometer studies of boron, cadmium and the energy distribution from paraffin.* Phys. Rev. *70,* 136 (1946).

Rainwater, Havens, Wu and Dunning, *Slow neutron velocity spectrometer studies. I. Cd, Ag, Sb, Ir, Mn.* Phys. Rev. *71,* 65 (1947).

Rasetti, *Excitation of neutrons from Be.* Z. Phys. *78,* 165 (1932).

Rasetti, Mitchell, Fink and Pegram, *On the absorption of slow neutrons in boron.* Phys. Rev. *49,* 777 (1936).

Rasetti, Segrè, Fink, Dunning and Pegram, *On the absorption law for slow neutrons.* Phys. Rev. *49,* 104 (1936).

Richards, *A photographic plate spectrum of the neutrons from the disintegration of lithium by deuterons.* Phys. Rev. *59,* 796 (1941).

Richards, *Angular distribution of the d-d neutrons.* Phys. Rev. *60,* 167 (1941).

Richards, Speck and Perlman, *Neutron spectra.* Phys. Rev. *70,* 118 (1946).

Riezler, *The effective cross section of krypton and xenon to slow neutrons.* Ann. Phys., Leipzig *41,* 193 (1942).

Riezler, *Absorption of C neutrons in rare earths.* Ann. Phys., Leipzig *41,* 476 (1942).

Ruben and Libby, *Width of iodine resonance neutron band.* Phys. Rev. *51,* 774 (1937).

Russinow, *Passage of fast neutrons through beryllium* (in English). Phys. Z. Sowjet, *10,* 219 (1936).

Sagane, *Minimum neutron energy to produce neutron loss process and its application to the measurement of Q values.* Phys. Rev. *53,* 492 (1938).

Salant, Horvath and Zagor, *Absorption of slow neutrons of different energies by boron, cobalt and manganese.* Phys. Rev. *55,* 111 (1939).

Sawyer, Wollan, Peterson and Bernstein, *Application of a bent crystal neutron spectrometer to measurements of reso-*

nance absorption. Phys. Rev. *70*, 791 (1946).

Schiff, *On the capture of thermal neutrons by deuterons.* Phys. Rev. *52*, 242 (1937).

Schultz and Goldhaber, *Capture cross section of hydrogen for slow neutrons.* Phys. Rev. *67*, 202 (1945).

Seren, Friedlander and Turkel, *Thermal neutron activation cross sections.* Phys. Rev. *71*, 463 (1947).

Sherr, *Collision cross sections for 25 Mev neutrons.* Phys. Rev. *68*, 240 (1945).

Sinma and Yamasaki, *Capture cross section for slow neutrons.* Phys. Rev. *59*, 402 (1941).

Sinma and Yamasaki, *Capture cross sections for slow neutrons.* Tokyo Inst. Chem. Res. *38*, 167 (1941).

Snell, Levinger, Wilkinson, Meiners and Sampson, *Chemical isolation of two of the delayed neutron activities: resolution of delayed neutron periods.* Phys. Rev. *70*, 111 (1946).

Snyder, Keuffel, Gilvarry and Way, *The temperature dependence of slow neutron resonance absorption in silver.* Phys. Rev. *61*, 390 (1942).

Staub and Stephens, *Neutrons from the breakup of He5.* Phys. Rev. *55*, 845 (1939).

Stephens, Djanab and Bonner, *Neutrons from the disintegration of nitrogen by deuterons.* Phys. Rev. *52*, 1079 (1937).

Stephens, *Neutrons from Li + deuterons.* Phys. Rev. *53*, 223 (1938).

Stetter and Jentschke, *Mass of the neutron.* Z. Phys. *110*, 214 (1938).

Stupp, *The filtering action and resonance absorption of thermal neutrons.* Ann. Phys., Leipzig (5) *43*, 630 (1943).

Sturm and Turkel, *Studies of neutron resonances with a crystal spectrometer.* Phys. Rev. *70*, 103 (1946).

Sturm and Arnold, *The total neutron cross section of dysprosium and neodymium.* Phys. Rev. *71*, 556 (1947).

Sundaracher, *The beta activity of the neutron.* Nature *157*, 268 (1946).

Sutton, McDaniel, Anderson and Lavatelli, *The capture cross section of boron for neutrons of energies from* 0.01 ev to 1000 ev. Phys. Rev. *71*, 272 (1947).

Szilard, *Absorption of residual neutrons.* Nature *136*, 950 (1935).

Tolman, *Static solutions of Einstein's field equations for spheres of fluid.* Phys. Rev. *55*, 364 (1939).

Torrey, *Analysis of neutron absorption in boron.* Phys. Rev. *53*, 266 (1938).

Turner, *Secondary neutrons from uranium.* Phys. Rev. *57*, 334 (1940).

Uehling, *The density distribution in the steady state diffusion of neutrons.* Phys. Rev. *59*, 137 (1941).

Van Vleck, *On the cross section of heavy nuclei for slow neutrons.* Phys. Rev. *48*, 367 (1935).

Verde, *Distribution of neutrons from a point source of fast neutrons, in an infinite hydrogeneous medium. I. Resonance neutrons.* Nuovo Cim (9) *3*, 116 (1946).

Volkoff, *On the equilibrium of massive spheres.* Phys. Rev. *55*, 413 (1939).

Volz, *Effective cross sections for the absorption of slow neutrons.* Z. Phys. *121*, 201 (1943).

Wang, *Neutron and negative proton.* Nature *157*, 549 (1946).

Wattenberg, *Photo-neutron sources and the energy of photo-neutrons.* Phys. Rev. *71*, 497 (1947).

Weekes, Livingston and Bethe, *A method for the determination of the selective absorption regions of slow neutrons.* Phys. Rev. *49*, 471 (1936).

Westcott and Bjerge, *Slowing down of neutrons by collisions with hydrogen nuclei.* Proc. Camb. Phil. Soc. *31*, 145 (1935).

Whitaker and H. G. Beyer, *Transmission of slow neutrons through crystals.* Phys. Rev. *55*, 1101 (1939).

Whitaker, Bright and Murphy, *Transmission of neutrons of different energies through quartz crystals.* Phys. Rev. *57*, 551 (1940).

Wick, *On the slowing down of neutrons.* Phys. Rev. *49*, 192 (1936).

Wick, *An application of the Fokker-Planck equation to the energy spectrum of thermal neutrons.* Phys. Rev. *70*, 103 (1946).

Wiedenbeck, *Neutron yields from the photo- and electro-disintegration of Be.* Phys. Rev. *69,* 235 (1946).

Wilson, *Delayed neutrons from Pu²³⁹.* Phys. Rev. *71,* 560 (1947).

Yalow and Goldhaber, *Wide and sharp neutron groups.* Phys. Rev. *69,* 47 (1946).

Yalow and Goldhaber, *Evidence of wide neutron groups.* Phys. Rev. *68,* 99 (1945).

Yalow, Yalow and Goldhaber, *Upper limit of the number of low energy neutrons from Ra-α-Be source.* Phys. Rev *69,* 253 (1946).

Yost and Dickinson, *The diffusion and absorption of neutrons in paraffin spheres.* Phys. Rev. *50,* 128 (1936).

Zahn, *Absorption coefficients for thermal neutrons.* Phys. Rev. *52,* 67 (1937).

Zimmer, *Ionization measurements on fast neutrons.* Phys. Z. *42,* 360 (1941).

Zinn and Seeley, *Production of neutrons with low voltage.* Phys. Rev. *50,* 1101 (1936).

Zinn, Seeley and Cohen, *Collision cross sections for D-D neutrons.* Phys. Rev. *56,* 260 (1939).

Zinn, *The Bragg reflection of neutrons by a single crystal.* Phys. Rev. *70,* 102 (1946).

Zwicky, *On the theory and observation of highly collapsed stars.* Phys. Rev. *55,* 726 (1939).

CHAPTER NINE

Theory of Nuclear Structure

Aten, *Nuclear packing effect* (in German). Physica *6,* 425 (1939).

Atkinson, *Evidence against He⁵.* Phys. Rev. *48,* 382 (1935).

Auluck and Kothari, *Distribution of energy levels for the liquid-drop nuclear model.* Nature *157,* 662 (1946).

Bagge, *Theory of heavy nuclei. Part I. Neutron excess in heavy nuclei.* Ann. Phys. Leipzig *33,* 359 (1938).

Bagge, *Theory of heavy nuclei. Part II. Decrease of Gamow barrier owing to excitation.* Ann. Phys. Leipzig *33,* 389 (1938).

Bagge, *Range of nuclear forces and disintegration by cosmic rays.* Ann. Phys. Leipzig *35,* 118 (1939).

Bagge, *Mass defect of the atomic nucleus and the multi-body problem.* Z. Naturforsch. *1,* 361 (1946).

Bardeen, *On the density of energy levels of heavy nuclei.* Phys. Rev. *51,* 799 (1937).

Bardeen and Feenberg, *Symmetry effects in the spacing of nuclear energy levels.* Phys. Rev. *54,* 809 (1938).

Barkas, *Interpretive bearing on the metrical field associated with heavy particles.* Phys. Rev. *52,* 1074 (1937).

Barkas, *The analysis of nuclear binding energies.* Phys. Rev. *55,* 691 (1939).

Bartlett, *Structure of atomic nuclei.* Phys. Rev. *41,* 370 (1932).

Bartlett, *Structure of atomic nuclei II.* Phys. Rev. *42,* 145 (1932).

Bartlett, *Negative protons in the nucleus?* Phys. Rev. *46,* 435 (1934).

Bartlett, *Exchange forces and the structure of the nucleus.* Phys. Rev. *49,* 102 (1936).

Beck, *Structure of proton and neutron.* C. R., Paris *208,* 332 (1939).

Beck and Tsien, *Nuclear levels of the compound Li.* Phys. Rev. *61,* 379 (1942).

Beck, *Remarks on the fine structure of "positronium".* Phys. Rev. *69,* 532 (1946).

Belinfante, *Under equation of the meson field* (in English). Physica *6,* 870 (1939).

Belinfante, *Spin of mesons* (in English). Physica *6,* 887 (1939).

Berenda, *Notes on the Wheeler-Feynman theory.* Phys. Rev. *71,* 850 (1947).

Bethe and Peierls, *Quantum theory of the diplon.* Proc. Roy. Soc. *148,* 146 (1935).

Bethe, *An attempt to calculate the number of energy levels of a heavy nucleus.* Phys. Rev. *50,* 332 (1936).

Bethe and Bacher, *Nuclear Physics. Part*

I. Stationary states of nuclei. Rev. Mod. Phys. *8*, 82 (1936).

Bethe and Rose, *Kinetic energy of the nuclei in the Hartree model.* Phys. Rev. *51*, 283 (1937).

Bethe and Placzek, *Resonance effects in nuclear processes.* Phys. Rev. *51*, 450 (1937).

Bethe, *The binding energy of the deuteron.* Phys. Rev. *53*, 313 (1938).

Bethe, *Coulomb energy of light nuclei.* Phys. Rev. *54*, 436 (1938).

Bethe, *The meson theory of nuclear forces.* Phys. Rev. *55*, 1261 (1939).

Bethe, *Meson theory of nuclear forces. I. General theory.* Phys. Rev. *57*, 260 (1940).

Bethe, *Meson theory of nuclear forces. Part II. Theory of the deuteron.* Phys. Rev. *57*, 390 (1940).

Bethe, *A new treatment of the compound nucleus.* Phys. Rev. *57*, 567 (1940).

Bethe, *Continuum theory of the compound nucleus.* Phys. Rev. *57*, 1125 (1940).

Bhabha, *Theory of heavy electrons and nuclear forces.* Proc. Roy. Soc. *166*, 501 (1938).

Bhabha, *Classical theory of mesons.* Proc. Roy. Soc. *172*, 384 (1939).

Bhabha, *Classical theory of spinning particles.* Proc. Indian Acad. Sci. *11A*, 247 (1940); Erratum *11A*, 467 (1940).

Bhabha, *Elementary heavy particles with any integral charge.* Proc. Indian Acad. Sci. *11A*, 347 (1940); Erratum *11A*, 468 (1940).

Bhabha, *Protons of double charge and the scattering of mesons.* Phys. Rev. *59*, 100 (1941).

Bhabha, *General classical theory of spinning particles in a meson field.* Proc. Roy. Soc. *178*, 314 (1941).

Bhabha and Basu, *The theory of particles of spin half and the Compton effect.* Proc. Indian. Acad. Sci. *15A*, 105 (1942); Erratum *15A*, 461 (1942).

Bhabha, *Relativistic wave equations for the proton.* Proc. Indian Acad. Sci. *21A*, 241 (1945).

Bhabha, *Relativistic wave equations for the elementary particles.* Rev. Mod. Phys. *17*, 200 (1945).

Bhabha and Harish-Chandra, *On the fields and equations of motion of point particles.* Proc. Roy. Soc. *185*, 250 (1946).

Blatt, *On the meson charge cloud around a proton.* Phys. Rev. *67*, 205 (1945).

Blatt, *On the heavy electron pair in the strong coupling limit.* Phys. Rev. *68*, 287 (1945).

Blatt, *Charge distribution around protons.* Phys. Rev. *65*, 352 (1944).

Bleuler, *A problem on the theory of the deuteron* (in German). Helv. Phys. Acta *17*, 405 (1944).

Bleuler, *A contribution to the two nucleon problem* (in German). Helv. Phys. Acta *18*, 317 (1945).

Bloch, *The principles of nuclear physics.* Rev. Gén Elect. *55*, 31 (1945).

Bohr, *Neutron capture and nuclear constitution.* Nature *137*, 344 (1936).

Bohr, *Conservation law in quantum theory.* Nature *138*, 25 (1936).

Bohr, *Quantum theory and the atomic nucleus.* Ann. Phys., Leipzig *32*, 5 (1938).

Bopp, *A linear theory of the electron.* Ann. Phys., Leipzig *38*, 345 (1940).

Born, *Application of "reciprocity" to nuclei.* Proc. Roy. Soc. *166*, 552 (1938).

Breit, *Nuclear stability and isotope shift.* Phys. Rev. *46*, 319 (1934).

Breit and Wigner, *Note on Majorana's exchange energy.* Phys. Rev. *48*, 918 (1935).

Breit and Feenberg, *The possibility of the same form of specific interaction for all nuclear particles.* Phys. Rev. *50*, 850 (1936).

Breit, *Approximately relativistic equations for nuclear particles.* Phys. Rev. *51*, 248 (1937). Addendum, *Proof of approximate invariance.* Phys. Rev. *51*, 778 (1937).

Breit and Stehn, *On the comparison of proton-proton and proton-neutron interactions.* Phys. Rev. *52*, 396 (1937).

Breit, *Approximately relativistic equations.* Phys. Rev. *53*, 153 (1938).

Breit and Wigner, *The saturation requirements for nuclear forces.* Phys. Rev. *53*, 998 (1938).

Breit, Hoisington, Share and Thaxton,

The approximate equality of the proton-proton and proton-neutron interactions for the meson potential. Phys. Rev. *55*, 1103 (1939).

Brillouin, *The undulatory mechanics of Schrödinger.* C. R., Paris *183*, 24 (1926).

de Broglie, *On the electrinos of Thibaud and the possible existence of a very small electric charge of neutrons.* C. R., Paris *224*, 615 (1947).

Brown and Inglis, *Coulomb energies and nuclear models.* Phys. Rev. *55*, 1182 (1939).

Brown, *The binding energy of H^3.* Phys. Rev. *56*, 1107 (1939).

Brulin and Hjalmars, *On the singlet and triplet state of the deuteron in the meson pair theory* (in English). Ark. Mat. Astr. Fys. *32A*, No. 7 (1945).

Brulin and Hjalmars, *On a modification of Klein's vector meson pair theory* (in English). Ark. Mat. Astr. Fys. *33B*, No. 4 (1946).

Bruno, *The Compton effect for the meson* (in English). Ark. Mat. Astr. Fys. *28B*, No. 5 (1942).

Bunge, *A new representation of types of nuclear forces.* Phys. Rev. *65*, 249 (1944).

Bunge, *A new representation of the types of nuclear forces.* Publ. Fac. Cienc. Fis.-Mat. La Plata *3*, 221 (1944).

Caldirola, *The meson field equations in five dimensional space.* Nuovo Cim. *19*, 25 (1942).

Carlson and Oppenheimer, *The impacts of fast electrons and magnetic neutrons.* Phys. Rev. *41*, 763 (1932).

Cassen and Condon. *On nuclear forces.* Phys. Rev. *50*, 847 (1936).

Cernuschi, *The neutron.* J. Phys. Rad. *8*, 273 (1937).

Chang, *The regularities in energy levels of light nuclei.* Phys. Rev. *65*, 352 (1944).

Cherdynzev, *On the frequency of even and odd atomic nuclei* (in English). C. R., URSS *33*, 22 (1941).

Coester, *Stability of heavy nuclei in the meson theory* (in German). Helv. Phys. Acta *17*, 35 (1944).

Condon, *Theory of nuclear structure.* J. Franklin Inst. *227*, 801 (1939).

Condon, *Note on electron-neutron interaction.* Phys. Rev. *49*, 459 (1936).

Cooper and Nelson, *Ground states of Be^{10} and C^{10}.* Phys. Rev. *58*, 1117 (1940).

Critchfield and Teller, *On the saturation of nuclear forces.* Phys. Rev. *53*, 812 (1938).

Critchfield, *Spin dependence in the electron-positron theory of nuclear forces.* Phys. Rev. *56*, 540 (1939).

Critchfield, *Excited states of nuclear particles in the meson pair theory.* Phys. Rev. *59*, 48 (1941).

Critchfield, *The electron waves in the magnetic dipole field of a neutron.* Phys. Rev. *70*, 793 (1946).

Critchfield, *Electron waves in the magnetic dipole field of a neutron.* Phys. Rev. *71*, 258 (1947).

Dancoff, *Virtual state of He^5 and meson forces.* Phys. Rev. *56*, 384 (1939).

Dancoff and Pauli, *Strong coupling mesotron theory of nuclear interactions.* Phys. Rev. *61*, 387 (1942).

Dancoff and Serber, *Nuclear forces in the strong coupling theory.* Phys. Rev. *61*, 395 (1942).

Dennison, *Specific heat of the hydrogen molecule.* Proc. Roy. Soc. *115*, 483 (1927).

Destouches, *Heavy electron.* C. R., Paris *206*, 1095 (1938).

Dolch, *Theory of the lightest nuclei.* Z. Phys. *100*, 401 (1936).

Duckworth, *New packing fractions and the packing fraction curve.* Phys. Rev. *62*, 19 (1942).

Eastman, *Energy and stability related to composition of atomic nuclei.* Phys. Rev. *46*, 1 (1946).

Eastman, *Heights of nuclear potential barriers.* Phys. Rev. *46*, 744 (1934).

Eckart, *A comparison of the nuclear theories of Heisenberg and Wigner. I.* Phys. Rev. *44*, 109 (1933).

Ehrenfest and Oppenheimer, *Note on the statistics of nuclei.* Phys. Rev. *37*, 333 (1931).

Eisenbud and Wigner, *Interaction between nuclear particles.* Proc. Nat. Acad. Sci. *27*, 281 (1941).

Elsasser, *Application of Pauli's principle to nuclei.* J. Phys. Rad. *4*, 549 (1933).

Elsasser, *Binding energies in the radioactive families U-Ra and Th.* C. R., Paris *199*, 46 (1934).

Elsasser, *Constitution of elementary particles and nuclear forces.* C. R., Paris *199*, 1213 (1934).

Elsasser, *Application of Pauli's principle to nuclei.* J. Phys. Rad. *5*, 389 (1934).

Elsasser, *Application of Pauli's principle to nuclei. Part III.* J. Phys. Rad. *5*, 635 (1934).

Eriksson, *Vector - pseudovector meson theory* (in English). Ark. Mat. Astr. Fys. *29A*, No. 10 (1943).

Eriksson, *On a generalization of Dirac's equations* (in English). Ark. Mat. Astr. Fys. *33B*, No. 6 (1946).

Euler, *Interaction of nuclear forces in heavy nuclei.* Z. Phys. *105*, 553 (1937).

Feenberg, *Neutron - proton interaction. Part I. The binding energies of the hydrogen and helium isotopes.* Phys. Rev. *47*, 850 (1935).

Feenberg and Knipp, *Intranuclear forces.* Phys. Rev. *48*, 906 (1935).

Feenberg and Share, *The approximate solution of the nuclear 3 and 4 particle eigenvalue problems.* Phys. Rev. *50*, 253 (1936).

Feenberg and Wigner, *On the structure of the nuclei between helium and oxygen.* Phys. Rev. *51*, 95 (1937).

Feenberg and Phillips, *On the structure of light nuclei.* Phys. Rev. *51*, 597 (1937).

Feenberg, *Further remarks on the saturation property of nuclear forces.* Phys. Rev. *52*, 667 (1937).

Feenberg, *A note on the Thomas-Fermi statistical method.* Phys. Rev. *52*, 758 (1937).

Feenberg, *On the shape and stability of heavy nuclei.* Phys. Rev. *55*, 504 (1939).

Feenberg, *A note on the density and compressibility of nuclear matter.* Phys. Rev. *59*, 149 (1941).

Feenberg, *Non-uniform particle density in nuclear structure.* Phys. Rev. *59*, 593 (1941).

Feenberg, *Theory of nuclear surface energy.* Phys. Rev. *60*, 204 (1941).

Feenberg, *Theory of nuclear coulomb energy.* Phys. Rev. *61*, 387 (1942).

Feenberg and Goertzel, *Theory of nuclear coulomb energy.* Phys. Rev. *70*, 597 (1946).

Feenberg and Primakoff, *Possibility of "conditional" saturation in nuclei.* Phys. Rev. *70*, 980 (1946).

Feenberg, *The nuclear energy surface.* Phys. Rev. *71*, 468 (1947).

Feinberg, *Angular distribution for the interaction of mesons with nuclei.* J. Phys., USSR *5*, 177 (1941).

Fierz, *Consequences of the assumption that heavy elementary particles may have a negative charge* (in German). Helv. Phys. Atta *14*, 105 (1941).

Fierz, *Meson theory of nuclear forces* (in German). Helv. Phys. Acta *15*, 329 (1942).

Fierz, *On the interaction between two nucleons in the meson theory* (in German). Helv. Phys. Acta *17*, 181 (1944).

Fierz and Wentzel, *On the deuteron problem. I* (in German). Helv. Phys. Acta *17*, 215 (1944).

Fierz, *On spin path coupling of two nucleons in meson theory* (in German). Helv. Phys. Acta *18*, 158 (1945).

Fisk, Schiff and Schockley, *On the binding of neutrons and protons.* Phys. Rev. *50*, 1090 (1936); Erratum, *50*, 1191 (1936).

Flint, *Survey of nuclear field theories.* Proc. Phys. Soc., Lond. *56*, 149 (1944).

Flint, *Quantum equations and nuclear field theories.* Phil. Mag. *36*, 635 (1945).

Flint, *A study of the nature of the field theories of the electron and positron and of the meson.* Proc. Roy. Soc. *185*, 14 (1946).

Flügge, *Structure of the light atomic nucleus.* Z. Phys. *96*, 459 (1935).

Flügge, *Mass defects of light nuclei from recent theories of nuclear forces.* Z. Phys. *105*, 522 (1937).

Flügge, *A note on the theory of the meson field.* Ann. Phys., Leipzig (5) *43*, 573 (1943).

Foldy, *On the meson theory of nuclear forces.* Phys. Rev. *71*, 276 (1947).

Fox, Creutz, White and Delsasso, *The difference in coulomb energy of light,* isobaric nuclei. Phys. Rev. *55*, 1106 (1939).

Frank, *Note on the Hartree and Hartree-Fock methods.* Phys. Rev. *51*, 577 (1937).

Fremberg, *Some applications of the Riesz potential to the theory of the electromagnetic field and the meson field.* Proc. Roy. Soc. *188*, 18 (1946).

Fröhlich, Heitler and Kahn, *Deviation from the Coulomb law for a proton.* Phys. Rev. *56*, 961 (1939).

Fröhlich, Heitler and Kahn, *Deviation from Coulomb law for proton.* Proc. Roy. Soc. *171*, 269 (1939).

Fuchs, *Statistical method in nuclear theory.* Proc. Roy. Soc. *174*, 509 (1940).

Furry, *Note on the theory of the neutral particle.* Phys. Rev. *54*, 56 (1938).

Galanin, *Movement of the meson in a homogeneous magnetic field.* J. Phys., USSR *6*, 27 (1942).

Gamow, *Quantum theory of the atomic nucleus.* Z. Phys. *51*, 204 (1928).

Gamow, *Stability of nuclei.* Int. Conf. Phys. London *1*, 60 (1934).

Gamow, *Isomeric Nuclei.* Nature *133*, 833 (1934).

Gamow, *Negative protons and nuclear structure.* Phys. Rev. *45*, 728 (1934).

Gamow, *Probability of nuclear meson absorption.* Phys. Rev. *71*, 550 (1947).

Gerjuoy and Schwinger, *The theory of light nuclei.* Phys. Rev. *60*, 158 (1941).

Gerjuoy and Schwinger, *On tensor forces and the theory of light nuclei.* Phys. Rev. *61*, 138 (1942).

Ginsburg, *Theory of interaction of mesons with the electromagnetic field.* J. Phys., USSR *5*, 47 (1941).

Ginsburg, *On the pseudoscalar theory of meson.* J. Phys., USSR *6*, 180 (1942).

Ginsburg, *On the theory of the particle of spin 3/2.* J. Phys., USSR *7*, 115 (1943).

Ginsburg, *On the theory of excited spin states of elementary particles.* J. Phys., USSR *8*, 33 (1944).

Godnew, *On the determination of symmetry of nuclear spin functions.* C. R., URSS *43*, 194 (1944).

Goeppert-Mayer, *Rare earth and trans-uranic elements.* Phys. Rev. *60*, 184 (1941).

Goudsmit, *Density of excited levels in heavy nuclei.* Phys. Rev. *51*, 64 (1937).

Graves, *Packing fraction differences among heavy elements.* Phys. Rev. *55*, 863 (1939).

Grönblom, *Variation of light atomic nuclei from the Hartree model* (in German). Acta Soc. Sci. Fennicae 2, No. 9 (1937).

Grönblom, *Deviation of the nuclei of $_2He^4$ and $_8O^{16}$ from the Hartree model.* Z. Phys. *110*, 37 (1938).

Grönblom and Marshak, *The binding energy of 4n nuclei on the alpha particle model.* Phys. Rev. *55*, 229 (1939).

Guggenheimer, *Constitution of atomic nuclei. Part I.* J. Phys. Rad. *5*, 253 (1934).

Guggenheimer, *Constitution of atomic nuclei. Part II.* J. Phys. Rad. *5*, 475 (1934).

Guggenheimer, *Series in nuclear energy levels.* Nature *145*, 104 (1940).

Guggenheimer, *On nuclear energy levels.* Proc. Roy. Soc. *181*, 169 (1942).

Gupta, *Electromagnetic field and the self-energy of the meson.* Proc. Nat. Inst. Sci. India *9*, 173 (1943).

Gupta, *Polarization of the vacuum in the meson theory.* Proc. Nat. Inst. Sci. India *9*, 337 (1943).

Gurevitsch, *Energy levels of a heavy nucleus.* Phys. Z. Sowjet. *12*, 489 (1937).

Gustafson, *Elimination of divergencies in quantum electrodynamics and in meson theory.* Nature *157*, 734 (1946).

Hafstad and Teller, *The alpha particle model of the nucleus.* Phys. Rev. *54*, 681 (1938).

Harish-Chandra, *On the equations of motion of point particles.* Proc. Roy. Soc. *185*, 269 (1946).

Harish-Chandra, *The correspondence between the particle and the wave aspects of the meson and the photon.* Proc. Roy. Soc. *186*, 502 (1946).

Harkin, *The nuclear exclusion principle and the neutron-proton pattern.* Phys. Rev. *52*, 39 (1937).

Hartree, *Wave mechanics of an atom with a non-Coulomb central field. Part*

I. Theory and Methods. Part II. Some results and discussion. Proc. Camb. Phil. Soc. *24*, 89, 111 (1928).

Hasse, *Polarizability of the helium atom and the lithium ion.* Proc. Camb. Phil. Soc. *26*, 542 (1930).

Heisenberg, *Structure of atomic nuclei.* Z. Phys. *77*, 1 (1932).

Heisenberg, *Structure of atomic nuclei. Part II.* Z. Phys. *78*, 156 (1932).

Heisenberg, *Structure of atomic nuclei.* Z. Phys. *80*, 587 (1932).

Heisenberg, *Structure of the light atomic nuclei.* Z. Phys. *96*, 473 (1935).

Heisenberg, *Limits of applicability of the present quantum theory.* Z. Phys. *110*, 251 (1938).

Heitler and Herzberg, *Do nitrogen nuclei obey Bose statistics?* Naturwiss. *17*, 673 (1929).

Heitler and Ma, *Inner excited states of proton and neutron.* Proc. Roy Soc. *176*, 368 (1940).

Heitler, *Physical concepts of the meson theory of the atomic nucleus.* Nature *148*, 680 (1941).

Heitler, *On the particle equation of the meson.* Proc. R. Irish. Acad. *49A*, 1 (1943).

Heitler and Peng, *On the production of mesons by proton-proton collisions.* Proc. R. Irish Acad. *49A*, 101 (1943).

Heitler, *Remarks on the self energy problem.* Proc. R. Irish Acad. *50A*, 1, (1944).

Heitler, *A theorem in the charge-symmetrical meson theory.* Proc. R. Irish Acad. *51A*, 33 (1946).

Hepner and Peierls, *Non-central forces in the nuclear two body problem.* Proc. Roy. Soc. *181*, 43 (1942).

Hönl, *Theory of charged elementary particles.* Ann. Phys., Leipzig *28*, 721 (1937).

Holmberg, *On nuclear forces described by a pseudo-vector-scalar neutral meson field* (in English). K. Fys. Sällsk, Lund Förh. *14*, 270 (1944).

Houriet, *Nuclear forces in the theory of pairs* (in French). Helv. Phys. Acta *16*, 529 (1943).

Houriet, *Structure of the nucleon according to the meson theory with strong coupling* (in French). Helv. Phys. Acta *18*, 473 (1945).

Houston, *A nuclear model.* Phys. Rev. *47*, 942 (1935).

Hoyt, *On the penetrability of a simple type of potential barrier.* Phys. Rev. *53*, 673 (1938).

Hu, *The relativistic correction in the meson theory of nuclear force.* Phys. Rev. *71*, 339 (1945).

Hulthén, *On the virtual state of the deuteron.* Phys. Rev. *61*, 671 (1942).

Hulthén, *Nuclear forces in a non-symmetrical scalar pseudoscalar meson field theory* (in English). K. Fys. Sällsk Lund. Förh. *14*, 18 (1944).

Hulthén, *Comments on the difficulties of the meson theory.* Rev. Mod. Phys. *17*, 263 (1945).

Hund, *Theoretical investigations of nuclear forces.* Phys. Z. *38*, 929 (1937).

Hund, *Symmetry properties of the forces in atomic nuclei.* Z. Phys. *105*, 202 (1937).

Hylleraas, *Binding forces between elementary nuclear particles.* Z. Phys. *107*, 258 (1937).

Hylleraas and Ore, *Stability of polyelectrons.* Phys. Rev. *71*, 468 (1947).

Hylleraas, *Electron affinity of positronium.* Phys. Rev. *71*, 491 (1947).

Hylleraas and Ore, *Binding energy of the positronium molecule.* Phys. Rev. *71*, 493 (1947).

Inglis, *Spin-orbit coupling in nuclei.* Phys. Rev. *50*, 783 (1936).

Inglis and Young, *Stable isobars.* Phys. Rev. *51*, 525 (1937).

Inglis, *Perturbation theory of light nuclei: He^4 and Li^6.* Phys. Rev. *51*, 531 (1937).

Inglis, *Angle dependence and range of nuclear forces.* Phys. Rev. *55*, 988 (1939).

Inglis, *Spin-orbit coupling in the alpha model of light nuclei.* Phys. Rev. *56*, 1175 (1939).

Inglis, *The alpha model of nuclear structure and nuclear moments.* Phys. Rev. *60*, 837 (1941).

Iskraut, *Observations on the energy impulse tensor of the field theories of matter.* Z. Phys. *119*, 659 (1942).

Iwanenko, *Constitution of atomic nuclei.* C. R., Paris *195*, 439 (1932).

Iwanenko, *The neutron hypothesis.* Nature *129*, 798 (1932).

Iwanenko and Rodichev, *Proca potential* (in Russian). J. Exp. Theor. Phys. USSR *9*, 526 (1939).

Iwanenko and Sokolow, *The dipole character of the meson and the polarization of vacuum.* J. Phys., USSR *6*, 175 (1942).

Iwanenko, *Further remarks on the difficulties of the meson theory.* Phys. Rev. *66*, 157 (1944).

Jánossy, *Exchange forces between three heavy particles due to the meson exchange field.* Proc. Camb. Phil. Soc. *35*, 616 (1939).

Jauch, *Nuclear forces in the electron pair theory* (in German). Helv. Phys. Acta *15*, 175 (1942).

Jauch and Hu, *On the mixed meson theory of nuclear forces.* Phys. Rev. *65*, 289 (1944).

Johnson and Primakoff, *Relations between the second and higher order processes in the neutrino-electron field theory.* Phys. Rev. *51*, 612 (1937).

Johnson, *Correction for nuclear motion in H_3.* Phys. Rev. *60*, 373 (1941).

Jordan, *Theory of atomic nuclei.* Ergeb. d. exakt. Naturwiss. *16*, 47 (1937).

Jordan, *Nuclear forces.* Naturwiss. *25*, 273 (1937).

Jordan, *Theory of elementary particles.* Z. Phys. *111*, 498 (1939).

Jost, *On the charge dependence of nuclear forces in the vector meson theory without neutral mesons* (in German). Helv. Phys. Acta *19*, 113 (1946).

Kahan, *Theory of the deuteron: proton-neutron interaction.* C. R., Paris *204*, 414 (1937).

Kahan, *Theory of the deuteron.* J. Phys. Rad. *8*, 281 (1937).

Kakinuma, *Coulomb force between electron and proton.* Proc. Phys.-Math. Soc. Jap. (3) *19*, 503 (1937).

Kar, *Proton-proton interaction and Yukawa particle.* Indian J. Phys. *16*, 187 (1942).

Kar and Roy, *A self-consistent method of determining the mass of the mesotron.* Indian J. Phys. *17*, 316 (1943).

Kemmer, *Neutron-proton interaction* (in German). Helv. Phys. Acta *10*, 47 (1937).

Kemmer, *Field theory of nuclear interaction.* Phys. Rev. *52*, 906 (1937).

Kemmer, *Change dependence of nuclear forces.* Proc. Camb. Phil. Soc. *34*, 354 (1938).

Kemmer, *Quantum theory of Einstein-Bose particles and nuclear interaction.* Proc. Roy. Soc. *166*, 127 (1938).

Kemmer, *Particle aspect of meson theory.* Proc. Roy. Soc. *173*, 91 (1939).

Kemmer, *The algebra of meson matrices.* Proc. Camb. Phil. Soc. *39*, 189 (1943).

Kikuchi, *A theory of the low mass neutral mesotron.* Phys. Rev. *69*, 125 (1946).

King and Van Vleck, *Dipole-dipole resonance forces.* Phys. Rev. *55*, 1165 (1939).

Kittel, *The fine structure of nuclear energy levels in the alpha model.* Phys. Rev. *62*, 109 (1942).

Knie, *Nuclear potential of the deuteron.* Publ. Fac. Cienc. Fis. Mat. La Plata No. 136, 329 (1940).

Kothari and Auluck, *Surface tension of nuclear matter and the enumeration of eigenfunctions of an enclosed particle.* Nature *159*, 204 (1947).

Kramers, *Wave mechanics and semi-numerical quantization.* Z, Phys. *39*, 828 (1926).

Kroeger, *The binding energy of O^{16}.* Phys. Rev. *54*, 1048 (1938).

Krogdahl, *The interaction of a proton and a helium atom in its excited states. II.* Astrophys. J. *102*, 64 (1945).

Kusaka, *Quantization of the wave field for particles with spin 3/2.* Phys. Rev. *61*, 205 (1942).

Kusaka, *Difficulties of the meson theories of nuclear forces.* Phys. Rev. *70*, 794 (1946).

Kwal, *On the wave mechanics of elementary particles.* C. R., Paris *218*, 548 (1944).

Kwal, *On the wave mechanics of elementary particles.* C. R., Paris *218*, 613 (1944).

Lamb and Schiff, *On the electromagnetic properties of nuclear systems.* Phys. Rev. *53*, 651 (1938).

Lamb, *Deviation from the Coulomb law for a proton.* Phys. Rev. *56*, 384 (1939).

Lamb, *Deviation from the Coulomb law for a proton.* Phys. Rev. *57*, 458 (1940).

Landau, *Statistical theory of nuclei.* Phys. Z. Sowjet. *11*, 556 (1937).

Landé, *Neutrons in the nucleus.* Phys. Rev. *43*, 620 (1933).

Landé and Thomas, *Finite self energies in radiation theory. Part II.* Phys. Rev. *60*, 514 (1941).

Langer, *Interaction between neutrons and protons.* Phys. Rev. *45*, 137 (1934).

Lattes and Wataghin, *On the abundance of nuclei in the universe.* Phys. Rev. *69*, 237 (1946).

Laue, *Quantum theory of the atomic nucleus.* Z. Phys. *52*, 726 (1929).

Lubanski, *Remark on the theory of mixed meson fields* (in English). Ark. Mat. Astr. Fys. *30B*, No. 7 (1944).

Ma, *Deviation from the Coulomb law for the proton.* Proc. Camb. Phil. Soc. *36*, 441 (1940).

Ma and Yu, *Electromagnetic properties of nuclei in the meson theory.* Phys. Rev. *62*, 118 (1942).

Majorana, *Nuclear theory.* Z. Phys. *82*, 137 (1933).

Majorana, *Symmetrical theory of electrons and protons.* Nuovo Cim. *14*, 171 (1937).

March, *Foundations and applications of the statistical metric.* Z. Phys. *105*, 620 (1937).

March, *Theory of nuclear forces.* Z. Phys. *106*, 532 (1937).

Margenau, *Possible models of an electrostatic neutron.* Phys. Rev. *45*, 559 (1934).

Margenau, *Complex neutron.* Phys. Rev. *46*, 107 (1934).

Margenau, *Nuclear energy levels and the model of a potential hole.* Phys. Rev. *46*, 613 (1934).

Margenau, *Relativity and nuclear forces.* Phys. Rev. *50*, 342 (1936).

Margenau, *Excited states of the alpha particle.* Phys. Rev. *53*, 198 (1938).

Margenau and Tyrrell, *Variational theory of the alpha particle.* Phys. Rev. *54*, 422 (1938).

Margenau and Carroll, *The bind energy of Li^6.* Phys. Rev. *54*, 705 (1938).

Margenau, *Binding energy of He^6 and nuclear forces.* Phys. Rev. *55*, 1173 (1939).

Margenau, *Interaction of alpha particles.* Phys. Rev. *59*, 37 (1941).

Margenau, *Statistics of excited states of nuclei.* Phys. Rev. *59*, 627 (1941).

Marshak, *Heavy electron pair theory of nuclear forces.* Phys. Rev. *57*, 1101 (1940).

Marvin, *Mutual magnetic interaction of electrons.* Phys. Rev. *71*, 102 (1947).

Massey and Mohr, *Interaction of light nuclei. Part II. Binding energies of $_1H^3$ and $_2H^3$.* Proc. Roy. Soc. *152*, 693 (1935).

Massey and Buckingham, *Collisions of neutrons with deuterons and the nature of nuclear forces.* Nature *146*, 776 (1940).

Matricon, *Application of the method of the self-consistent field to atomic nuclei.* C. R., Paris *206*, 651 (1938).

Matricon, *Application of method of self consistent field to atomic nuclei.* C. R., Paris *206*, 1809 (1938).

Møller, *The theory of nuclear forces.* Nature *142*, 290 (1938).

Møller and Rosenfeld, *The electric quadrupole moment of the deuteron and the field theory of nuclear forces.* Nature *144*, 476 (1939).

Møller, *Distance of energy levels of atomic nuclei.* Phys. Z. *40*, 615 (1939).

Motz and Feenberg, *The spacing of energy levels in light nuclei.* Phys. Rev. *54*, 1055 (1938).

Murard, *Behavior of particles in an exterior field: application to the nucleon.* C. R., Paris *221*, 547 (1945).

Murard, *Properties of some types of particles. Application to the nucleon.* C. R., Paris *219*, 577 (1944).

Murard, *On the various types of elementary particles.* C. R., Paris *221*, 607 (1945).

Nagakura, *Reaction between very light nuclei.* Proc. Phys.-Math. Soc. Jap (3) *20*, 977 (1938).

Nakabayasi, *Nuclear forces* (in German). Tohoku Univ. Sci. Rep. *25*, 1141 (1937).

Nakabayasi, *Exchange character of nuclear forces from the beta decay theory* (in German). Tohoku Univ. Sci. Rep. *26*, 261 (1937).

Nagamiya and Noguchi, *A liquid model of atomic nuclei.* Proc. Phys.-Math. Soc. Jap. (3) *24*, 306 (1942).

Oppenheimer and Serber, *The density of nuclear levels.* Phys. Rev. *50*, 391 (1936).

Oppenheimer and Schwinger, *On the interaction of mesotrons and nuclei.* Phys. Rev. *60*, 150 (1941).

Ore, *Binding energy of polyelectrons.* Phys. Rev. *70*, 90 (1946).

Pais, *Lifetime of the neutral meson.* Nature *156*, 715 (1945).

Pais, *On the theory of the electron and of the nucleon.* Phys. Rev. *68*, 227 (1945).

Pais, *On the self energy of mesons* (in English). Physica *12*, 81 (1946).

Pais, *On the theory of elementary particles* (in English). K. Ned. Akad. Wet. Verh. *19*, 1 (1947).

Pauli, *On the question of the theoretical interpretation of the satellites of some spectral lines and the effect of magnetic fields on them.* Naturwiss. *12*, 741 (1924).

Pauli and Dancoff, *The pseudoscalar meson field with strong coupling.* Phys. Rev. *62*, 85 (1942).

Pauli, *On strong coupling and weak coupling theories of the meson field.* Phys. Rev. *63*, 221 (1943).

Pauli and Kusaka, *On the theory of a mixed pseudoscalar and a vector meson field.* Phys. Rev. *63*, 400 (1943).

Pauli, *On applications of the λ-limiting process to the theory of the meson field.* Phys. Rev. *64*, 332 (1943).

Pauli and Hu, *On the strong coupling case for spin-dependent interactions in scalar and vector pair theories.* Rev. Mod. Phys. *17*, 267 (1945).

Peierls, *Theory of nuclear forces.* Nature *145*, 687 (1940).

Peierls, *Fundamental particles.* Nature *158*, 773 (1946).

Peng, *On the divergence difficulty of quantized field theories and the rigorous treatment of radiation reaction.* Proc. Roy. Soc. *186*, 119 (1946).

Perrin, *Neutral particles of intrinsic mass zero.* C. R., Paris *197*, 1625 (1933).

Pétiau, *Wave equation of a corpuscle with two mass states capable of representing the proton-neutron.* C. R., Paris *209*, 194 (1939).

Pétiau, *On the theory of the particle of spin 1.* J. Phys. Rad. *6*, 62 (1945).

Pétiau, *On the interactions between material particles exerted by means of a particle of spin 2 $h/2_\pi$.* J. Phys. Rad. *6*, 115 (1945).

Pétiau, *The wave equations of particles of arbitrary spin. I.* J. Phys. Rad. *7*, 124 (1946).

Pétiau, *On the wave equations of particles of arbitrary spin. II.* J. Phys. Rad. *7*, 181 (1946).

Pétiau, *Matrix theory of the representations of particles of spin $h/2_\pi$.* Rev. Sci., Paris *83*, 67 (1946).

Phillips and Feenberg, *Coulomb exchange energy in light nuclei.* Phys. Rev. *59*, 400 (1941).

Platt, *A suggestion concerning fundamental particles.* Phys. Rev. *70*, 110 (1946).

Plesset, *Note on neutron-proton exchange interaction.* Phys. Rev. *49*, 551 (1936).

Pollard, *Indications of a simple rule relating nuclear resonance levels with atomic number.* Phys. Rev. *45*, 218 (1934).

Pollard, *Evidence for a resonance level in the B^{10} nucleus.* Phys. Rev. *45*, 555 (1934).

Pollard, *Nuclear potential barriers; experiment and theory.* Phys. Rev. *47*, 611 (1935).

Present, *Must neutron-neutron forces exist in the $_1H^3$ nucleus?* Phys. Rev. *48*, 640 (1935).

Present, *Proton-proton forces in anomalous scattering and in nuclear binding.* Phys. Rev. *48*, 919 (1935).

Present, *Proton-proton forces in anomalous scattering and nuclear binding.* Phys. Rev. *49*, 201 (1936).

Present, *Must neutron-neutron forces ex-*

ist in the H³ nucleus? Phys. Rev. *50*, 635 (1936).

Present, *A note on nuclear radii.* Phys. Rev. *60*, 28 (1941).

Preston, *Radii of the nuclei of natural alpha emitters.* Phys. Rev. *69*, 535 (1946).

Primakoff and Holstein, *Many body interactions in atomic and nuclear systems.* Phys. Rev. *55*, 1218 (1939).

Racah, *Symmetry between particles and anti-particles.* Nuovo Cim. *14*, 322 (1937).

Racah, *Theory of complex spectra. I.* Phys. Rev. *61*, 186 (1942).

Radzinski, *Binding energy of N^{16}* (in German). Acta Phys. Polonica 7, 231 (1938).

Rarita and Present, *On the nuclear 2, 3 and 4 body problems.* Phys. Rev. *51*, 788 (1937).

Rarita and Slawsky, *Nuclear two-body variational problem.* Phys. Rev. *54*, 1053 (1938).

Rarita and Schwinger, *On the neutron-proton interaction.* Phys. Rev. *59*, 436 (1941).

Rarita and Schwinger, *On the exchange properties of the neutron-proton interaction.* Phys. Rev. *59*, 556 (1941).

Rarita and Schwinger, *On a theory of particles with half-integral spin.* Phys. Rev. *60*, 61 (1941).

Renner, *Representation of elementary particles by models.* Naturwiss. *26*, 736 (1938).

Rogers and Rogers, *An independent determination of the binding energy of the deuteron.* Phys. Rev. *55*, 263 (1939).

Rosenfeld, *Meson theories in five dimensions* (in English). Proc. K. Akad. Amsterdam *45*, 155 (1942).

Rutherford, *Structure of atomic nuclei.* Proc. Roy. Soc. *123*, 373 (1929).

Sachs and Goeppert-Mayer, *Calculations on a new neutron-proton interaction potential.* Phys. Rev. *53*, 991 (1938).

Saha, Sirkar and Mukherji, *Structure of atomic nuclei.* Proc. Nat. Inst. Sci. India *6*, 45 (1940).

Scherrer, *Interaction between elementary particles* (in German). Helv. Phys. Acta *14*, 81, 130 (1941).

Schiff, *Excited state of He³.* Phys. Rev. *54*, 92 (1938).

Schiff, Snyder and Weinberg, *On the existence of stationary states of the mesotron field.* Phys. Rev. *57*, 315 (1940).

Schiff, *Field theories for charged particles of arbitrary spin.* Phys. Rev. *57*, 903 (1940).

Schoenberg, *On the theory of integer spin mesons.* Phys. Rev. *60*, 468 (1941).

Schoenberg, *The electron's self energy.* Phys. Rev. *67*, 193 (1945).

Schrödinger, *Systematics of meson matrices.* Proc. R. Irish Acad. *49A*, 29 (1943).

Schüler, *On the deviation from spherical symmetry in several atomic nuclei.* Phys. Z. *36*, 812 (1935).

Schwinger, *On the charged scalar mesotron field.* Phys. Rev. *60*, 159 (1941).

Schwinger, *On a field theory of nuclear forces.* Phys. Rev. *61*, 387 (1942).

Sen, *Nuclear structure of light atoms.* Indian J. Phys. *11*, 427 (1938).

Serber and Dancoff, *Strong coupling mesotron theory of nuclear forces.* Phys. Rev. *63*, 143 (1943).

Serpe, *The theory of the meson and the electromagnetic properties of atomic nuclei* (in French). Physica *11*, 495 (1946).

Sexl, *Choice of nuclear statistics.* Naturwiss. *25*, 153 (1937).

Sexl, *On a simple derivation of nuclear dispersion formula.* Z. Phys. *115*, 571 (1940).

Share and Breit, *Relativistic effects for the deuteron.* Phys. Rev. *52*, 546 (1937).

Share, *The excited states of the H³ and He⁴ nuclei.* Phys. Rev. *53*, 875 (1938).

Siegert, *Note on the interaction between nuclei and electromagnetic radiation.* Phys. Rev. *52*, 787 (1937).

Sizoo, *Energetic stability of isobaric nuclei* (in English). Physica *4*, 467 (1937).

Slater, *The theory of complex spectra.* Phys. Rev. *34*, 1293 (1929).

Snyder and Weinberg, *Stationary states of scalar and vector fields.* Phys. Rev. *57*, 307 (1940).

Solomon, *Statistical theory of nuclei.* C. R., Paris *207*, 910 (1938).

Solomon, *Neutral mesotrons and electron pairs.* C. R., Paris *209*, 678 (1939).

Solomon, *Theory of the deuteron.* C. R., Paris *210*, 477 (1940).

Sommerfeld, *Higher ionization potentials of atoms according to the Thomas-Fermi model.* Z. Phys. *80*, 415 (1933).

Soonawala, *The structure of atomic nuclei.* Indian J. Phys. *16*, 291 (1942).

Stephens, *Effect of the Coulomb force on binding energies of light nuclei.* Phys. Rev. *57*, 938 (1940).

Stueckelberg, *Interaction between elementary particles.* C. R., Paris *207*, 387 (1938).

Stueckelberg, *Interaction energy in electrodynamics and in the field theory of nuclear forces* (in German). Helv. Phys. Acta *11*, 225 (1938).

Stueckelberg, *Interaction forces in electrodynamics and in the field theory of nuclear forces* (in German). Helv. Phys. Acta *11*, 299 (1938).

Stueckelberg, *Rigorous theory of interaction between nuclear particles.* Phys. Rev. *54*, 889 (1938).

Stueckelberg and Patry, *Interaction between nuclear particles* (in German). Helv. Phys. Acta *12*, 300 (1939).

Stueckelberg and Patri, *Classical theory of exchange forces* (in German). Helv. Phys. Acta *13*, 167 (1940).

Stueckelberg, *Influence of the pseudoscalar field on the classical theory of exchange forces* (in French). Helv. Phys. Acta *13*, 347 (1940).

Stueckelberg, *Difficulties in the field theory of exchange forces.* Phys. Z. *41*, 523 (1940).

Stueckelberg, *The difficulty in the field theory of exchange forces.* Z. techn. Phys. *21*, 275 (1940).

Stueckelberg, *New model of the point electron of classical theory* (in French). Helv. Phys. Acta *14*, 51 (1941).

Stueckelberg, *A principle connecting the theory of relativity and quantum theory* (in French). Helv. Phys. Acta *16*, 173 (1943).

Stueckelberg, *Present state of the theories of elementary particles and of quanta* (in French). Experientia *1*, 33 (1945).

Svartholm, *The binding energies of the lightest atomic nuclei with an application of the theory of integral equations to the eigenvalue problems.* Univ. Lund (Thesis Phys. Inst.) (1945).

Taketani and Sakata, *Wave equation of the meson.* Proc. Phys.-Math. Soc. Jap. (3) *22*, 757 (1940).

Taub, *Solutions of equations for particles of spin zero or one when no field is present.* Phys. Rev. *57*, 807 (1940).

Teller and Wheeler, *On the rotation of the atomic nucleus.* Phys. Rev. *53*, 778 (1938).

Thomas, *The interaction between a neutron and a proton and the structure of H^3.* Phys. Rev. *47*, 903 (1935).

Tobias and Segrè, *High energy carbon nuclei.* Phys. Rev. *70*, 89 (1946).

Tomonaga, *Kinetic energy of the nucleus in the Hartree-Fock model.* Tokyo Inst. Phys. Chem. Res. *32*, 229 (1937).

Tonnelat, *A new type of unitary theory —study of the particle of spin 2.* Ann. Phys., Paris *17*, 158 (1942).

Turner, *The missing heavy nuclei.* Rev. Mod. Phys. *17*, 292 (1945).

Tyrrell, Carroll and Margenau, *Binding energies of light nuclei.* Phys. Rev. *55*, 790 (1939).

Tyrrell, *The nuclear five-body problem.* Phys. Rev. *56*, 250 (1939).

Umeda, Tomonaga and Ono, *Mutual potential energy between two deuterons* (in German). Tokyo Inst. Phys. Chem. Res. *32*, 87 (1937).

Umeda and Ono, *Dynamic fluid model of atomic nucleus* (in German). Tokyo Inst. Phys. Chem. Res. *32*, 120 (1937).

Umeda, *Density of nuclear energy levels on the oscillator model* (in German). Tokyo Inst. Phys. Chem. Res. *33*, 197 (1938).

Umeda, *Contribution of fluctuation linkage to nuclear excitation energy* (in German). Tokyo Inst. Phys. Chem. Res. *36*, 57 (1939).

Vand, *A note on the analysis of the nuclear binding energies.* Phil. Mag. (7) *34*, 280 (1943).

Van Isacker, *On the decomposition of the equations for a particles of ar-*

bitrary spin. C. R., Paris *219*, 51 (1944).

Van Lier and Uhlenbeck, *Statistical calculation of density of energy levels of nuclei* (in English). Physica *4*, 531 (1937).

Véronnet, *Constitution of atomic nuclei.* C. R., Paris *207*, 42 (1938).

Villars, *A contribution to the deuteron problem* (in German). Helv. Phys. Acta *19*, 323 (1946).

Volkoff, *Tensor forces and heavy nuclei.* Phys. Rev. *62*, 126 (1942).

Volkoff, *A note on exchange tensor forces in heavy nuclei.* Phys. Rev. *62*, 134 (1942).

Volkoff, *On the lack of saturation of exchange tensor forces.* Phys. Rev. *62*, 304 (1942).

Volz, *Magnitude of nuclear forces.* Z. Phys. *105*, 537 (1937).

Walsh, *The point singularity in a nonlinear meson theory.* Proc. R. Irish Acad. *50A*, 167 (1945).

Wang, *On the binding energy of deuteron and the neutron-proton scattering by a new potential.* Phys. Rev. *66*, 103 (1944).

Wang and Tsao, *An attempt at finding the relationship between nuclear forces and the gravitational force.* Phys. Rev. *66*, 155 (1944).

Wang and Tsao, *Relationship between nuclear force and gravitational force.* Nature *155*, 512 (1945).

Wang, *Calculation of the binding energy of the deuteron and the neutron-proton scattering by a new potential.* Phys. Rev. *68*, 163 (1945).

Wang, *A calculation of the binding energies of H³ and He⁴ with a new potential.* Phys. Rev. *70*, 492 (1946).

Wang, *Neutron and negative proton.* Nature *157*, 549 (1946).

Warren and Margenau, *Note on the validity of methods used in nuclear calculations.* Phys. Rev. *52*, 1027 (1937).

Wataghin, *On the abundance of nuclei in the universe.* Phys. Rev. *70*, 430 (1946).

Watanabe, *Light atomic nuclei and the Hartree model.* Z. Phys. *112*, 159 (1939).

Watanabe, *Application of thermodynamical notions to normal state of the nucleus.* Z. Phys. *113*, 482 (1939).

Way, *Photoelectric cross section of the deuteron.* Phys. Rev. *51*, 552 (1937).

Way, *A simple picture of the binding energies of H³ and He⁵.* Phys. Rev. *56*, 556 (1939).

Wefelmeier, *Geometrical model of the atomic nucleus.* Z. Phys. *107*, 332 (1937).

Wefelmeier, *Nuclear properties of Sm and its structural isomerism.* Ann. Phys., Leipzig *36*, 373 (1939).

Wefelmeier, *A model of the transuranic elements.* Naturwiss. *27*, 110 (1939).

Weisskopf and Wigner, *Calculation of the natural breadth of spectral lines on the basis of Dirac's theory.* Z. Phys. *63*, 54 (1930).

Weisskopf, *Exchange forces between elementary particles.* Phys. Rev. *50*, 1187 (1936).

Weisskopf, *Statistics and nuclear reactions.* Phys. Rev. *52*, 295 (1937).

Weizsäcker, *The forces valid for the structure of the nucleus.* Phys. Z. *36*, 779 (1935).

Weizsäcker, *Theory of nuclear masses.* Z. Phys. *96*, 431 (1935).

Weizsäcker, *Nuclear structure models.* Naturwiss. *36*, 209, 225 (1938).

Weizsäcker, *On the Wefelmeier model of the transuranic elements.* Naturwiss. *27*, 133 (1939).

Wentzel, *A generalization of the quantum conditions for the purposes of wave mechanics.* Z. Phys. *38*, 518 (1926).

Wentzel, *Heavy electrons and theories of nuclear phenomena.* Naturwiss. *26*, 273 (1938).

Wentzel, *Hypothesis of higher proton isobars* (in German). Helv. Phys. Acta *14*, 2 (1941).

Wentzel, *Pair theory of nuclear forces* (in German). Helv. Phys. Acta *15*, 111 (1942).

Wentzel, *Saturation character of nuclear forces and mesons* (in German). Helv. Phys. Acta *15*, 685 (1942).

Wentzel, *Theory of meson fields with strong coupling to nucleons* (in German). Helv. Phys. Acta *16*, 222 (1943).

Wentzel, *Vector meson theory* (in German). Helv. Phys. Acta *16*, 551 (1943).

Wentzel, *On the deuteron problem II* (in German). Helv. Phys. Acta *17*, 252 (1944).

Wentzel, *Recent research in meson theory.* Rev. Mod. Phys. *19*, 1 (1947).

Wergeland, *Meson theory and nuclear forces.* Phys. Rev. *60*, 835 (1941).

Wheeler, *The dependences of nuclear forces on velocity.* Phys. Rev. *50*, 643 (1936).

Wheeler, *Molecular viewpoints in nuclear structure.* Phys. Rev. *52*, 1083 (1937).

Wheeler, *On the mathematical description of light nuclei by the method of resonating group structure.* Phys. Rev. *52*, 1107 (1937).

Wheeler, *Wave functions for large arguments by the amplitude phase methol.* Phys. Rev. *52*, 1125 (1937).

Wheeler, *The alpha particle model and the properties of the nucleus Be*[8]. Phys. Rev. *59*, 27 (1941).

Wheeler and Ladenburg, *Mass of the meson by the method of momentum loss.* Phys. Rev. *60*, 754 (1941).

Wheeler, *The stability of polyelectrons.* Ann. N. Y. Acad. Sci. *46*, 221 (1946).

White, Creutz, Delsasso and Wilson, *Positrons from light nuclei.* Phys. Rev. *59*, 63 (1941).

Wick, *On the interaction between neutrons and protons.* Z. Phys. *84*, 799 (1933).

Wick, *Properties of the nucleus.* Nuovo Cim. *11*, 227 (1934).

Wightman, *Pais' f-field theory and the mirror nuclei.* Phys. Rev. *71*, 447 (1947).

Wigner, *On the mass defect of helium.* Phys. Rev. *43*, 252 (1933).

Wigner, *Saturation of exchange forces.* Proc. Nat. Acad. Sci. *22*, 662 (1936).

Wigner, *On the consequences of the symmetry of the nuclear Hamiltonian on the spectroscopy of nuclei.* Phys. Rev. *51*, 106 (1937).

Wigner, *On the structure of the nuclei beyond oxygen.* Phys. Rev. *51*, 947 (1937).

Wigner, Critchfield and Teller, *The electron-positron field theory of nuclear forces.* Phys. Rev. *56*, 530 (1939).

Wigner and Feenberg, *Symmetry properties of nuclear levels.* Rep. Phys. Soc. Prog. Phys. *8*, 274 (1942).

Wilson, *The calculations of processes involving mesons by matrix methods.* Proc. Camb. Phil. Soc. *35*, 363 (1940).

Wilson, *A spherical shell nuclear model.* Phys. Rev. *69*, 538 (1946).

Winans, *A classical model for the nucleus.* Phys. Rev. *71*, 379 (1947).

Wisniewski, *Model of atomic nuclei* (in German). Acta Phys. Polonica *6*, 125 (1937).

Wisniewski, *Attempt at a theory of the nucleus* (in French). Acta Phys. Polonica *6*, 335 (1937).

Wisniewski, *Structure of atomic nuclei* (in French). Acta Phys. Polonica *7*, 5 (1938).

Wisniewski, *Theory of proton* (in French). Acta Phys. Polonica *7*, 310 (1939).

Yamanouchi, *Binding energies of atomic nuclei. Part I.* Proc. Phys.-Math. Soc. Jap. (3) *19*, 557 (1937).

Yamanouchi, *Binding energies of atomic nuclei. Part II.* Proc. Phys.-Math. Soc. Jap. (3) *19*, 790 (1937).

Yamasaki, *On the meson theory involving a mixture of scalar and pseudoscalar fields. Part I.* Proc. Phys.-Math. Soc. Jap. (3) *25*, 659 (1943).

Yamasaki *On the meson theory involving a mixture of scalar and pseudoscalar fields. Part II.* Proc. Phys.-Math. Soc. Jap. (3) *26*, 32 (1944).

Yost, Wheeler and Breit, *Coulomb functions in repulsive fields.* Phys. Rev. *49*, 174 (1936).

Young, *Note on the interaction of nuclear particles.* Phys. Rev. *48*, 913 (1935).

Young, *On the shape and stability of heavy nuclei.* Phys. Rev. *55*, 1102 (1939).

Yukawa, *Interaction of elementary particles. Part I.* Proc. Phys.-Math. Soc. Jap. (3) *17*, 48 (1935).

Yukawa and Sakata, *Interaction of elementary particles. Part II.* Proc. Phys.-Math. Soc. Jap (3), *19*, 1084 (1937).

Yukawa, Sakata and Taketani, *Interaction of elementary particles. Part III.* Proc. Phys. Math. Soc. Jap. (3) *20,* 319 (1938).

Yukawa, Sakata, Kobayasi and Taketani, *Interaction of elementary particles. Part IV.* Proc. Phys.-Math. Soc. Jap. (3) *20,* 720 (1938).

Yukawa and Sagata, *Mass and mean life time of the meson.* Nature *143,* 761 (1939).

Yukawa, *Remarks on the nature of the mesotron.* Z. Phys. *119,* 201 (1942).

CHAPTER TEN

Theories of Disintegration Processes

Alichanian and Berestezky, *On the interpretation of beta disintegration data.* Phys. Rev. *55,* 978 (1939).

Alvarez, *Electron capture and internal conversion in* Ga^{67}. Phys. Rev. *53,* 606 (1938).

Alvarez, Helmholz and Nelson, *Isomeric silver and the Weizsäcker theory.* Phys. Rev. *57,* 660 (1940).

Badarau, *Passage of particles through Coulomb potential barriers.* C. R., Paris *207,* 842 (1938).

Badarau, *Passage and diffusion of particles across Coulomb potential barriers.* C. R., Paris *207,* 1030 (1938).

Badarau, *Passage of particles through Coulomb potential barriers.* C. R., Paris *209,* 89 (1939).

Bardham, *Nuclear structure of radioactive substances.* Phil. Mag. (7) *25,* 1033 (1938).

Barnóthy and Forró, *Proper lifetime of the mesotron.* Phys. Rev. *60,* 154 (1941).

Bell, *Bond energies and repulsive energies in proton transfer reactions.* Trans. Faraday Soc. *37,* 493 (1941).

Bessey, *On the theory of the internal conversion of gamma radiation.* Phys. Rev. *61,* 544 (1942).

Bethe and Peierls, *Photoelectric disinte-gration of the diplon.* Int. Conf. Phys., London *1,* 93 (1934).

Bethe and Peierls, *The "neutrino."* Nature *133,* 532 (1934).

Bethe and Peierls, *The neutrino.* Nature *133,* 689 (1934).

Bethe, *Theory of disintegration of nuclei by neutrons.* Phys. Rev. *47,* 747 (1935).

Bethe and Heitler, *Stopping power of fast particles and the creation of electron pairs.* Proc. Roy. Soc. *146,* 83 (1934).

Bethe, *Nucleus radius and the many-body problem.* Phys. Rev. *50,* 977 (1936).

Bethe, *Disintegrations of very heavy nuclei by deuterons.* Phys. Rev. *51,* 1004 (1937).

Bethe, *Nuclear Physics. Part II. Nuclear dynamics.* Rev. Mod. Phys. *9,* 69 (1937).

Bethe, *The Oppenheimer-Phillips process.* Phys. Rev. *53,* 39 (1938).

Bethe and Critchfield, *The formation of deuterons by proton combination.* Phys. Rev. *54,* 248 (1938).

Bethe and Critchfield, *On the formation of deuterons by proton combination.* Phys. Rev. *54,* 862 (1938).

Bethe, *Energy production in stars.* Phys. Rev. *55,* 103 (1939).

Bethe, *Energy production in stars.* Phys. Rev. *55,* 434 (1939).

Bethe and Nordheim, *On the theory of meson decay.* Phys. Rev. *57,* 998 (1940).

Bethe, *Multiple production of mesons by protons.* Phys. Rev. *70,* 787 (1946).

Bhabha, *Creation of electron pairs by fast charged particles.* Nature *134,* 934 (1934).

Bloch and Gamow, *On the probability of gamma ray emission.* Phys. Rev. *50,* 260 (1936).

Bloch, *On the continuous gamma radiation accompanying the beta decay.* Phys. Rev. *50,* 272 (1936).

Bohr and Kalckar, *On the transmutation of atomic nuclei by impact of material particles. I. General theoretical remarks.* Kgl. Danske Vid. Sels. *14,* No. 10 (1936).

Bohr, *Nuclear photo-effects.* Nature *141,* 326 (1938).

Bohr, Peierls and Placzek, *Nuclear reactions in the continuous energy region.* Nature *144*, 200 (1939).

Booth and Wilson, *Radiative processes involving fast mesons.* Proc. Roy. Soc. *175*, 483 (1940).

Born, *Theory of nuclear disintegration.* Z. Phys. *58*, 306 (1929).

Borsellino and Ghizzetti, *Formation of electron-pairs by gamma rays in the field of an electron.* Ric. Sci. Recostruz. *16*, 625 (1946).

Bramley, *K-A transformation.* Science *86*, 424 (1937).

Breit and Yost, *Capture of charged particles by nuclei due to emission of gamma radiation.* Phys. Rev. *46*, 1110 (1934).

Breit and Yost, *Radiative capture of protons by C.* Phys. Rev. *48*, 203 (1935).

Breit and Condon, *The photoelectric effect of the deuteron.* Phys. Rev. *49*, 904 (1936).

Breit and Wigner, *The disintegration of Li^8.* Phys. Rev. *51*, 593 (1937).

Breit and Knipp, *Note on K-electron capture in Be^7.* Phys. Rev. *54*, 652 (1938).

Breit, *Schematic treatment of nuclear resonances.* Phys. Rev. *69*, 472 (1946).

Breit and Darling, *Calculation of angular distributions in resonance reactions.* Phys. Rev. *71*, 141 (1947).

Camp, *Heavy particle interactions from beta decay theory.* Phys. Rev. *51*, 1046 (1937).

Carlson and Oppenheimer, *On multiplicative showers.* Phys. Rev. *51*, 220 (1937).

Chang and Wang, *Analysis of beta disintegration data. II. The probability of beta disintegration and the complexity of the atomic nuclei* (in English). Sci. Rec. Acad. Sinica *1*, 103 (1942).

Cherdynzev, *Binding energy of the atomic nucleus and alpha decay.* Phys. Z. Sowjet. *13*, 170 (1938).

Colby and Little, *Possible results of a new reaction.* Phys. Rev. *70*, 437 (1946).

Cooper, *On the separation of nuclear isomers.* Phys. Rev. *61*, 1 (1942).

Courant, *On the excitation functions of proton induced reactions in Li.* Phys. Rev. *63*, 219 (1943).

Critchfield and Teller, *On the angular distribution of alpha particles produced in the Li^7-proton reaction.* Phys. Rev. *60*, 10 (1941).

Critchfield and Wigner, *The anti-symmetrical interaction in beta decay theory.* Phys. Rev. *60*, 412 (1941).

Critchfield, *On the theoretical half-lives of beta activities.* Phys. Rev. *61*, 249 (1942).

Critchfield, *Theoretical half-lives of beta transitions.* Phys. Rev. *61*, 387 (1942).

Critchfield, *The anti-symmetrical interaction in beta decay theory.* Phys. Rev. *63*, 417 (1943).

Curie, *Empirical determination of the atomic number Z_A corresponding to the maximum stability of atoms of mass number A.* J. Phys. Rad. *6*, 209 (1945).

Dancoff and Morrison, *Note on internal conversion of arbitrary multipole order.* Phys. Rev. *58*, 149 (1938).

Dancoff and Morrison, *The calculation of internal conversion coefficients.* Phys. Rev. *55*, 122 (1939).

Dancoff, *Short range alpha's in natural radioactivity.* Phys. Rev. *70*, 116 (1946).

Delbruck and Gamow, *Transition probabilities of excited nuclei.* Z. Phys. *72*, 492 (1931).

Dennison, *Excited states of the O^{16} nucleus.* Phys. Rev. *57*, 454 (1940).

Diemer and de Vries, *On the activities caused by nearly thermal neutrons* (in English). Physica *11*, 345 (1946).

Dube and Jha, *On the theory of emissions of alpha particles from radioactive nuclei.* Indian J. Phys. *17*, 344 (1943).

Eastman, *Energy of removal of neutrons and alpha particles from nuclei and alpha instability below the radio-elements.* Phys. Rev. *46*, 238 (1934).

Eisner, *Interpretation of the angular distribution of alpha's from Li^7 $(p, \alpha)\alpha$.* Phys. Rev. *65*, 85 (1944).

Ellickson, *A method of calculating fluctuations.* Phys. Rev. *58*, 572 (1938).

Ellis and Mott, *Internal conversion of gamma rays and nuclear level systems* of Th B and C. Proc. Roy. Soc. *139*, 369 (1933).

Ellis, Cockcroft, Peierls and Richardson, *Discussion on beta type of nuclear transmutation*. Proc. Roy. Soc. *161*, 447 (1937).

Feather, *The probability of excitation of the nucleus Pb206 in the alpha disintegration of Po*. Phys. Rev. *70*, 88 (1946).

Feenberg, *Does the alpha particle possess excited states?* Phys. Rev. *49*, 328 (1936).

Feinberg, *Ionization of the atom due to beta decay*. J. Phys. USSR *4*, 423 (1941).

Ferber, *Theoretical method in radioactivity*. J. Phys. Rad. *9*, 337 (1938).

Fermi and Uhlenbeck, *On the recombination of electrons and positrons*. Phys. Rev. *44*, 510 (1933).

Fermi, *A theory of the emission of beta rays*. Ric. Sci. *2*, 12 (1933).

Fermi, *Theory of beta rays*. Z. Phys. *88*, 161 (1934).

Fierz, *Quantization of theories of beta disintegration* (in German). Helv. Phys. Acta *10*, 123 (1937); Erratum *10*, 284 (1937).

Fierz, *Fermi theory of beta disintegration*. Z. Phys. *104*, 553 (1937).

Fisk and Taylor, *The internal conversion of gamma rays*. Proc. Roy. Soc. *146*, 178 (1934).

Flammersfeld, *Isomers of the stable nuclei rhodium and silver*. Z. Naturforsch. *1*, 3 (1946).

Flügge and Krebs, *Nuclear Physics*. Phys. Z. *38*, 13 (1937).

Fowler, *Matrix elements for positron decay*. Phys. Rev. *69*, 134 (1946).

Frame, *Collision of a slow alpha particle with lithium*. Proc. Camb. Phil. Soc. *33*, 115 (1937).

Frenkel, *Solid body model of heavy nuclei* (in English). Phys. Z. Sowjet. *9*, 533 (1936).

Frenkel, *On the spectroscopy of heavy nuclei. II. Rotation and magnetic excitation levels of heavy nuclei in conjunction with soft gamma rays*. J. Phys., USSR *4*, 493 (1941).

Fröhlich, Heitler and Kahn, *The photodisintegration of the deuteron in the meson theory*. Proc. Roy. Soc. *174*, 85 (1940).

Fuchs, *Stability of nuclei against beta emission*. Proc. Camb. Phil. Soc. *35*, 242 (1939).

Furry, *On transition probabilities in double beta disintegration*. Phys. Rev. *56*, 1184 (1939).

Gamow, *Quantum theory of nuclear disintegration*. Nature *122*, 805 (1929).

Gamow and Houtermans, *Quantum mechanics of the radioactive nucleus*. Z. Phys. *52*, 496 (1929).

Gamow, *Quantum theory of atomic disruption*. Z. Phys. *52*, 510 (1929).

Gamow, *Fine structure of alpha rays*. Nature *126*, 397 (1930).

Gamow, *Theory of radioactive alpha disintegration. Nuclear disruption and excitation by alpha rays*. Phys. Z. *32*, 651 (1931).

Gamow and Rosenblum. *The effective diameters of radioactive nuclei*. C. R., Paris *197*, 1620 (1933).

Gamow and Teller, *Selection rules for the beta disintegration*. Phys. Rev. *49*, 895 (1936).

Gamow, *Beta decay*. Phys. Z. *38*, 800 (1937).

Gamow and Teller, *The rate of selective thermo-nuclear reactions*. Phys. Rev. *53*, 608 (1938).

Goeppert-Mayer, *Double beta disintegration*. Phys. Rev. *48*, 512 (1935).

Goertzel and Lowen, *Magnetic multipile conversion for K electrons*. Phys. Rev. *67*, 203 (1945).

Goertzel and Lowen, *Angular correlation of successive gamma rays*. Phys. Rev. *69*, 533 (1946).

Goertzel, *Angular correlation of gamma rays*. Phys. Rev. *70*, 897 (1946).

Goldhaber, *Probability of artificial nuclear transformation and its connection with the vector model of the nucleus*. Proc. Camb. Phil. Soc. *30*, 561 (1934).

Goldhaber, *Mechanism of the neutron loss of reaction produced by deuterons*. Nature *146*, 167 (1940).

Goldhaber, *Radiation widths of highly*

excited nuclei. Phys. Rev. *67,* 59 (1945).

Goldhaber, *Width of nuclear levels.* Phys. Rev. *68,* 99 (1945).

Goldstein, *Theory of alpha disintegration.* J. Phys. Rad. 7, 527 (1936).

Goldstein, *Theory of nuclear reactions.* C. R., Paris *206,* 52 (1938).

Goldstein, *Theory of showers of protons and neutrons.* C. R., Paris *206,* 488 (1938).

Goldstein, *Excitation mechanism in atomic nuclei.* C. R., Paris *206,* 1880 (1938).

Goldstein, *Recombination of atomic nuclei.* C. R., Paris *207,* 965 (1938).

Goldstein, *Recoil atoms of radioactive bodies.* J. Phys. Rad. *8,* 316 (1937).

Goldstein, *Highly activated atomic nuclei.* J. Phys. Rad. *9,* 262 (1938).

Gora, *Fermi's theory of beta decay.* Proc. Indian Acad. Sci. *4A,* 551 (1936).

Greuling, *Theoretical half-lives of forbidden beta transitions.* Phys. Rev. *61,* 568 (1942).

Grönblom, *Beta decay and spin of light nuclei.* Phys. Rev. *56,* 508 (1939).

Grosev and Frank, *Nuclear impulse in pair creation.* C. R., URSS *19,* 239 (1938).

Gueutal and Daudel, *On the stability of heavy nuclei.* Rev. Sci., Paris *84,* 109 (1946).

Gumbel, *Intervals between radioactive emissions.* J. Phys. Rad. *8,* 321 (1937).

Gurney and Condon, *Wave mechanics and radioactive disintegration.* Nature *122,* 439 (1928).

Gurney and Condon, *Wave mechanics and radioactive disintegration.* Phys. Rev. *33,* 127 (1929).

Gurney, *Nuclear levels and artificial disintegration.* Nature *123,* 565 (1929).

Guth, *Theory of electro-disintegration of Be.* Phys. Rev. *55,* 411 (1939).

Guth, *Radiative transition probabilities in heavy nuclei. Excitation of nuclei by X-rays.* Phys. Rev. *59,* 325 (1941).

Guthrie and Sachs, *On the excited states of nuclei beyond oxygen.* Phys. Rev. *62,* 8 (1942).

Hamilton, *Angular distribution of gamma quanta emitted following beta decay.* Phys. Rev. *60,* 168 (1941).

Hamilton, Heitler and Peng, *Theory of cosmic-ray mesons.* Phys. Rev. *64,* 78 (1943).

Harkins and Madorsky, *A graphical study of the stability relations of atomic nuclei.* Phys. Rev. *19,* 135 (1922).

Harkins and Gans, *Inelastic collisions with changes of mass and the problem of nuclear disintegration with capture or non-capture of a neutron or another nuclear projectile.* Phys. Rev. *46,* 397 (1934).

Hazen, *The multiple production of penetrating particles by cosmic-ray protons and neutrons.* Phys. Rev. *64,* 257 (1943).

Hebb and Uhlenbeck, *Stability of nuclear isomers* (in English). Physica 5, 605 (1938).

Hebb, *Theory of beta decay* (in English). Physica *5,* 701 (1938).

Heitler, *Remarks on nuclear disintegrations by cosmic rays.* Phys. Rev. *54,* 873 (1938).

Heitler, *On the production of mesons by proton-proton collisions. II.* Proc. R. Irish Acad. *50A,* 155 (1945).

Houtermans and Jordan, *The hypothesis of the variation in time of beta decay and the possibility of its experimental proof.* Z. Naturforsch. *1,* 125 (1946).

Hoyle, *Generalized Fermi interaction.* Proc. Camb. Phil. Soc. *33,* 277 (1937).

Hoyle, *Beta transitions in a Coulomb field.* Proc. Roy. Soc. *166,* 249 (1938).

Hull, *Calculation of activities of radioactive substances in series disintegrations.* J. Phys. Chem. *45,* 1305 (1941).

Hulme, *The internal conversion coefficient for Ra C.* Proc. Roy. Soc. *138,* 643 (1932).

Hulme, McDougall, Buckingham and Fowler, *Photoelectric absorption of gamma rays in heavy elements.* Proc. Roy. Soc. *149,* 131 (1935).

Hulme, Mott, Oppenheimer and Taylor, *Internal conversion coefficient for gamma rays.* Proc. Roy. Soc. *155,* 315 (1936).

Isidu, *On the disintegration of deuterons*

by neutrons. Proc. Phys.-Math. Soc. Jap. (3) *24*, 828 (1942).

Itoh, *Analysis of experimental data for beta disintegration*. Proc. Phys. Math. Soc. Jap. (3) *22*, 531 (1940).

Jaeger and Hulme, *The internal conversion of X-rays with positrons*. Proc. Roy. Soc. *148*, 708 (1935).

Jauch, *On the inelastic photo-dissociation of the deuteron*. Phys. Rev. *69*, 275 (1946).

Jauncey, *Heavy particles and the neutrino*. Phys. Rev. *53*, 106 (1938).

Kahan, *Nuclear disintegrations*. C. R., Paris *206*, 1289 (1938); Erratum *206*, 1420 (1938).

Kahan, *Transmutation of light nuclei and their theoretical interpretation*. C. R., Paris *208*, 649 (1939).

Kalckar, Oppenheimer and Serber, *Note on nuclear photo-effect at high energies*. Phys. Rev. *52*, 273 (1937).

Kalckar, Oppenheimer and Serber, *Note on resonances in transmutations of light nuclei*. Phys. Rev. *52*, 279 (1937).

Kapur, *Transmutation functions for deuterons*. Proc. Roy. Soc. *163*, 553 (1937).

Kapur and Peierls, *Dispersion formula for nuclear reactions*. Proc. Roy. Soc. *166*, 277 (1938).

Klein, *On the meson-pair theory of nuclear interaction* (in English). Ark. Mat. Astr. Fys. *30A*, No. 3 (1944).

Knipp and Teller, *On the energy loss of heavy ions*. Phys. Rev. *59*, 659 (1941).

Kobayasi and Utiyama, *Interaction of mesons with radiation fields*. Proc. Phys.-Math. Soc. Jap. (3) *22*, 882 (1940).

Kobayasi, *Distinction between longitudinal and transverse mesons*. Phys. Rev. *59*, 843 (1941).

Kofoed-Hansen, *Maximum beta energies and the mass of the neutrino*. Phys. Rev. *71*, 451 (1947).

Kompanejez, *Induced beta disintegration by central collision of an electron with a heavy nucleus*. Phys. Z. Sowjet. *12*, 138 (1937).

Konopinski and Uhlenbeck, *On the Fermi theory of beta radioactivity*. Phys. Rev. *48*, 7 (1935).

Konopinski and Bethe, *The evaporation model of nuclear disintegrations*. Phys. Rev. *51*, 1004 (1937).

Konopinski and Bethe, *The theory of excitation functions on the basis of the many body model*. Phys. Rev. *54*, 130 (1938).

Konopinski and Uhlenbeck, *On the Fermi theory of beta radioactivity. II. The "forbidden" spectra*. Phys. Rev. *60*, 308 (1941).

Kreisler, *Disintegration of deuterons by deuterons* (in German). Acta Phys. Polonica *6*, 327 (1937).

Kurie, Richardson and Paxton, *On the shape of the distribution curves of electrons emitted from artificially produced radioactive substances*. Phys. Rev. *48*, 167 (1935).

Kusaka, *Beta decay with neutrino of spin 3/2*. Phys. Rev. *60*, 61 (1941).

Kusaka, *The effect of radiation damping on burst production*. Phys. Rev. *64*, 256 (1943).

Landau, *Stability of neon and carbon with respect to alpha particle disintegration*. Phys. Rev. *52*, 1251 (1937).

Langer, *Excitation and disintegration of protons and the neutret*. Phys. Rev. *45*, 495 (1934).

Lifschitz, *Collisions of deuterons with heavy nuclei*. Phys. Z. Sowjet. *13*, 224 (1938).

Lowen, *Lifetimes of nuclear levels with respect to electric multipole radiation*. Phys. Rev. *59*, 835 (1941).

Lowen, *K-L conversion ratio*. Phys. Rev. *67*, 203 (1945).

Lubanski and Rosenfeld, *On the disintegration of the deuteron by electron impact* (in German). Experientia *1*, 198 (1945).

Marshak, *Beta ray spectrum of K^{40} and theory of beta decay*. Phys. Rev. *70*, 980 (1946).

Martins, *Influence of "Schwinger" forces on nuclear processes* (in French). Portugaliae Physica *2*, 99 (1946).

Massey, *Creation of electron pairs by nuclear capture of neutrons*. Proc. R. Irish Acad. *44A*, 77 (1938).

Massey and Burhop, *Probability of annihilation of positrons without emis-*

244

sion of radiation. Proc. Roy. Soc. *167*, 53 (1938).

McConnell, *Production and annihilation of negative protons.* Proc. R. Irish Acad. *50A*, 189 (1945).

Mercier, *Theory of beta radioactivity.* C. R., Paris *204*, 1117 (1937).

Mercier, *A note on the theory of beta radioactivity.* Nature *139*, 797 (1937).

Migdal, *Quadrupole and dipole gamma radiation of nuclei.* J. Phys., USSR *8*, 331 (1944).

Møller, *On the capture of orbital electrons by nuclei.* Phys. Rev. *51*, 84 (1937).

Møller, *Fermi theory of positron emission.* Phys. Z. Sowjet. *11*, 9 (1937).

Mott, *Theory of excitation by collision with heavy particles.* Proc. Camb. Phil. Soc. *27*, 553 (1931).

Motz and Schwinger, *Beta radioactivity of neutrons.* Phys. Rev. *48*, 704 (1935).

Nelson and Oppenheimer, *Multiple production of mesotrons by protons.* Phys. Rev. *60*, 159 (1941).

Nelson, *On the pseudoscalar mesotron theory of beta decay.* Phys. Rev. *60*, 830 (1941).

Nordheim, *Probability of radiative processes for very high energies.* Phys. Rev. *49*, 189 (1936).

Nordheim and Nordheim, *On the production of heavy electrons.* Phys. Rev. *54*, 254 (1938).

Nordheim and Yost, *On the matrix element in Fermi's theory of beta decay.* Phys. Rev. *51*, 942 (1937).

Nordheim, Nordheim, Oppenheimer and Serber, *The disintegration of high energy protons.* Phys. Rev. *51*, 1037 (1937).

Oppenheimer, *The disintegration of lithium by protons.* Phys. Rev. *43*, 380 (1933).

Oppenheimer and Plesset, *On the production of the positive electron.* Phys. Rev. *44*, 53 (1933).

Oppenheimer, *The disintegration of the deuteron by impact.* Phys. Rev. *47*, 845 (1935).

Oppenheimer and Phillips, *Note on the transmutation functions for deuterons.* Phys. Rev. *48*, 500 (1935).

Oppenheimer and Serber, *Note on boron plus proton reactions.* Phys. Rev. *53*, 636 (1938).

Oppenheimer and Schwinger, *On pair emission in the proton bombardment of fluorine.* Phys. Rev. *56*, 1066 (1939).

Oppenheimer, *On the selection rules in beta decay.* Phys. Rev. *59*, 908 (1941).

Ortner, *Nuclear disintegration in photographic films caused by cosmic rays.* S. B. Akad. Wiss. Wien *149*, 231 (1940).

Ostrofsky, Breit and Johnson, *The excitation function of Li under proton bombardment.* Phys. Rev. *49*, 22 (1936).

Ostrofsky, Bleick and Breit, *Effects of exchange forces on the excitation function of Li^7 under proton bombardment.* Phys. Rev. *49*, 352 (1936).

Pauli and Jauch, *On the application of Dirac's method of field-quantization to the problem of emission of low frequency photons.* Phys. Rev. *65*, 255 (1944).

Peierls, *The Bohr theory of nuclear reactions.* Rep. Phys. Soc. Progr. Phys. *7*, 87 (1941).

Peng, *Cascade productions of mesons.* Proc. R. Irish Acad. *49A*, 245 (1944).

Perrin, *Dissymmetry of positive and negative beta spectra and intrinsic mass of neutrino or ergon.* C. R., Paris *198*, 2086 (1934).

Peters and Richman, *Deuteron disintegration by electron.* Phys. Rev. *59*, 804 (1941).

Pollard, *Potential barrier of nitrogen nucleus.* Proc. Leeds Phil. Lit. Soc. *2*, 324 (1932).

Pontecorvo, *Probability of radiative transitions in nuclei.* C. R., Paris *207*, 230 (1938).

Pool and Campbell, *Note on K electron capture and isomerism in radiosilver.* Phys. Rev. *53*, 272 (1938).

Predmestnikow, *Directional distribution in beta decay.* Phys. Z. Sowjet. *13*, 32 (1938).

Racah, *Pair production by collision of charged particles.* Nuovo Cim. *14*, 93 (1937).

Rarita, Schwinger and Nye, *The photodisintegration of the deuteron.* Phys. Rev. *59*, 209 (1941).

Reinsberg, *Angular distribution in nuclear transitions.* Z. Phys. *108,* 189 (1938).

Richardson, *Relations in beta ray transformations and the neutrino theory.* Nature *139,* 505 (1937).

Richman, *Angular distribution in deuteron disintegration by electrons.* Phys. Rev. *61,* 203 (1942).

Rogers and Rogers, *The energy-range relations for deuterons, protons and alpha particles.* Phys. Rev. *53,* 713 (1938).

Rose and Uhlenbeck, *The formation of electron-positron pairs by internal conversion of gamma radiation.* Phys. Rev. *48,* 211 (1935).

Rose, *A note on the possible effect of screening in the theory of beta disintegration.* Phys. Rev. *49,* 727 (1936).

Rozental, *Meson lifetime and radioactive beta decay.* Phys. Rev. *60,* 612 (1941).

Rüchardt, *On the transformations of hydrogen canal rays.* Ann. Phys., Leipzig *71,* 377 (1923).

Sachs, *Internal conversion in 6.4 hr Cd.* Phys. Rev. *57,* 159 (1940).

Sachs, *A note on nuclear isomerism.* Phys. Rev. *57,* 194 (1940).

Sakata and Tanikawa, *Capture of mesotrons by atomic nucleus.* Proc. Phys. Math. Soc. Jap. (3) *21,* 58 (1939).

Sakata and Tanikawa, *Spontaneous disintegration of the neutral mesotron (neutretto).* Phys. Rev. *57,* 548 (1940).

Sakata, *Theory of the meson decay.* Proc. Phys.-Math. Soc. Jap. (3) *23,* 283 (1941).

Sakata, *Yukawa's theory of the beta disintegration and the lifetime of meson.* Proc. Phys.-Math. Soc. Jap. (3) *23,* 291 (1941).

Schiff, *Resonance fluorescence of nuclei.* Phys. Rev. *70,* 761 (1946).

Schiff, *Discussion of the fluorine-proton resonances.* Phys. Rev. *70,* 891 (1946).

Schönberg, *Beta ray selective rules and meson decay.* Phys. Rev. *56,* 612 (1939).

Sedgwick, *On the theory of successive radiative transformations.* Proc. Camb. Phil. Soc. *38,* 280 (1942).

Segrè, *A paradox on nuclear isomerism.* Phys. Rev. *70,* 445 (1946).

Segrè, *Possibility of altering the decay rate of a radioactive substance.* Phys. Rev. *71,* 274 (1947).

Sen Chowdhury, *Sargent's curves for artificially beta active nuclei.* Indian J. Phys. *17,* 262 (1943).

Serber, *Beta decay and mesotron lifetime.* Phys. Rev. *56,* 1065 (1939).

Sexl, *Wave mechanics derivation of radioactive disintegration constants.* Z. Phys. *56,* 62 (1929).

Sexl, *Differential equation in wave mechanics. Theory of radioactive alpha particle emission.* Z. Phys. *56,* 72 (1929).

Sexl, *Quantum mechanics of alpha radiation.* Z. Phys. *59,* 579 (1929).

Sexl, *Quantitative theory of radioactive alpha emission.* Z. Phys. *81,* 163 (1933).

Siegert, *On the derivation of the dispersion formula for nuclear reactions.* Phys. Rev. *56,* 750 (1939).

Tamm, *Beta radioactivity and nuclear forces.* Phys. Z. Sowjet. *10,* 567 (1936).

Taylor and Mott, *A theory of the internal conversion of gamma rays.* Proc. Roy. Soc. *138,* 665 (1932).

Taylor and Mott, *Internal conversion of gamma rays.* Proc. Roy. Soc. *152,* 215 (1933).

Thibaud, *Systematic time distribution of alpha disintegration.* Ann. Phys., Paris *15,* 225 (1941).

Tisza, *Pair production at beta disintegration.* Phys. Z. Sowjet. *11,* 425 (1937).

Tsien, *The characteristics of the nuclei composed of $_3Li^5$ and the resonance mixture.* C. R., Paris *218,* 996 (1944).

Uhlenbeck and Kulper, *Law of Sargent* (in German). Physica *4,* 601 (1937).

Umeda, *Amplitude factor in beta decay* (in German). Tokyo Inst. Phys. Chem. Res. *33,* 137 (1938).

Umeda, *Nuclear excitation and partitio numerorum* (in German). Tokyo Inst. Phys. Chem. Res. *33,* 629 (1938).

Urry, *The radio-elements in non-equilibrium systems.* Amer. J. Sci. *240,* 426 (1942).

Valadares, *Einstein's photoelectric law and the phenomenon of internal conversion* (in French). Portugaliae Physica *1,* 35 (1943).

Véronnet, *Theory of radioactivity, na-*

tural and artificial, based on nuclear constitution. C. R., Paris 207, 121 (1938).

Volkoff, The Oppenheimer-Phillips process. Phys. Rev. 57, 866 (1940).

Wang, A suggestion on the detection of the neutrino. Phys. Rev. 61, 97 (1942).

Wang and Chang, Analysis of beta disintegration data. I. The Sargent curves and Fermi and K. U. theories of beta radioactivity (in English). Sci. Rec. Acad. Sinica 1, 98 (1942).

Wataghin, On the nuclear photo-effect. Phys. Rev. 69, 33 (1946).

Watase, Disintegration of N^{13} nucleus. Proc. Phys.-Math. Soc. Jap (3) 22, 639 (1940).

Weisskopf, Excitation of nuclei by bombardment with charged particles. Phys. Rev. 53, 1018 (1938).

Weisskopf and Ewing, On the yield of nuclear reactions with heavy elements. Phys. Rev. 57, 472 (1940).

Weisskopf, Note on the radiation properties of heavy nuclei. Phys. Rev. 59, 318 (1941).

Weizsäcker, Metastable states of atomic nuclei. Naturwiss. 24, 813 (1936).

Weltin, On the branching ratio of Na^{22}. Phys. Rev. 64, 128 (1943).

Wentzel, Theory of beta disintegration and nuclear forces. Z. Phys. 104, 34 (1936).

Wentzel, Beta interaction (in German). Helv. Phys. Acta 10, 107 (1937).

Wentzel, Theory of beta disintegration and nuclear forces. Z. Phys. 105, 738 (1937).

Wick, Excitation of nuclei by fast electrons. Ric. Sci. 11, 49 (1940).

Wigner and Breit, The beta ray spectrum of Li^8. Phys. Rev. 50, 1191 (1936).

Wigner, On coupling conditions in light nuclei and the lifetimes of beta radioactivities. Phys. Rev. 56, 519 (1939).

Wilhelmy, Resonance emergence of protons and alpha particles. Z. Phys. 107, 769 (1937).

Wísniewski, Excited states of nuclei (in French). Acta Phys. Polonica 7, 245 (1938).

Wu, Remark on the energy distribution of neutrons from fluorine. Phys. Rev. 45, 846 (1934).

Yukawa, Interaction of elementary particles. Proc. Phys.-Math. Soc. Jap. (3) 17, 48 (1935).

Yukawa and Sakata, On the nuclear transformation with the absorption of the orbital electron. Phys. Rev. 51, 677 (1937).

Zavelsky, On the upper limit of the beta spectrum of Th C" and Th B. Phys. Rev. 55, 317 (1939).

CHAPTER ELEVEN

Nuclear Fission

Abelson, Cleavage of the uranium nucleus. Phys. Rev.55, 418 (1939).

Abelson, Further products of U decay. Phys. Rev. 55, 670 (1939).

Abelson, The identification of some of the products of U cleavage. Phys. Rev. 55, 876 (1939).

Abelson, An investigation of the products of the disintegration of U by neutrons. Phys. Rev. 56, 1 (1939).

Adler, Growth of chain reaction in mass containing U. C. R., Paris 209, 301 (1939).

Adler and von Halban, Control of the chain reaction involved in fission of the U nucleus. Nature 143, 793 (1939).

Ageno, Amaldi, Bocciarelli, Cacciapuoti and Trabacchi, Fission yield by fast neutrons. Phys. Rev. 60, 67 (1941).

Anderson, Booth, Dunning, Fermi, Glasoe and Slack, The fission of U. Phys. Rev. 55, 511 (1939).

Anderson, Fermi and Hanstein, Production of neutrons in U bombarded by neutron:. Phys. Rev. 55, 797 (1939).

Anderson and Fermi, Simple capture of neutrons by U. Phys. Rev. 55, 1106 (1939).

Anderson, Fermi and Szilard, Neutron production and absorption in uranium. Phys. Rev. 56, 284 (1939).

Anderson, Fermi and Grosse, Branching ratios in the fission of U^{235}. Phys. Rev. 59, 52 (1941).

Arakatsu, Uemura, Sonoda, Shimizu, Kimura and Muraoka, *Photo-fission of U and Th produced by the gamma rays of Li and F bombarded with high speed protons*. Proc. Phys. Math. Soc. Jap. (3) *23*, 440 (1941).

Arakatsu, Sonoda, Uemura and Shimizu, *Range of the photo-fission fragments of U produced by the gamma rays of Li bombarded with protons*. Proc. Phys.-Math. Soc. Jap. (3) *23*, 633 (1941).

Aten, Bakker and Heyn, *Transmutation of thorium by neutrons*. Nature *143*, 679 (1939).

Bakker, *Splitting of heavy nuclei by neutrons*. Ned. T. Natuurk. *6*, 333 (1939).

Baldwin and Klaiber, *Photo-fission in heavy elements*. Phys. Rev. *71*, 3 (1947).

Beck and Havas, *Dissymmetry of rupture of U*. C. R., Paris *208*, 1084 (1939).

Bernstein, Preston, Wolfe and Slattery, *Yield of photo-neutrons from U^{235} fission products in heavy watar*. Phys. Rev. *71*, 140 (1947).

Bjerge, Brostrøm and Koch, *Decay curves of U and Th fission products*. Nature *143*, 794 (1939).

Bøggild, Brostrøm and Lauritsen, *Range and straggling of fission fragments*. Phys. Rev. *59*, 275 (1941).

Bøggild, *Range velocity relation for fission fragments in He*. Phys. Rev. *60*, 827 (1941).

Bøggild, Arroe and Sigurgeirsson, *Cloud chamber studies of electronic and nuclear stopping of fission fragments in different gases*. Phys. Rev. *71*, 281 (1947).

Bohr, *Disintegration of heavy nuclei*. Nature *143*, 330 (1939).

Bohr, *Resonance in U and Th disintegration and the phenomenon of nuclear fission*. Phys. Rev. *55*, 418 (1939).

Bohr and Wheeler, *The mechanism of nuclear fission*. Phys. Rev. *56*, 426 (1939).

Bohr and Wheeler, *The fission of protactinium*. Phys. Rev. *56*, 1065 (1939).

Bohr, *Velocity-range relation for fission fragments*. Phys. Rev. *59*, 270 (1941).

Bohr, *Mechanism of deuteron induced fission*. Phys. Rev. *59*, 1042 (1941).

Bohr, *Disintegration of heavy nuclei*. Nature *143*, 330 (1943).

Booth, Dunning and Slack, *Delayed neutron emission from U*. Phys. Rev. *55*, 876 (1939).

Booth, Dunning and Slack, *Energy distribution of U fission fragments*. Phys. Rev. *55*, 981 (1939).

Booth, Dunning and Glasoe, *Range distribution of the uranium fission fragments*. Phys. Rev. *55*, 982 (1939); Erratum *55*, 1273 (1939).

Booth, Dunning, Grosse and Nier, *Neutron capture by uranium*. Phys. Rev. *58*, 475 (1940).

Borst and Floyd, *Fissionability studies*. Phys. Rev. *70*, 107 (1946).

Bothe and Gentner, *The energy limit of the neutrons produced by the fission of U*. Z. Phys. *119*, 568 (1942).

Braun, Preiswerk and Scherrer, *Detection of alpha particles in the disintegration of Th*. Nature *140*, 682 (1937).

Bretscher and Cook, *Transmutations of uranium and thorium nuclei by neutrons*. Nature *143*, 559 (1939).

Broda, *Determination of the upper limits of the fission cross sections of lead and bismuth for Li-D neutrons by a track count method*. Nature *158*, 872 (1946).

Broda and Wright, *Determination of the upper limits of the fission cross sections of lead and bismuth for Li-D neutrons by a chemical method*. Nature *158*, 871 (1946).

Brostrøm, Koch and Lauritsen, *Delayed neutron emission accompanying uranium fission*. Nature *144*, 830 (1939).

Burgy, Pardue, Willard and Wollan, *Energy of delayed neutrons from U^{235} fissions*. Phys. Rev. *70*, 104 (1946).

Chatterjee, *On the spontaneous emission of neutrons from uranium nucleus*. Indian J. Phys. *19*, 211 (1945).

Chlopin, Passwik-Chlopin and Wolkov, *A particular mode of fission of the U nucleus*. Nature *144*, 595 (1939).

Cocconi, *Fission of heavy nuclei*. Nuovo Cim. *16*, 417 (1939).

Corson and Thornton, *Disintegration of U*. Phys. Rev. *55*, 509 (1939).

Crane, Delsasso, Fowler and Lauritsen, *The emission of negative electrons*

248

from lithium and fluorine bombarded with deuterons. Phys. Rev. *47*, 971 (1935).

Curie, von Halban and Preiswerk, *Artificial production of elements of an unknown radioactive family by the irradiation of thorium by neutrons*. C. R., Paris *200*, 1841 (1935).

Curie, von Halban and Preiswerk, *Radioactive elements produced on the irradiation of thorium by neutrons*. C. R., Paris *200*, 2079 (1935).

Curie, von Halban and Preiswerk, *On the artificial creation of elements belonging to an unknown radioactive family upon irradiation of Th by neutrons*. J. Phys. Rad. *6*, 361 (1935).

Curie and Savitch, *Radioactive elements from U by bombardment with neutrons*. J. Phys. Rad. *8*, 385 (1937).

Curie and Savitch, *On the radio-element of period 3.5 hours found in uranium irradiated by neutrons*. C. R. *206*, 906 (1938).

Curie and Savitch, *Radio-element of period 3.5 hr formed in Th irradiated with neutrons*. C. R., Paris *206*, 1643 (1938).

Curie and Savitch, *Radio-elements formed by the action of neutrons on U. Part II*. J. Phys. Rad. *9*, 355 (1938).

Curie, Savitch and da Silva, *On the radiation of a body of period 3.5 hours formed by uranium irradiated by neutrons*. J. Phys. Rad. *9*, 440 (1938).

Curie and Savitch, *Radio-elements formed in U and Th irradiated by neutrons*. C. R., Paris *208*, 343 (1939).

Demers, *Pairs of fission fragments from U235*. Phys. Rev. *70*, 974 (1946).

Demers, *Pairs of fission fragments*. Phys. Rev. *71*, 483 (1947).

Dessauer and Hafner, *Proton-induced fission*. Phys. Rev. *59*, 840 (1941).

Dodé, von Halban, Joliot and Kowarski, *Energy of neutrons liberated by division of U nucleus*. C. R., Paris *208*, 995 (1939).

Dodson and Fowler, *Products of U fission. Radioactive isotopes of iodine and xenon*. Phys. Rev. *57*, 967 (1940).

Dodson and Fowler, *Radioactive halogens produced by the neutron bombardment of U and Th*. Phys. Rev. *55*, 880 (1939).

Fabre, Magnan and Muraour, *Detonation of nitrogen iodide by fragments of U bombarded by neutrons*. C. R., Paris *209*, 436 (1939).

Fano, *Possibility of disintegration of very heavy nuclei into two nuclei of medium weight*. J. Phys. Rad. *10*, 229 (1939).

Farwell, Segrè and Wiegand, *Long range alpha particles emitted in connection with fission. Preliminary report*. Phys. Rev. *71*, 327 (1947).

Fearon, Engle, Thayer, Swift and Johnson, *Results of atmosphere analyses done at Tulsa, Okla. during the period neighboring the time of the second Bikini atomic bomb test*. Phys. Rev. *70*, 564 (1946).

Feather and Bretscher, *Atomic numbers of the so-called transuranic elements*. Nature *143*, 516 (1939).

Feather, *The time involved in the process of nuclear fission*. Nature *143*, 597 (1939).

Feather, *Fission of heavy nuclei*. Nature *143*, 1027 (1939).

Feenberg, *The detonation of nitrogen iodide by nuclear fission*. Phys. Rev. *55*, 980 (1939).

Fermi, *Possible production of elements of atomic number higher than 92*. Nature *133*, 898 (1934).

Fermi and Segrè, *Fission of uranium by alpha particles*. Phys. Rev. *59*, 680 (1941).

Ferretti, *On a new type of artificial disintegration*. Ric. Sci. *10*, 332 (1939).

Flammersfeld, *The partition conditions and energies in uranium fission*. Z. Phys. *120*, 450 (1943).

Flügge, *Prospects of technical utilization of energy-content of atomic nuclei*. Naturwiss. *27*, 402 (1939).

Flügge and von Droste, *Energy considerations in the production of barium in neutron irradiation of uranium*. Z. Phys. Chem. *B42*, 274 (1939).

Flügge, *On the spontaneous fission of uranium and its neighboring elements*. Z. Phys. *121*, 298 (1943).

Fowler and Dodson, *Intensely ionizing*

particles produced by neutron bombardment of U and Th. Phys. Rev. *55,* 417 (1939).

Frenkel, *Electro-capillary theory of splitting of heavy nuclei by slow neutrons* (in Russian). J. Exp. Theor. Phys. USSR *9,* 641 (1939).

Frenkel, *On the splitting of heavy nuclei by slow neutrons.* Phys. Rev. *55,* 987 (1939).

Frenkel, *On some features of the process of fission of heavy nuclei.* J. Phys., USSR *10,* 533 (1940).

Frisch, *Physical evidence for the division of heavy nuclei under neutron bombardment.* Nature *143,* 276 (1939).

Frisch, *Statistical calculation of composite decay curves.* Nature *143,* 852 (1939).

Frisch, *The distribution in energy of fission fragments near the fast fission threshold.* Phys. Rev. *71,* 478 (1947).

Gant, *Fission of uranium under deuteron bombardment.* Nature *144,* 707 (1939).

Gant and Krishnan, *Deuteron-induced fission in U and Th.* Proc. Roy. Soc. *178,* 474 (1941).

Gibbs and Thomson, *Possible delay in the emission of neutrons from U.* Nature *144,* 202 (1939).

Glasoe and Steigman, *Fission products from U.* Phys. Rev. *55,* 982 (1939).

Glasoe and Steigman, *Radioactive products from the gases produced in U fission.* Phys. Rev. *57,* 566 (1940).

Glasoe and Steigman, *Radioactive products from gases produced in uranium fission.* Phys. Rev. *58,* 1 (1940).

Götte, *A nuclear isomer of xenon appearing in U fission.* Naturwiss. *28,* 449 (1940).

Goldstein, Rogozinski and Walen, *Interaction of fast neutrons with U nuclei.* J. Phys. Rev. *10,* 477 (1939).

Goldstein, Rogozinski and Walen, *The scattering by U nuclei of fast neutrons and the possible neutron emission resulting from fission.* Nature *144,* 201 (1939).

Green and Alvarez, *Heavy ionizing particles from U.* Phys. Rev. *55,* 417 (1939).

Green and Livesey, *Fission fragment tracks in photographic plates.* Nature *158,* 272 (1946).

Grosse and Agruss, *The identity of Fermi's reactions of element 93 with element 91.* J. Amer. Chem. Soc. *57,* 438 (1935).

Grosse, Booth and Dunning, *The fission of protactinium.* Phys. Rev. *56,* 382 (1939).

Grosse and Booth, *Radioactive zirconium and columbium from U fission.* Phys. Rev. *57,* 664 (1940).

Grummitt and Wilkinson, *Fission products of U^{235}.* Nature *158,* 163 (1946).

Haenny and Rosenberg, *Emission of neutrons on rupture of U nucleus. Possible chain reaction.* C. R., Paris *208,* 898 (1939).

Hagiwara, *Liberation of fast neutrons in the nuclear explosions of U irradiated by thermal neutrons.* Mem. Coll. Sci. Kyoto Imp. Univ. *23,* 19 (1940).

Hahn, *Artificial radio-elements by neutron irradiation. Elements beyond uranium.* Ber. Deutsch. Chem. Ges. *69A,* 217 (1935).

Hahn, Meitner and Strassmann, *New transformation processes in neutron irradiation of uranium. Elements beyond uranium.* Ber. Deutsch. Chem. Ges. *69B,* 905 (1935).

Hahn and Meitner, *On the artificial transformation of uranium by neutrons. I.* Naturwiss. *23,* 37 (1935).

Hahn and Meitner, *On the artificial transformation of uranium by neutrons. II.* Naturwiss. *23,* 230 (1935).

Hahn, Meitner and Strassmann, *Some further observations on the artificial transformation products of uranium.* Naturwiss. *23,* 544 (1935).

Hahn and Meitner, *New transformation products in the irradiation of uranium with neutrons.* Naturwiss. *24,* 158 (1936).

Hahn, Meitner and Strassmann, *On the transuranic elements and their chemical behavior.* Ber. Deutsch. Chem. Ges. *70B,* 1374 (1937).

Hahn, Meitner and Strassmann, *New transformation product of long life in the trans-uranium series.* Naturwiss. *26,* 475 (1938).

Hahn and Strassmann, *Neutron induced Ra isotopes from U.* Naturwiss. *26*, 755 (1938).

Hahn, *Fission products of U and Th.* Ann. Phys., Leipzig *36*, 368 (1939).

Hahn and Strassmann, *Neutron induced radioactivity of U.* Naturwiss. *27*, 11 (1939).

Hahn and Strassmann, *Active Ba isotopes from U and Th.* Naturwiss. *27*, 89 (1939).

Hahn and Strassmann, *Disintegration of uranium.* Naturwiss. *27*, 163 (1939).

Hahn and Strassmann, *Further disintegration product from bombardment of U with neutrons.* Naturwiss. *27*, 529 (1939).

Hahn, Strassmann and Flügge, *Disintegration of Th.* Naturwiss. *27*, 544 (1939).

Hahn and Strassman, *Collapse of U and Th nuclei into lighter atoms.* Phys. Z. *40*, 673 (1939).

Hahn and Strassmann, *Application of the emanation capacity of a uranium compound to the extraction of fission products of U. Two short lived alkali metals.* Naturwiss. *28*, 54 (1940).

Hahn and Strassmann, *Separation of the Kr and Xe isotopes arising in U fission.* Naturwiss. *28*, 455 (1940).

Hahn and Strassmann, *Some additional products of U fission.* Naturwiss. *28*, 543 (1940).

Hahn and Strassmann, *New method for separation of active fission products from U and Th.* Naturwiss. *28*, 54 (1940).

Hahn and Strassmann, *On the Mo isotopes from the fission of U.* Naturwiss. *29*, 369 (1941).

Hahn and Strassmann, *On the strontium and yttrium isotopes resulting from the fission of U.* Z. Phys. *121*, 729 (1943).

von Halban, Joliot, Kowarski and Perrin, *Evidence for a chain nuclear reaction in the middle of a uraniferous mass.* J. Phys. Rad. *10*, 428 (1939).

von Halban, Joliot and Kowarski, *Liberation of neutrons in the nuclear explosion of uranium.* Nature *143*, 470 (1939).

von Halban, Joliot and Kowarski, *Number of neutrons liberated in the nuclear fission of U.* Nature *143*, 680 (1939).

von Halban, Joliot and Kowarski, *Energy of neutrons liberated in the nuclear fission of uranium induced by thermal neutrons.* Nature *143*, 939 (1939).

Haxby, Shoupp, Stephens and Wells, *Photo-fission of uranium and thorium.* Phys. Rev. *58*, 92 (1940).

Haxby, Shoupp, Stephens and Wells, *Fast neutron threshold for uranium fission.* Phys. Rev. *58*, 1035 (1940).

Haxby, Shoupp, Stephens and Wells, *Photo-fission of uranium and thorium.* Phys. Rev. *59*, 57 (1941).

Haxel, *Energy and range of heavy disintegration products of U.* Z. Phys. *112*, 681 (1939).

Haynes, *Effect of neutron energy on the total decay curves of fission products.* Phys. Rev. *59*, 834 (1941).

Henderson, *The heat of fission in U.* Phys. Rev. *56*, 703 (1939).

Herzog, *Gamma ray anomaly following the atomic bomb test of July 1, 1946.* Phys. Rev. *70*, 227 (1946).

Hess and Luger, *The ionization of the atmosphere in the N. Y. area before and after the Bikini atom bomb test.* Phys. Rev. *70*, 564 (1946).

Heyn, Aten and Bakker, *Transmutation of U and Th by neutrons.* Nature *143*, 516 (1939).

Jacobsen and Lassen, *Fission cross section in U and Th for deuteron impact.* Phys. Rev. *59*, 1043 (1939).

Jacobsen and Lassen, *Deuteron induced fission in uranium and thorium.* Phys. Rev. *58*, 867 (1940).

Jacobsen and Lassen, *Fission cross section in U and Th for deuteron impact.* Nature *148*, 230 (1941).

Jentschke and Prankl, *Investigation of the heavy fragments in the decay of uranium and thorium by neutron bombardment.* Naturwiss. *27*, 134 (1939).

Jentschke, Prankl and Herneger, *The fission of ionium under neutron bombardment.* Naturwiss. *28*, 315 (1940).

Jentschke and Prankl, *Energies and masses of the U fragments on irradiation chiefly with thermal neutrons.* Z. Phys. *119*, 696 (1942).

Joliot, *Explosive rupture of U and Th nuclei by neutrons.* C. R., Paris *208*, 341 (1939).

Joliot, *Trajectories of products of explosion of U nuclei.* C. R., Paris *208*, 647 (1939).

Joliot, *Explosive rupture of nuclei of U and Th by neutrons.* J. Phys. Rad. *10*, 159 (1939).

Joliot, *On a physical method of extraction of radio-elements from fission and evidence of a radio-praseodymium (13 days).* C. R., Paris *218*, 733 (1944).

Kanner and Barschall, *Distribution in energy of the fragments from U fission.* Phys. Rev. *57*, 372 (1940).

Kennedy and Seaborg, *Search for beta particles emitted during U fission process.* Phys. Rev. *55*, 877 (1939).

Kennedy and Wahl, *Search for spontaneous fission in 94²³⁹.* Phys. Rev. *69*, 367 (1946).

Kingdon, Pollock, Booth and Dunning, *Fission of the separated isotopes of U.* Phys. Rev. *57*, 749 (1940).

Krishnan and Banks, *Fission of uranium and thorium under deuteron bombardment.* Nature *145*, 860 (1940).

Kwal, *Several phenomena accompanying nuclear fission of U.* C. R., Paris *224*, 563 (1947).

Ladenburg, Kanner, Barschall and Van Voorhis, *Study of U and Th fission produced by fast neutrons of nearly homogeneous energy.* Phys. Rev. *56*, 168 (1939).

Langer and Stephens, *Radioactive barium and strontium from photo-fission of uranium.* Phys. Rev. *58*, 759 (1940).

Langsdorf, *Fission products of thorium.* Phys. Rev. *56*, 205 (1939).

Lark-Horovitz, *Uranium fission with Li-D neutrons. Energy distribution of the fission fragments.* Phys. Rev. *60*, 156 (1941).

Lassen, *On the variation along range of the Hρ distribution and the charge of the fission fragments of the light group.* Phys. Rev. *69*, 137 (1946).

Lassen, *Specific ionization by fission fragment.* Phys. Rev. *70*, 577 (1946).

Lassen, *Hρ distribution of fission fragments.* Phys. Rev. *68*, 142 (1945).

Lassen, *Ionization by fission fragments in nitrogen, argon and xenon.* Phys. Rev. *68*, 230 (1945).

Libby, *Stability of uranium and thorium for natural fission.* Phys. Rev. *55*, 1269 (1939).

Lieber, *The fission products from the irradiation of uranium with neutrons. The Sr isotope.* Naturwiss. *27*, 421 (1939).

Magnan, *Rupture of elements lighter than U by neutron bombardment.* C. R., Paris *208*, 742 (1939).

Magnan, *Neutrons emitted during splitting of U by neutrons.* C. R., Paris *208*, 1218 (1939).

Maurer and Pose, *Neutron emission of the U nucleus as the result of its spontaneous fission.* Z. Phys. *121*, 285 (1943).

McMillan, *Radioactive recoils from U activated by neutrons.* Phys. Rev. *55*, 510 (1939).

Meitner and Hahn, *New transmutation processes in the irradiation of uranium with neutron:.* Naturwiss. *24*, 158 (1936).

Meitner, *Beta and gamma rays of elements beyond U.* Ann. Phys., Leipzig *29*, 246 (1937).

Meitner, Hahn and Strassmann, *Disintegration of U by neutrons.* Z. Phys. *106*, 249 (1937).

Meitner, Strassmann and Hahn, *Artificial disintegration of Th with neutrons.* Z. Phys. *109*, 538 (1938).

Meitner and Frisch, *Disintegration of U by neutron:. A new type of nuclear reaction.* Nature *143*, 239 (1939).

Meitner and Frisch, *Products of the fission of the U nucleus.* Nature *143*, 471 (1939).

Meitner, *New products of the thorium nucleus.* Nature *143*, 637 (1939).

Meitner, *Attempt to single out some fission processes of U by using the differences in their energy release.* Rev. Mod. Phys. *17*, 287 (1945).

Michaels, Parry and Thomson, *Production of neutrons by the fission of uranium.* Nature *143*, 760 (1939).

Moussa and Goldstein, *Radioactive bromine isotopes from uranium fission.* Phys. Rev. *60*, 534 (1941).

Mouzon and Park, *Delayed gamma rays from U activated by neutrons.* Phys. Rev. *56*, 238 (1939).

Myssowsky and Jdanoff, *Tracks on photographic plates of the recoil nuclei of disintegration of U.* Nature *143*, 794 (1939).

Nier, Booth, Dunning and Grosse, *Nuclear fission of separated U isotopes.* Phys. Rev. *57*, 546 (1940).

Nier, Booth, Dunning and Grosse, *Further experiments on the fission of .eparated uranium isotopes.* Phys. Rev. *57*, 748 (1940).

Nishina, Yasaki, Ezoe, Kimura and Ikawa, *Fission of thorium by neutrons.* Nature, *144*, 547 (1939).

Nishina, Yasaki, Ezoe, Kimura and Ikawa, *Fission products of U produced by fast neutrons.* Nature *146*, 24 (1940).

Nishina, Yasaki, Kimura and Ikawa, *Fission products of uranium by fast neutrons.* Phys. Rev. *58*, 660 (1940).

Nishina, Yasaki, Kimura and Ikawa, *Fission products of U by fast neutrons.* Phys. Rev. *59*, 323 (1941).

Nishina, Yasaki, Kimura and Ikawa, *Fission products of U by fast neutrons.* Phys. Rev. *59*, 677 (1941).

Nishina, Kimura, Yasaki and Ikawa, *Some fission products from the irradiation of U with fast neutrons.* Z. Phys. *119*, 195 (1942).

Noddack, *On element 93.* Z. angew. Chemie *47*, 653 (1934).

Nordheim, *Pile kinetics.* Phys. Rev. *70*, 115 (1946).

Oliphant, *The release of atomic energy.* Nature *157*, 5 (1946).

Peierls, *Critical conditions in neutron multiplication.* Proc. Camb. Phil. Soc. *35*, 610 (1939).

Perfilov, *Registration of U fragments with removal of background due to alpha particles emitted by U.* C. R., URSS *47*, 623 (1945).

Perrin, *Possibility of multiple disintegration.* C. R., Paris *208*, 1394 (1939).

Perrin, *Calculation of possible conditions for chain reactions in U.* C. R., Paris *208*, 1573 (1939).

Plesset, *On the classical model of nuclear fission.* Amer. J. Phys. *9*, 1 (1941).

Plutonium Project, *Nuclei formed in fission: decay characteristics, fission yields and chain relationship.*. Rev. Mod. Phys. *18*, 513 (1946).

Polessitsky, Orbeli and Nemewsky, *Chemical nature of the radioactive fragments of Th fission. Radioactive halogens* (in English). C. R., URSS *28*, 215 (1940).

Present and Knipp, *On the dynamics of complex fission.* Phys. Rev. *57*, 751 (1940).

Present, Reines and Knipp, *The liquid drop model for nuclear fission.* Phys. Rev. *70*, 557 (1946).

Reddemann and Bamke, *Absorption of slow neutrons in uranium.* Naturwiss. *27*, 518 (1939).

Richards and Speck, *Range distribution of U235 fission fragments in photographic emulsion.* Phys. Rev. *71*, 141 (1947).

Roberts, Meyer and Hafstad, *Droplet fission of U and Th nuclei.* Phys. Rev. *55*, 416 (1939).

Roberts, Meyer and Wang, *Further observations on the splitting of U and Th.* Phys. Rev. *55*, 510 (1939).

Roberts, Hafstad, Meyer and Wang, *The delayed neutron emission which accompanies fission of U and Th.* Phys. Rev. *55*, 664 (1939).

Rona and Neuninger, *Further notes on artificial activity of Th due to neutrons.* Naturwiss. *24*, 491 (1936).

Rotblat, *Emission of neutrons accompanying the fission of U nuclei.* Nature *143*, 852 (1939).

Savitch, *Gaseous radio-element formed by bombardment of U by neutrons.* C. R., Paris *208*, 646 (1939).

Scalettar and Nordheim, *Theory of pile control rods.* Phys. Rev. *70*, 115 (1946).

Scharff-Goldhaber and Klaiber, *Spontaneous emission of neutrons from uranium.* Phys. Rev. *70*, 229 (1946).

Schrödinger, *Probability problems in nuclear chemistry.* Proc. R. Irish Acad. *51A*, 1 (1945).

Seaborg, Gofman and Stoughton, *Nuclear properties of U233. A new fissionable isotope of uranium.* Phys. Rev. *71*, 378 (1947).

Seelmann-Eggebert, *Direct measurement of the inert gases appearing in U fission*. Naturwiss. *28*, 451 (1940).

Segrè, *An unsuccessful search for transuranic elements*. Phys. Rev. *55*, 1104 (1939).

Segrè and Wu, *Some fission products of U*. Phys. Rev. *57*, 552 (1940).

Segrè and Seaborg, *Fission products of U and Th produced by high energy neutron*. Phys. Rev. *59*, 212 (1941).

Segrè and Wiegand, *Stopping powers of various substances for fission fragments*. Phys. Rev. *70*, 808 (1946).

Smythe, *Atomic energy for military purposes*. Rev. Mod. Phys. *17*, 351 (1945).

Solomon, *Rupture of radioactive nuclei by neutrons*. C. R., Paris *208*, 570 (1939).

Solomon, *Idea of surface tension in nuclear physics*. C. R., Paris *208*, 896 (1939).

Soodak, *Pile perturbation*. Phys. Rev. *70*, 115 (1946).

Strassmann and Hahn, *Short lived bromine and iodine isotopes in U fission*. Naturwiss. *28*, 817 (1940).

Szilard and Zinn, *Instantaneous emission of fast neutrons in the interaction of slow neutrons with U*. Phys. Rev. *55*, 799 (1939).

Thibaud and Moussa, *Disintegration of U nuclei by neutrons*. C. R., Paris *208*, 652 (1939).

Thibaud and Moussa, *Rupture of U nuclei by neutrons and liberation of the resultant energy*. C. R., Paris *208*, 744 (1939).

Thibaud and Moussa, *Degradation of the uranium nucleus*. J. Phys. Rad. *10*, 388 (1939).

Tsien, Chastel, Zah-Wei and Vigneron, *On the tri-partition of U caused by neutron capture*. C. R., Paris *223*, 986 (1946).

Tsien, Zah-Wei, Chastel and Vigneron, *Energies and frequencies of the phenomena of tri-partition and quadri-partition*. C. R., Paris *224*, 272 (1947).

Tsien, Zah-Wei and Faruggi, *The fission energy of thorium*. C. R., Paris *224*, 825 (1947).

Tsien, Zah-Wei, Chastel and Vigneron, *On the new fission processes of uranium nuclei*. Phys. Rev. *71*, 382 (1947).

Turner, *Nuclear fission*. Rev. Mod. Phys. *12*, 1 (1940).

Turner, *Atomic energy from U^{238}*. Phys. Rev. *69*, 366 (1946).

Von Droste, *Investigation of production of alpha rays during the irradiation of thorium and uranium by radium plus beryllium neutrons*. Z. Phys. *110*, 84 (1938).

Von Droste, *On the energy division in the fragments resulting upon irradiation of uranium with neutrons*. Naturwiss. *27*, 198 (1939).

Von Droste and Reddemann, *On the neutrons resulting from fission of the uranium nucleus*. Naturwiss. *27*, 371 (1939).

Way and Wigner, *Radiation from fission products*. Phys. Rev. *70*, 115 (1946).

Weekes and Weekes, *Effort to observe anomalous gamma rays connected with atomic bomb test of July 1, 1946*. Phys. Rev. *70*, 565 (1946).

Weil, *Beta spectra of some U fission products*. Phys. Rev. *60*, 167 (1941).

Weinberg, *Calculation of critical size of heterogeneous slow neutron chain reactors*. Phys. Rev. *70*, 115 (1946).

Wertenstein, *Radioactive gases evolved in U fission*. Nature *144*, 1045 (1939).

Whitaker, Barton, Bright and Murphy, *The cross sections of metallic uranium for slow neutrons*. Phys. Rev. *55*, 793 (1939).

Wick, *Stability of elongated drop as model for nucleus*. Nuovo Cim. *16*, 229 (1939).

Wu, *Identification of two radioactive xenons from uranium fission*. Phys. Rev. *58*, 926 (1940).

Yasaki, *Fission products and induced beta ray radioactivity of U by fast neutrons*. Tokyo Inst. Phys. Chem. Res. *37*, 457 (1940).

Zah-Wei, Tsien, Vigneron and Chastel, *Experimental proof of the quadri-partition of U*. C. R., Paris *223*, 1119 (1946).

Zinn and Szilard, *Emission of neutrons by U*. Phys. Rev. *56*, 619 (1939).

CHAPTER TWELVE

Methods and Apparatus

Aebersold, *The production of a beam of fast neutrons.* Phys. Rev. *56,* 714 (1939).

Aiya, *A self-stabilized high-voltage source for Geiger counters.* Curr. Sci. *12,* 227 (1943).

Alaoglu and Smith, *Statistical theory of a scaling circuit.* Phys. Rev. *53,* 832 (1938).

Alfvin, *Simple scale-of-two counter.* Proc. Phys. Soc., Lond. *50,* 358 (1938).

Allison, Skaggs and Smith, *Correction to electrostatic analyzer measurements.* Phys. Rev. *57,* 550 (1940).

Alvarez, McMillan and Snell, *The removal of the ion beam of the cyclotron from the magnet field.* Phys. Rev. *51,* 148 (1937).

Alvarez, *The production of collimated beams of monochromatic neutrons in the temperature range 300° - 10° K.* Phys. Rev. *54,* 609 (1938).

Alvarez, *The design of a proton linear accelerator.* Phys. Rev. *70,* 799 (1946).

Amaldi and Ferretti, *Two modifications of the induction accelerator (betatron).* Nuovo Cim. (9) *3,* 22 (1946).

Anderson and Feld, *Preparation of pressed Ra + Be neutron sources.* Rev. Sci. Instr. *18,* 186 (1947).

Aoki, Narimatu and Siotani, *Features of Geiger Müller counters.* Proc. Phys.-Math. Soc. Jap (3) *22,* 746 (1940).

Aoki, Yamane and Kawanoue, *On some features of Geiger Müller counters.* Proc. Phys.-Math. Soc. Jap (3) *23,* 861 (1941).

Ardenne, *On a system of ion sources with mass monochromator for generators of neutrons.* Phys. Z. *43,* 91 (1942).

Ardenne, *On a million volt installation for the transmutation of atoms.* Z. Phys. *121,* 236 (1943).

Asana, *An experimental possibility of nuclear spectrum analysis by the angular distribution experiments on nuclear disintegration* (in German). Proc. Phys.-Math. Soc. Jap. (3) *25,* 276 (1943).

Aston, *Mass Spectra. Part II. Accelerated anode rays.* Phil. Mag. (6) *47,* 385 (1924).

Aston, *Second-order focusing mass spectrograph and isotope weights by the doublet method.* Proc. Roy. Soc. *163,* 391 (1937).

Bacher and Swanson, *On the collimation of fast neutrons.* Phys. Rev. *56,* 483 (1939).

Bacher, Baker and McDaniel, *Experiments with a slow neutron velocity spectrometer. II.* Phys. Rev. *69,* 443 (1946).

Bainbridge and Jordan, *Mass spectrum analysis. 1. The mass spectrograph. 2. The existence of isobars of adjacent elements.* Phys. Rev. *50,* 282 (1936).

Baker and Bacher, *Experiments with a slow neutron velocity spectrometer.* Phys. Rev. *59,* 332 (1941).

Baldwin, Klaiber and Hartzler, *Synchronization of auxiliary apparatus with a betatron.* Rev. Sci. Instr. *18,* 121 (1947).

Bale and Bonner, *A combined voltage-regulating and quenching circuit for the Geiger-Müller counter.* Rev. Sci. Instr. *14,* 222 (1943).

Bardeen, *Concentration of isotopes by thermal diffusion. Rate of approach to equilibrium.* Phys. Rev. *57,* 35 (1940).

Barschall and Bethe, *Energy sensitivity of fast neutron counters.* Rev. Sci. Instr. *18,* 147 (1947).

Bartlett, *The shape of betatron pole faces.* Phys. Rev. *64,* 185 (1943).

Barwick, Schütze, *Enrichment of the light argon isotope by diffusion.* Z. Phys. *105,* 395 (1937).

Beams and Trotter, *Acceleration of electrons to high energies.* Phys. Rev. *45,* 849 (1934).

Beams and Haynes, *The separation of isotopes by centrifuging.* Phys. Rev. *50,* 491 (1936).

Beams and Skarstrom, *The concentration of isotopes by the evaporative centrifuge method.* Phys. Rev. *56,* 266 (1939).

Beers, *A precision method of measuring*

Geiger counter resolving times. Rev. Sci. Instr. *13*, 72 (1942).

Bell and Veksler, *A flat proportional counter*. J. Phys., USSR *10*, 386 (1946).

Bergstrand, *A scale-of-N counter and the sensitiveness towards light of a Geiger-Müller tube* (in English). Ark. Mat. Astr. Fys. *29A*, No. 31 (1943).

Berlovich, *Statistics of misses in Geiger Müller counters* (in Russian). J. Exp. Theor. Phys. *16*, 543 (1945).

Berlovich, *The theory of misses in an electromagnetic counter at the output of recording sets with Geiger-Müller counters* (in Russian). J. Exp. Theor. Phys. *16*, 547 (1946).

Bethe and Rose, *The maximum energy obtainable from the cyclotron*. Phys. Rev. *52*, 1254 (1937).

Bethe, *Influence of multiple scattering on curvature measurements*. Phys. Rev. *69*, 689 (1946).

Bijl, *Separation of isotopes by thermal diffusion*. Ned. T. Natuurk 7, *147* (1940).

Blau and Dreyfus, *The mutiplier phototube in radioactive measurements*. Rev. Sci. Instr. *16*, 245 (1945).

Bleakney, *The ionization potential of molecular hydrogen*. Phys. Rev. *40*, 496 (1932).

Bleakney and Hipple, *A new mass spectrometer with improved focusing properties*. Phys. Rev. *53*, 521 (1938).

Blewett and Jones, *Filament sources of positive ions*. Phys. Rev. *50*, 464 (1936).

Blewett, *Radiation losses in the induction electron accelerator*. Phys. Rev. *69*, 87 (1946).

Blewett, *The transition from betatron to synchrotron operation*. Phys. Rev. *70*, 798 (1946).

Bohm and Foldy, *The theory of the synchrotron*. Phys. Rev. *70*, 249 (1946).

Bosley, Craggs and McEwan, *Measurement of out-of-phase magnetic fields in betatrons*. Nature *159*, 229 (1947).

Bosley, *The betatron*. J. Sci. Instr. *23*, 277 (1946).

Bothe, *On the simplication of coincidence counting*. Z. Phys. *59*, 1 (1929).

Bothe and Baeyer, *Examination of nuclear processes by coincidence methods*. Gött. Nachr. *1*, 195 (1935).

Bothe and Gentner, *Fast corpuscular-ray apparatus and transmutation results*. Z. Phys. *104*, 685 (1937).

Bothe, *On the principle of the method of neutron probes*. Z. Phys. *120*, 437 (1943).

Bouwers, Heyn and Kuntke, *Neutron generator* (in English). Physica *4*, 153 (1937).

Bradt and Scherrer, *Apparatus for coincidence measurements of very high resolving power* (in German). Helv. Phys. Acta *16*, 251 (1943).

Bradt, Gugelot, Huber, Medicus, Preiswerk and Scherrer, *Sensitivity of counter tubes with lead, brass and aluminum cathodes for gamma rays in the energy range 0.1 to 3 Mev* (in German). Helv. Phys. Acta *19*, 77 (1946).

Bramley, *Theory of the separation of isotopes by thermal or centrifugal methods*. Science. *92*, 427 (1940).

Brobeck, Lawrence, MacKenzie, McMillan, Serber, Sewell, Simpson and Thornton, *Initial performance of the 184 inch cyclotron of the University of California*. Phys. Rev. *71*, 449 (1947).

Brown, Irvine and Livingston, *Cyclotron targets: preparation and radiochemical separation. II. Krypton*. J. Chem. Phys. *12*, 132 (1934).

Brown, Mitchell and Fowler, *The construction of a mass spectrometer for isotope analysis*. Rev. Sci. Instr. *12*, 435 (1941).

Brown and Curtiss, *Thin walled Al beta ray tube counters*. J. Res. Nat. Bur. Stand. *35*, 147 (1945).

Buechner, Lamar and Van der Graaf, *Experimental investigation of ion beam focusing* J. Appl. Phys. *12*, 141 (1941).

Buechner, Van de Graaff, Sperduto, Burrell, McIntosh and Urquhart, *A compact high voltage electrostatic generator using sulphur hexafluoride insulation*. Phys. Rev. *69*, 692 (1946).

Capron, Stokkink and van Meerssche, *Separation of nuclear isomers in the electric field*. Nature *157*, 806 (1946).

Chadwick, May, Pickavance and Powell, *An investigation of the scattering of* high-energy particles from the cyclotron by the photographic method. I. The experimental method. Proc. Roy. Soc. *183*, 1 (1944).

Chagas, Gonçalves and Machado, *On the use of a multivibrator as a Geiger-Müller counter external quenching circuit.* Rev. Brasil Biol. *4*, 123 (1944).

Chang, *A study of the alpha ray spectra by the cyclotron magnet.* Phys. Rev. *67*, 58 (1945).

Chang, *A study of the alpha particles with a cyclotron magnet alpha ray spectrograph.* Phys. Rev. *69*, 60 (1946).

Chang and Rosenblum, *A simple counting system for alpha ray spectra and the energy distribution of Po alpha particles.* Phys. Rev. *67*, 222 (1947).

Clark, *Magnetic fields due to dee structures in a synchrotron.* Phys. Rev. *70*, 444 (1946).

Clark, Getting and Thomas, *A new method for displacing the electron beam in a synchrotron.* Phys. Rev. *70*, 562 (1946).

Clusius and Dickel, *Thermal diffusion tube for separation of isotopes.* Z. Phys. Chem. *44*, 397 (1939).

Clusius, Staveley and Dickel, *On separation experiments of the xenon isotopes by rectification; the triple point of xenon.* Z. Phys. Chem. *50*, 403 (1941).

Coates, *Production of X-rays by swiftly moving mercury ions.* Phys. Rev. *40*, 542 (1934).

Cockcroft and Walton, *Experiments with high velocity positive ions.* Proc. Roy. Soc. *129*, 477 (1930).

Cockcroft and Walton, *Production of high velocity positive ions.* Proc. Roy. Soc. *136*, 619 (1932).

Coggeshall and Jordan, *An experimental mass spectrometer.* Rev. Sci. Instr. *14*, 125 (1943).

Coggeshall and Muskat, *The paths of ions and electrons in non-uniform magnetic fields.* Phys. Rev. *66*, 187 (1944).

Coggeshall, *The paths of ions and electrons in crossed, non-uniform electric and magnetic fields.* Phys. Rev. *68*, 98 (1945).

Coggeshall, *The paths of ions and elec-trons in non-uniform crossed electric and magnetic fields.* Phys. Rev. *70*, 270 (1946).

Coggeshall, *Fringing flux corrections for magnetic focusing devices.* Phys. Rev. *71*, 482 (1947).

Colby and Hatfield, *Apparatus for measurement of alpha particle range and relative stopping power of gases.* Rev. Sci. Instr. *12*, 62 (1941).

Condit, *A cloud chamber study of heavy particles accelerated in the cyclotron.* Phys. Rev. *62*, 301 (1942).

Cooksey and Lawrence, *6 Mev magnetic resonance accelerator with emergent beam.* Phys. Rev. *49*, 866 (1936).

Coon and Nobles, *Hydrogen recoil proportional counter for neutron detection.* Rev. Sci. Instr. *18*, 44 (1947).

Copp and Greenberg, *A mica-window Geiger counter tube for measuring soft radiation.* Rev. Sci. Instr. *14*, 205 (1943).

Cosslett, *Magnetic lens for beta rays of high energy.* J. Sci. Instr. *17*, 259 (1940).

Courant and Bethe, *Theoretical considerations on bringing the beam out of a betatron.* Phys. Rev. *70*, 798 (1946).

Cowie, *Geiger-counter characteristics with applied potentials reversed.* Phys. Rev. *48*, 833 (1935).

Cowie and Ksanda, *Arc-ion source with direct current filament supply for 60 inch cyclotron.* Rev. Sci. Instr. *16*, 224 (1945).

Craggs, *Absolute sensitivity of Geiger counters.* Nature *148*, 661 (1941).

Craggs, *The electrostatic focusing of high speed ion and electron beams.* J. Appl. Phys. *13*, 772 (1942).

Craggs, *A high voltage apparatus for atomic disintegration experiments.* Proc. Phys. Soc., Lond. *54*, 439 (1942).

Craggs, *A high voltage apparatus for atomic disintegration experiments.* Metrop.-Vick. Gaz. *20*, 289 (1944).

Crane, *High potential apparatus for nuclear disintegration experiments.* Phys. Rev. *52*, 11 (1937).

Crane and Mouzon, *Simple design for a cloud chamber.* Rev. Sci. Instr. *8*, 351 (1937).

Crane, *The racetrack: a proposed modification of the synchrotron.* Phys. Rev. *69,* 542 (1946).

Crane, *The racetrack.* Phys. Rev. *70,* 800 (1946).

Curtis and Bender, *The use of elkonite for cyclotron ion source.* Rev. Sci. Instr. *13,* 266 (1942).

Curtiss, *Integrating circuit for vapor type Geiger-Müller counters.* J. Res. Nat. Bur. Stand. *25,* 369 (1940).

Curtiss, *Miniature Geiger-Müller counter.* J. Res. Nat. Bur. Stand. *30,* 157 (1943).

Curtiss and Davis, *A counting method for the determination of small amounts of radium and of radon.* J. Res. Nat. Bur. Stand. *31,* 181 (1943).

Curtiss and Brown, *An arrangement with small solid angle for measurement of beta rays.* J. Res. Nat. Bur. Stand. *37,* 91 (1946).

Dahl, Hafstad and Tuve, *Technique and design of Wilson cloud chambers.* Rev. Sci. Instr. *4,* 373 (1933).

Davis and Curtiss, *Interval selector for random pulses.* J. Res. Nat. Bur. Stand. *29,* 405 (1942).

Debye, *Theory of thermal diffusion method of separation.* Ann. Phys., Leipzig *36,* 284 (1939).

Dempster, *New method of positive ray analysis.* Phys. Rev. *11,* 316 (1918).

Dempster, *Electric and magnetic focusing in mass spectroscopy.* Phys. Rev. *51,* 67 (1937).

Den Hartog and Muller, *New apparatus for measurements with Geiger-Müller counter* (in English). Physica *11,* 441 (1946).

Dennison, *The stability of orbits in the racetrack.* Phys. Rev. *69,* 542 (1946).

Dennison, *The stability of synchrotron orbits.* Phys. Rev. *70,* 58 (1946).

Dennison and Berlin, *Racetrack stability.* Phys. Rev. *70,* 764 (1946).

Deutsch, *A magnetic lens beta ray spectrometer.* Phys. Rev. *59,* 684 (1941).

Deutsch, Elliott and Evans, *Theory, design and applications of a short magnetic lens electron spectrometer.* Rev. Sci. Instr. *15,* 178 (1944).

De Vault, *Vacuum tube scaling circuit.* Rev. Sci. Instr. *12,* 83 (1941).

De Vos and du Toit, *A copper evaporation method of Geiger-Müller tube construction.* Rev. Sci. Instr. *16,* 270 (1945).

Dittrich, *Investigations on double coincidence arrangement* . Phys. Z. *41,* 256 (1940).

Dodd, *The reversed cyclotron.* Phys. Rev. *65,* 353 (1944).

Dodd, *The spirotron.* Phys. Rev. *66,* 160 (1944).

Dodd, *Frequency elimination in spirotron systems for accelerating ions and electrons.* Phys. Rev. *67,* 65 (1945).

Drehmann, *Experiments on the enrichment of radioactive manganese (Mn^{56}).* Naturwiss. *29,* 708 (1941).

Driscoll, Hodge and Ruark, *Interval meter and its application to studies of Geiger counter statistics.* Rev. Sci. Instr. *11,* 241 (1940).

Du Bridge and Brown, *Improved D. C. Amplifier.* Rev. Sci. Instr. *4,* 532 (1933).

Dunning, *Amplifier systems for the measurement of ionization by single particles.* Rev. Sci. Instr. *5,* 387 (1934).

Dunworth, *Application of method of coincidence counting to experiments in nuclear physics.* Rev. Sci. Instr. *11,* 167 (1940).

Eckart and Shonka, *Accidental coincidences in counter circuits.* Phys. Rev. *53,* 752 (1938).

Edwards, Pool and Blake, *The adaptation of the Cauchois spectrograph to artificial radioactive sources.* Phys. Rev. *67,* 150 (1945).

Eklund, *An automatic cloud chamber for nuclear disintegration studies and a determination of the beta ray spectrum of activated dysprosium* (in English). Ark. Mat. Astr. Fys. *28A,* No. 3 (1942).

Eklund, *A simple neutron generator using the D-D process* (in English). Ark. Mat. Astr. Fys. *29A,* No. 7 (1943).

Erbacher and Philipp, *Separation of radioactive atoms from isotopic stable atoms.* Z. phys. Chem. *176,* 169 (1936).

Erbacher and Philipp, *Extraction of artificially radioactive phosphorus from a mixture with the stable isotope.* Z. phys. Chem. *179*, 263 (1937)

Estermann and Stern, *Intensity measurements of molecular rays.* Z. Phys. *85*, 135 (1933).

Ewald, *A new Mattauch-Herzog double focusing mass spectrograph. The masses of C^{13} and N^{15}.* Z. Naturforsch. *1*, 131 (1946).

Fay and Paneth, *Concentration of artificially produced elements by means of an electric field.* J. Chem. Soc. 384 (1936).

Feather, *Application of method of coincidence counting in beta particle spectrograph of novel design.* Proc. Camb. Phil. Soc. *36*, 224 (1940).

Feld, *Proposed neutron spectrometer in the 10-1000 kev range.* Phys. Rev. *70*, 429 (1946).

Feld, Scalettar and Szilard, *Use of threshold detectors for fast neutron studies.* Phys. Rev. *71*, 464 (1947).

Feldenkreis, *"Industrial" separation of isotopes.* Nature *157*, 481 (1946).

Fertel, Gibbs, Moon, Thomson and Wynn-Williams. *Experiments with velocity spectrometer for slow neutrons.* Proc. Roy. Soc. *175*, 316 (1940).

Finney and Evans, *The radioactivity of solids determined by alpha ray counting.* Phys. Rev. *48*, 503 (1935).

Fleischmann, *Enrichment of N^{15} by the separation column method of Clusius and Dickel.* Phys. Z. *41*, 14 (1940).

Foldy and Bohm, *Efficiency of the frequency modulated cyclotron.* Phys. Rev. *70*, 445 (1946).

Fowler, Dudley and Gibson, *Production of intense beams of positive ions.* Phys. Rev. *46*, 1075 (1934).

Fox, Hipple, Williams, *Mass spectrometer with a small magnet.* Phys. Rev. *65*, 353 (1944).

Franck, *Proposal of a method for the separation of He^3 from He^4.* Phys. Rev. *70*, 561 (1946).

Frank, *Phase stability of synchrotron orbits.* Phys. Rev. *69*, 689 (1946).

Frank, *The stability of electron orbits in the synchrotron.* Phys. Rev. *70*, 177 (1946).

Freed, Jaffey and Schultz, *High centrifugal fields and radioactive decay.* Phys. Rev. *63*, 12 (1943).

Freundlich, Hincks and Ozeroff, *Pulse analyser for nuclear research.* Rev. Sci. Instr. *18*, 90 (1947).

Friedland, *A mass spectroscopic analysis of the polyatomic gases in a fast counter.* Phys. Rev. *71*, 377 (1947).

Furry, Jones and Onsager, *On the theory of isotope separation by thermal diffusion.* Phys. Rev. *55*, 1083 (1939).

Furry and Jones, *Isotope separation by thermal diffusion. The cylindrical case.* Phys. Rev. *69*, 459 (1946).

Gabor, *Stabilizing linear particle accelerator by means of grid lenses.* Nature *159*, 303 (1947).

Gaerttner and Crane, *An improved method for the measurement of high energy gamma rays.* Phys. Rev. *51*, 58 (1937).

Gautier and Ruark, *Composition of mixed vapors in the cloud chamber.* Phys. Rev. *57*, 1040 (1940).

Gentner and Segrè, *Appendix on the calibration of the ionization chamber.* Phys. Rev. *55*, 814 (1939).

Getting, *Multivibrator Geiger counter circuit.* Phys. Rev. *53*, 103 (1938).

Getting, Fisk and Vogt, *Some features of an electrostatic generator and ion source for high voltage research.* Phys. Rev. *56*, 1098 (1939).

Getting, *A proposed detector for high energy electrons and mesons.* Phys. Rev. *71*, 123 (1947).

Gideon and Kurbatov, *Correction of Geiger-Müller counter data for high counting rates. III.* Phys. Rev. *71*, 140 (1947).

Gingrich, Evans and Edgerton, *A direct reading counting rate meter for random pulses.* Rev. Sci. Instr. 7, 441 (1936).

Good, Kip and Brown, *Design of beta ray and gamma ray Geiger-Müller counters.* Rev. Sci. Instr. *17*, 262 (1946).

Götte, *A separation method for the rare earths appearing in nuclear fission, by* the process of Szilard and Chalmers. Z. Naturforsch. *1*, 377 (1946).

Govaerts, *A photographic method of studying beta ray absorption by* $_{15}P^{32}$. Nature *145*, 624 (1940).

Grahame and Seaborg, *The separation of radioactive substances without the use of a carrier.* Phys. Rev. *54*, 240 (1938).

Graves and Walker, *A method for measuring half-lives.* Phys. Rev. *71*, 1 (1947).

Gray, *The ionization method of measuring neutron energy.* Proc. Camb. Phil. Soc. *40*, 72 (1944).

Greanias and Wukasch, *Electron injection technique for betatrons.* Phys. Rev. *70*, 797 (1946).

Greinacher, *A new method for measuring elementary rays.* Z. Phys. *36*, 364 (1926).

Hafstad, *The application of the FP-54 pliotron to atomic disintegration studies.* Phys. Rev. *44*, 201 (1933).

Haworth and Gillette, *A slow neutron velocity spectrometer.* Phys. Rev. *69*, 254 (1946).

Haxby, Akeley, Ginzbarg, Smith, Welch and Whaley, *Preliminary studies on the Purdue microwave electron accelerator.* Phys. Rev. *70*, 797 (1946).

Hayes, *Radio frequency heating cyclotron filaments.* Phys. Rev. *70*, 220 (1946).

Hazen, *Some operating characteristics of the Wilson cloud chamber.* Rev. Sci. Instr. *13*, 247 (1942).

Heine, *Wilson cloud-chamber investigation of the emission of light positive particles by beta radioactive bodies* (in German). Helv. Phys. Acta *17*, 273 (1944).

Hellund, *Isotope separation by transient pressure diffusion.* Phys. Rev. *57*, 743 (1940).

Helmholtz, Franck and Peterson, *Synchrotron radiofrequency systems.* Phys. Rev. *70*, 448 (1946).

Henderson and White, *Design and operation of a large cyclotron.* Rev. Sci. Instr. *9*, 19 (1938).

Herb, Parkinson and Kerst, *The development and performance of an electrostatic generator operating under high air pressure.* Phys. Rev. *51*, 75 (1937).

Herb and Bernet, *Maximum voltage of Wisconsin electrostatic generator as a function of air pressure.* Phys. Rev. *52*, 379 (1937).

Herb, Turner, Hudson and Warren, *Electrostatic generator with concentric electrodes.* Phys. Rev. *58*, 519 (1940).

Hertz, *A method for the separation of gaseous isotope mixtures and its application to the isotopes of neon.* Z. Phys. *79*, 108 (1932).

Herzog, *Report on a very sensitive counter for gamma rays and its technical applications* (in German). Helv. Phys. Acta *19*, 414 (1946).

Hill and Dunworth, *Rate of spread of discharge along the wire of a Geiger counter,* Nature *158*, 833 (1946).

Hipple and Condon, *Detection of metastable ions with the mass spectrometer.* Phys. Rev. *68*, 54 (1945).

Hoffmann, *Methods and results of recent research in nuclear physics.* Phys. Z. *41*, 514 (1940).

Hole, *On the statistical treatment of counting experiments in nuclear physics* (in English). Ark. Mat. Astr. Fys, *33A*, No. 11 (1946).

Houtermans, *The energy consumed in the separation of isotopes.* Ann. Phys. Leipzig *40*, 493 (1941).

Huber and Metzger, *Investigations on an ion source of 60KV operating potential* (in German). Helv. Phys. Acta *19*, 200 (1946).

Huber, Alder and Baldinger, *A method for the measurement of velocity of the ion tube in a fast counter* (in German). Helv. Phys. Acta *19*, 204 (1946).

Hudspeth, *Wave guide acceleration of particles.* Phys. Rev. *69*, 671 (1946).

Hull, *Calibration of Neher-Harper counter circuit for accurate comparison of beta or gamma sources.* Rev. Sci. Instr. *11*, 404 (1940).

Humphreys, *Separation of Br isotopes by centrifugation.* Phys. Rev. *56*, 684 (1939).

Huntoon and Ellett, *The ionization*

gauge for atomic beam measurements. Phys. Rev. *49*, 381 (1936).

Hutchison, Stewart and Urey, *The concentration of C^{13}*. J. Chem. Phys. *8*, 532 (1940).

Hutter, *The electron optics of mass spectrographs and velocity focusing devices*. Phys. Rev. *67*, 248 (1945).

Imre, *Physical chemical separation of isomeric atomic types*. Naturwiss. *28*, 158 (1940).

Irvine, *Concentrating the uranium isotope of 23 minute half life*. Phys. Rev. *55*, 1105 (1939).

Jaeger and Zimmer, *Radiation protection and protective measurements*. Phys. *42*, 25 (1941).

Jánossy and Ingleby, *Circuit for self-recording Geiger-Müller counters*. J. Sci. Instr. *19*, 30 (1942).

Jesse, Hannum, Forstat and Hart, *Ionization chamber techniques in measurement of C^{14}*. Phys. Rev. *71*, 478 (1947).

Johnson and Johnson, *A theoretical analysis of ionization of chambers and pulse amplifiers*. Phys. Rev. *50*, 170 (1936).

Johnston and Hutchison, *Efficiency of the electrolytic separation of Cl isotopes*. J. Chem. Phys. *10*, 469 (1942).

Jones, *On the theory for the thermal diffusion coefficient for isotopes*. Phys. Rev. *59*, 1019 (1941).

Jones and Furry, *Separation of isotopes by thermal diffusion*. Rev. Mod. Phys. *18*, 151 (1946).

Jordan and Coggeshall, *Measurement of relative abundance with the mass spectrometer*. J. Appl. Phys. *13*, 539 (1942).

Kaiser and Greanias, *The synchro-betatron electron accelerator guide fields*. Phys. Rev. *69*, 536 (1946).

Kaiser and Greanias. *The synchro-betatron electron accelerator guide fields*. Phys. Rev. *70*, 797 (1946).

Kanne, *On the preparation of Po sources*. Phys. Rev. *52*, 380 (1937).

Kapur, Sarna and Charanjit, *Helium-filled Geiger-Müller counters*. Curr. Sci. *10*, 521 (1941).

Keevil and Grasham, *Theory of alpha ray couting from solid sources*. Canad. J. Res. *21*, 21 (1943).

Kerst, *The acceleration of electrons by magnetic induction*. Phys. Rev. *60*, 47 (1941).

Kerst and Serber, *Electronic orbits in the induction accelerator*. Phys. Rev. *60*, 53 (1941).

Kerst, *A new induction accelerator generating 20 Mev*. Phys. Rev. *61*, 93 (1942).

Kerst, *Method of increasing betatron energy*. Phys. Rev. *68*, 233 (1945).

Keston, *Stable self-quenching Geiger-Müller counters containing organic metallic compounds*. Rev. Sci. Instr. *14*, 293 (1943).

Klemm, *Theory of mass spectrographs which are double focusing for all masses*. Z. Naturforsch. *1*, 137 (1946).

Klemm, *The phenomenology of two processes for isotope separation*. Z. Naturforsch. *1*, 252 (1946).

Koch, *A device producing ion beams homogeneous as to mass and energy*. Phys. Rev. *69*, 238 (1946).

Kohman, *Counter corrections at high counting rates*. Phys. Rev. *65*, 63 (1944).

Kopfermann and Walcher, *Separation of the thallium isotope II. Optical investigation of various thallium mixtures*. Z. Phys. *122*, 465 (1946).

Korff and Danforth, *Neutron measurements with boron trifluoride counters*. Phys. Rev. *55*, 980 (1939).

Korff, *Fast neutron measurements with recoil counters*. Phys. Rev. *56*, 1241 (1939).

Korff, *Operation of proportional counters*. Rev. Sci. Instr. *12*, 94 (1941).

Korff, *Operation of proportional counters*. Rev. Mod. Phys. *14*, 1 (1942).

Korff and Present, *The quenching mechanism in "self-quenching" counters*. Phys. Rev. *65*, 253 (1944).

Korff, *Experiments on counters with grid*. Phys. Rev. *68*, 53 (1945).

Korff, *Experiments with triode counters*. Phys. Rev. *68*, 284 (1945).

Korsunsky, *On the four pole beta spectrograph*. J. Phys., USSR *9*, 14 (1945).

Kosten, *On the frequency distribution of the number of discharges counted by a Geiger-Müller counter in a constant interval* (in English). Physica *10*, 749 (1943).

Krasny-Ergen, *Separation of U isotopes.* Nature *145*, 742 (1940).

Krüger, *Separation and spectroscopic investigation of N¹⁵.* Z. Phys. *111*, 467 (1939).

Kruger, Green and Stallmann, *Cyclotron operation without filaments.* Phys. Rev. *51*, 291 (1937).

Kruger and Green, *The construction and operation of a cyclotron to produce 1 MV deuterons.* Phys. Rev. *51*, 699 (1937).

Kruger, Richardson, Groetzinger, Lyman, Ogle, Nelson, Schwarz, Greene, Anderson, Colbey, Lee, McClellan, Scag, Smith and Tallmadge, *A cyclotron which allows the accelerated particles to emerge in a direction parallel to the dee interface.* Phys. Rev. *65*, 62 (1944).

Kruger, Groetzinger, Richarson, Lyman, Ogle, Nelson, Schwarz, Greene, Colby, Lee, McClellan, Scag, Smith and Tallmadge, *Cyclotron which allows the accelerated particles to emerge in a direction parallel to the dee surface.* Rev. Sci. Instr. *15*, 365 (1944).

Kurbatov and Mann, *Correction of Geiger-Müller counter data.* Phys. Rev. *68*, 40 (1945).

Kurbatov and Groetzinger, *Correction of Geiger-Müller counter data. II.* Phys. Rev. *69*, 253 (1946).

Kurie, *Use of the Wilson cloud chamber for measuring the range of alpha particles from weak sources.* Rev. Sci. Instr. *2*, 655 (1932).

Kurie, *Present day design and technique of cyclotron.* J. Appl. Phys. *9*, 691 (1938).

Kurie, *Technique of high intensity bombardment with fast particles.* Rev. Sci. Instr. *10*, 199 (1939).

Lamar and Luhr, *Convenient proton source.* Phys. Rev. *45*, 287 (1934).

Lamar and Luhr, *Proton currents from a low voltage arc.* Phys. Rev. *45*, 745 (1934).

Lamar and Luhr, *Proton production in the low voltage arc.* Phys. Rev. *46*, 87 (1934).

Lamar, Samson and Compton, *High current ion sources for nuclear investigations.* Phys. Rev. *48*, 886 (1935).

Lamar, Buechner and Compton, *Low voltage proton sources.* Phys. Rev. *51*, 936 (1937).

Lamar, Buechner and Van de Graaff, *Production of proton beams.* J. Appl. Phys. *12*, 132 (1940).

Langendijk and Ornstein, *The application of the photographic method in beta ray spectroscopy* (in English). Physica 7, 475 (1940).

Laslett, *A nomogram to facilitate the determination of the resolving time of counter equipment.* Phys. Rev. *71*, 144 (1947).

Lauritsen and Bennett, *A new high potential X-ray tube.* Phys. Rev. *32*, 850 (1928).

Lauritsen and Lauritsen, *A simple quartz fiber electrometer.* Rev. Sci. Instr. *8*, 438 (1937).

Lauritsen, Lauritsen and Fowler, *Application of a pressure electrostatic generator to the transmutation of light elements by protons.* Phys. Rev. *59*, 241 (1941).

Lauritsen, Fowler and Lauritsen, *Device for introducing short-lived radioactive samples into a cloud chamber.* Phys. Rev. *71*, 275 (1947).

Lauterjung and Neuert, *After-effect with counter tubes.* Z. Phys. *122*, 266 (1944).

Lawrence and Livingston, *The production of high speed light ions without the use of high voltages.* Phys. Rev. *40*, 19 (1932).

Lawrence and Livingston, *Multiple acceleration of ions to very high speeds.* Phys. Rev. *45*, 608 (1934).

Lawrence and Cooksey, *On the apparatus for the multiple acceleration of light ions to high speeds.* Phys. Rev. *50*, 1131 (1936).

Lawrence, Alvarez, Brobeck, Cooksey, Corson, McMillan, Salisbury and Thornton, *Initial performance of the 60 inch cyclotron of the William H. Crocker Radioation laboratory, University of California.* Phys. Rev. *56*, 124 (1939).

Lawson and Tyler, *An improvement in*

Geiger-Müller tube design. Phys. Rev. *53,* 605 (1938).

Lawson and Tyler, *Design of magnetic beta ray spectrometer.* Rev. Sci. Instr. *11,* 6 (1940).

Lewis and McDonald, *Concentration of* H^2 *isotope.* J. Chem. Phys. *1,* 341 (1933).

Lewis and Hayden, *A mass spectrograph for the study of fission product mixtures.* Phys. Rev. *70,* 111 (1946).

Lifschutz and Duffendack, *The counting losses in Geiger-Müller counter circuits and recorders.* Phys. Rev. *54,* 714 (1938).

Lifschutz and Duffendack, *The advantages of scaling circuits in recording random counts.* Phys. Rev. *55,* 412 (1939).

Lifschutz, *New vacuum tube scaling circuits of arbitrary integral or fractional scaling rates.* Phys. Rev. *57,* 243 (1940).

Linder, *Nuclear electrostatic generator.* Phys. Rev. *71,* 129 (1947).

Livingston and Lawrence, *The production of 4,800,000 volt hydrogen ions.* Phys. Rev. *43,* 212 (1933).

Livingston, *Magnetic resonance accelerator.* Rev. Sci. Instr. *7,* 55 (1936).

Livingston, *Ion sources for cyclotrons.* Rev. Mod. Phys. *18,* 293 (1946).

Lofgren and Peters, *Description of a frequency modulated cyclotron and a discussion of the deflector problem.* Phys. Rev. *70,* 444 (1946).

Lu, *A low temperature thermal source of Li ions.* Phys. Rev. *53,* 845 (1938).

Luhr, *A source of doubly ionized helium.* Phys. Rev. *49,* 317 (1936).

MacDonald, *The thermionic amplification of direct currents.* Physics *7,* 265 (1936).

MacKenzie and Schmiddt, *Frequency modulation for Berkeley 37 inch cyclotron.* Phys. Rev. *70,* 445 (1946).

Maier-Liebnitz, *"Slow" Wilson chamber experiments.* Z. Phys. *112,* 569 (1939).

Maier-Leibnitz, *The coincidence method and its application to nuclear physics problems.* Phys. Z. *43,* 333 (1942).

Manley, Berger and Gillette, *Experiments with a slow neutron velocity spectrometer.* Phys. Rev. *69,* 254 (1946).

Mann and Kurbatov, *A correction of Geiger-Müller counter data.* Phys. Rev. *68,* 101 (1945).

Marschall, *Basis of the electron optic theory of a mass spectrograph.* Phys. Z. *45,* 1 (1944).

Martelly, *Method of separating isotopes, based on the use of rotating high frequency electric fields.* C. R., Paris *215,* 106 (1942).

der Mateosian and Friedman, *Organic vapors for self-quenched Geiger-Müller counters.* Phys. Rev. *69,* 689 (1946).

Mattauch, *A double focusing mass spectrograph and the masses of* N^{15} *and* O^{18}. Phys. Rev. *50,* 617 (1936); Erratum *50,* 1089 (1936).

Mattauch, *Mass spectrography and nuclear structure.* Phys. Z. *38,* 951 (1937).

Mattauch, *Correction to measurements with the electrostatic analyzer.* Phys. Rev. *57,* 549 (1940).

McCusker, *A new type of counter for ionizing particles.* J. Sci. Instr. *21,* 120 (1944).

McGee, *A high voltage D.C. regulator entirely A.C. operated.* Phys. Rev. *66,* 160 (1944).

McKibben, *Nomographic charts for nuclear reactions.* Phys. Rev. *70,* 101 (1946).

McKibben, Frisch and Hush, *Control equipment for 2.5Mev van de Graaff giving an ion beam constant to* $\pm$ *1.5 Kev.* Phys. Rev. *70,* 117 (1946).

McMillan and Salisbury, *A modified arc source for the cyclotron.* Phys. Rev. *56,* 836 (1939).

McMillan, *The synchroton—a proposed high energy particle accelerator.* Phys. Rev. *68,* 143 (1945).

McMillan, *Resonance acceleration of charged particles.* Phys. Rev. *70,* 800 (1946).

Meaker and Roberts, *Some phase effects with coincidence proportional counters.* Rev. Sci. Instr. *15,* 149 (1944).

Mellen, *A radium source ion gauge.* Phys. Rev. *69,* 691 (1946).

Members of Nuclear Physics Laboratory of Osaka Imperial University. *Separation of isotopes by thermal diffusion. I. Preliminary test of the separation* column. *II. Separation of Cl isotopes.* Proc. Phys.-Math. Soc. Jap. (3) *23,* 590 (1941).

Mikhaleva, *The Geiger-Müller counter with a hollow anode.* J. Phys., USSR *10,* 296 (1946).

Miller, *Geiger-Müller counter pulse size.* Rev. Sci. Instr. *14,* 68 (1943).

Miller, *A new type of electrostatic generator.* Phys. Rev. *69,* 666 (1946).

Montgomery and Montgomery, *Geiger-Müller counters.* J. Franklin Inst. *231,* 447, 509 (1941).

Montgomery and Montgomery, *Time lags in Geiger-Müller counter discharges.* Phys. Rev. *59,* 1045 (1941).

Muehlhause and Friedman, *Geiger-Müller counter technique for high counting rates.* Phys. Rev. *69,* 46 (1946).

Muehlhause and Friedman, *Measurement of betatron radiation with Geiger-Müller counter.* Phys. Rev. *69,* 691 (1946).

Murray, *Recording negative ions and electrons in a gas with a Geiger-Müller counter.* Phys. Rev. *65,* 63 (1944).

Nawijn, *The mechanism of the Geiger-Müller counter* (in English). Physica *9,* 481 (1942).

Nawijn and Mulder, *The mechanism of the Geiger-Müller counter* (in English). Physica *10,* 531 (1943).

Neher and Harper, *A high speed Geiger counter circuit.* Phys. Rev. *49,* 940 (1936).

Neher and Pickering, *Modified high speed Geiger counter circuit.* Phys. Rev. *53,* 316 (1938).

Neher and Pickering, *Light weight high voltage supply for Geiger counters.* Rev. Sci. Instr. *12,* 140 (1941).

Ney and Mann, *Mass measurement with a single field mass spectrograph.* Phys. Rev. *69,* 239 (1946).

Ney and Armistead, *The self-diffusion coefficient of uranium hexifluoride.* Phys. Rev. *71,* 14 (1947).

Nielson, *Efficiency of positive and negative ions as condensation nuclei in the Wilson cloud chamber.* Phys. Rev. *61,* 202 (1942).

Nier, *The coefficient of thermal diffusion of methane.* Phys. Rev. *56,* 1009 (1939).

Nier, *Concentration of* C^{13} *by thermal diffusion.* Phys. Rev. *57,* 30 (1940).

Nier, *A mass spectrometer.* Rev. Sci. Instr. *11,* 212 (1940).

Nier, Ney and Inghram, *A null method for the comparison of two ion currents in a mass spectrometer.* Phys. Rev. *70,* 116 (1946).

Norris, Snell, Meiners and Slotin, *Manufacture of* C^{14} *in the chain reacting pile.* Phys. Rev. *70,* 110 (1946).

Oppenheimer, Johnston and Richman, *Drift tubes for linear proton accelerator.* Phys. Rev. *70,* 447 (1946).

Ortner and Stretter, *On the electrical measurement of particular corpuscular rays.* Z. Phys. *54,* 449 (1929).

Parkinson, Herb, Bernet and McKibben, *Electrostatic generator operating under high air pressure—operational experience and acces ory apparatus.* Phys. Rev. *53,* 642 (1938).

Parsegian, *Use of 6AK5 and 954 tubes in ionization chamber pulse amplifiers.* Rev. Sci. Instr. *17,* 39 (1946).

Peacock, *The measurement of absolute efficiencies of gamma ray counters.* Phys. Rev. *66,* 160 (1944).

Peacock and Osborne, *Control and recording circuits for a magnetic beta ray spectrograph.* Phys. Rev. *69,* 679 (1946).

Peacock and Good, *An automatic sample changer to be used for measuring radioactive samples.* Rev. Sci. Instr. *17,* 255 (1946).

Peck, *Photographic neutron detection.* Phys. Rev. *71,* 464 (1947).

Pegram, Urey and Huffman, *Distilling apparatus for separation of isotopes.* Phys. Rev. *49,* 883 (1936).

Peierls, *Statistical error in counting experiments.* Proc. Roy. Soc. *149,* 467 (1935).

Penick, *D.C. amplifier circuits for use with the electrometer tube.* Rev. Sci. Instr. *6,* 115 (1935).

Perfilov, *A new method of recording of particles of the type of U fragments by means of a photographic plate* (in English). J. Phys. USSR *10,* 1 (1946).

Pi, *Magnesium ion source for high in-*

tensity mass spectrograph. Phys. Rev. 67, 65 (1945).

Pickering and Snowden, A voltage regulator for high voltages. Phys. Rev. 65, 151 (1944).

Plesset, Studies with a permanent magnet beta ray spectrograph. Phys. Rev. 59, 936 (1941).

Plesset, Harnwell and Seidl, A permanent beta ray spectrograph. Rev. Sci. Instr. 13, 351 (1942).

Pollard, Use of a proportional counter for meson detection. Phys. Rev. 69, 689 (1946).

Pollock, Combination of betatron and synchrotron for electron. Phys. Rev. 69, 125 (1946).

Pollock, Langmuir, Elder, Blewett, Gurewitsch and Watters, Design of a 70 Mev synchrotron. Phys. Rev. 70, 798 (1946).

Post, A proposed high energy particle accelerator. Phys. Rev. 69, 126 (1946).

Potter, A four tube counter decade. Electronics 17, 110, 358, 360 (1944).

Powell, Further applications of the photographic method in nuclear physics. Nature 145, 155 (1940).

Powell, The protographic plate in nuclear physics. Endeavour 1, 151 (1942).

Powell, Applications of the photographic method to problems in nuclear physics. I. (a) the determination of the energy of homogeneous groups of alpha particles. (b) The determination of the energy of fast neutrons. Proc. Roy. Soc. 181, 344 (1943).

Powell, Occhialini, Livesey and Chilton, A new photographic emulsion for the detection of fast charged particles. J. Sci. Instr. 23, 102 (1946).

Purcell, The focusing of charged particles by a spherical condenser. Phys. Rev. 54, 818 (1938).

Rabinovich, Investigation of the phasing properties of the relativistic resonance accelerator. I. Synchrotron. II. Cyclotron with varying frequency of dee voltage (phasotron). J. Phys. USSR 10, 523 (1946).

Ramsey, Measurements of discharge characteristics of Geiger-Müller counters. Phys. Rev. 57, 1022 (1940).

Ramsey and Lees, Further confirmation of the Montgomery theory of counter discharge. Phys. Rev. 60, 411 (1941).

Ramsey and Hudspeth, Some discharge characteristics of self-quenching counters. Phys. Rev. 61, 95 (1942).

Ramsey, Directional properties of self-quenching counters. Phys. Rev. 61, 96 (1942).

Rayton and Wilkins, A Wilson cloud chamber investigation of the alpha particles from uranium. Phys. Rev. 51, 818 (1937).

Regener, Decade counting circuits. Phys. Rev. 69, 46 (1946).

Regener, Improved decade counting circuit. Phys. Rev. 69, 689 (1946).

Regener, Multiple master coincidence method for cosmic ray studies. Phys. Rev. 70, 449 (1946).

Reuterswärd, A new mass spectrograph (in German). Ark. Mat. Astr. Fys. 30A, No. 7 (1944).

Richardson, MacKenzie, Lofgren and Wright, Frequency modulated cyclotron. Phys. Rev. 69, 669 (1946).

Ringo, A magnetic alpha particle spectrograph. Phys. Rev. 59, 107 (1941).

Roberts, Interval selector: a device for measuring time distribution of pulses. Rev. Sci. Instr. 12, 71 (1941).

Rochester and Jánossy, The preparation and efficiency of the fast Geiger-Müller counter. Phys. Rev. 63, 52 (1943).

Rogers, Theory of the electrostatic beta particle energy spectrograph. Rev. Sci. Instr. 8, 22 (1937).

Rogers, Theory of electrostatic beta particle energy spectrograph. Rev. Sci. Instr. 11, 19 (1940).

Rogers and Horton, On the theory of the electrostatic beta-particle energy spectrograph. Rev. Sci. Instr. 14, 216 (1943).

Rogers, On the theory of the electrostatic beta particle energy spectrograph. IV. Phys. Rev. 69, 537 (1937).

Rose, Focusing and maximum energy of ions in the cyclotron. Phys. Rev. 53, 392 (1938).

Rose, Magnetic field corrections in the cyclotron. Phys. Rev. 53, 715 (1938).

Rose and Ramsey, On time lags in coincidence discharges of Geiger-Müller counters. Phys. Rev. *59*, 616 (1941).

Rose and Korff, *Properties of proportional counters.* Phys. Rev. *29*, 850 (1941).

Rose and Ramsey, *The behavior of proportional counter amplification at low voltages.* Phys. Rev. *61*, 198 (1942).

Rose and Ramsey, *Proportional counter amplification at low voltages.* Phys. Rev. *61*, 504 (1942).

Rossi, *Method of registering multiple simultaneous impulses of several Geiger's counters.* Nature *125*, 636 (1930).

Rotblat, *Application of the coincidence method for measurements of short life periods.* Proc. Roy. Soc. *177*, 260 (1941).

Ruark and Brammer, *The efficiency of counters and counter circuits.* Phys. Rev. *52*, 322 (1937).

Ruark, *Multivibrator Geiger counter circuit.* Phys. Rev. *53*, 316 (1938).

Rubin, *Particle scattering camera.* Phys. Rev. *70*, 447 (1946).

Rukavichnikov, *The first USSR cyclotron.* Phys. Rev. *52*, 1077 (1937).

Rumbaugh, *The isolation of Li isotopes with a mass spectrometer.* Phys. Rev. *49*, 882 (1936).

Rutherford, Wynn-Williams, Lewis and Bowden, *Analysis of alpha rays by an annular magnetic field.* Proc. Roy. Soc. *139*, 617 (1933).

Sagane, Miyamoto, Nakamura and Taketi, *Preliminary tests on the spiral orbit gamma ray spectrometer.* Proc. Phys.-Math. Soc. Jap. (3) *25*, 274 (1943).

Saha, *The theory of the 180° magnetic focusing type of beta ray spectrometer.* Indian J. Phys. *19*, 97 (1945).

Saxon and Schwinger, *Electron orbits in the synchrotron.* Phys. Rev. *69*, 702 (1946).

Schiff, *Statistical analysis of counter data.* Phys. Rev. *50*, 89 (1936).

Schiff, *On the paths of ions in the cyclotron.* Phys. Rev. *54*, 1114 (1938).

Schiff, *Energy angle distribution of betatron target radiation.* Phys. Rev. *70*, 87 (1946).

Schultz, *System for high speed counting of nuclear particles.* Phys. Rev. *69*, 689 (1946).

Schultz, *The determination of small time intervals between related nuclear events.* Phys. Rev. *71*, 134 (1947).

Schwarz, *A voltage stabilizer for a D. C. generator.* Phys. Rev. *61*, 544 (1942).

Schwarz and Pool, *Use of a mass spectroscope in artificial radioactive assignments.* Phys. Rev. *64*, 43 (1943).

Schwartz, *A note on the separation of gases by diffusion into a fast streaming vapor.* Phys. Rev. *68*, 145 (1945).

Schwinger, *Electron radiation in high energy accelerators.* Phys. Rev. *70*, 798 (1946).

Scott, *Focused beam source of hydrogen and helium ions.* Phys. Rev. *55*, 954 (1939).

Seaborg and Kennedy, *Nuclear isomerism and the chemical separation of isomers in tellurium.* Phys. Rev. *55*, 410 (1939).

Segrè, Halford and Seaborg, *Chemical separation of nuclear isomers.* Phys. Rev. *55*, 321 (1939).

Segrè and Wiegand, *Boron trifluoride detector for low neutron intensities.* Rev. Sci. Instr. *18*, 86 (1947).

Serber, *Orbits of particles in the racetrack.* Phys. Rev. *70*, 434 (1946).

Seth and Rao, *The linearity of the mass scale of Aston's mass spectrograph.* Indian J. Phys. *16*, 219 (1947).

Shapiro, *Tracks of nuclear particles in photographic emulsions.* Rev. Mod. Phys. *13*, 58 (1941).

Shaw and Rall, *An A. C. operated Mattuch type mass spectrograph.* Phys. Rev. *70*, 117 (1946).

Shaw, *An intense positive ion source for solids.* Phys. Rev. *71*, 277 (1947).

Shepherd and Haxby, *A scale-of-eight impulse counter.* Rev. Sci. Instr. *7*, 425 (1936).

Sherr, *Separation of gaseous isotopes by diffusion.* J. Chem. Phys. *6*, 251 (1938).

Sherwin, *Delays in firing time of Geiger-Müller counters.* Phys. Rev. *71*, 479 (1947).

Shive, *Practice and theory of the modulation of Geiger counters.* Phys. Rev. *56*, 579 (1939).

Shrader, *Partial separation of the iso-*

topes of chlorine by thermal diffusion. Phys. Rev. *69*, 439 (1946).

Siegbahn, *Investigations on the use of magnetic lenses for beta spectroscopy* (in German). Ark. Mat. Astr. Fys. *28A*, No. 17 (1942).

Siegbahn, *Studies in beta spectroscopy* (in English). Ark. Mat. Astr. Fys. *30A*, No. 20 (1944).

Siegbahn, *A magnetic lens of special field form for beta and gamma ray investigations, designs and applications.* Phil. Mag. (7) *37*, 162 (1946).

Simoda, *On the motion of positive ions in the Geiger-Müller counter.* Proc. Phys.-Math. Soc. Jap. (3) *25*, 445 (1943).

Simon, *Isotope separation factor of a thermal diffusion column.* Phys. Rev. *61*, 388 (1942).

Simon, *Performance of a hot wire Clusius and Dickel column.* Phys. Rev. *69*, 596 (1946).

Simpson, *Reduction of the natural insensitive time in Geiger-Müller counters.* Phys. Rev. *66*, 39 (1944).

Simpson, *Four-pi solid angle Geiger-Müller counters.* Rev. Sci. Instr. *15*, 119 (1944).

Simpson, *A precision alpha proportional counter.* Phys. Rev. *70*, 117 (1946).

Skaggs, Almy, Kerst and Langl, *Removal of the electron beam from the betatron.* Phys. Rev. *70*, 95 (1946).

Slack and Ehrcke, *200KV neutron source.* Rev. Sci. Instr. *8*, 193 (1937).

Slätis, *On a photographic method for investigating the optical properties of magnetic lenses and for recording beta ray lines* (in English). Ark. Mat. Astr. Fys. *32A*, No. 20 (1946).

Slater, *The design of linear accelerators.* Phys. Rev. *70*, 799 (1946).

Sloan and Coates, *Recent advances in the production of heavy high speed ions without the use of high voltages.* Phys. Rev. *46*, 539 (1934).

Sloan, *A radiofrequency high voltage generator.* Phys. Rev. *47*, 62 (1935).

Smith, Lozier and Bleakney, *Automatic recording mass-spectrographs.* Phys. Rev. *45*, 761 (1934).

Smith and Scott, *Conditions for producing intense ionic beams.* Phys. Rev. *55*, 946 (1939).

Smythe and Mattauch, *A new mass spectrometer.* Phys. Rev. *40*, 429 (1932).

Smythe, Rumbaugh and West, *High intensity mass spectrometer.* Phys. Rev. *45*, 220, 724 (1934).

Spatz, *Factors influencing the plateau characteristics of self-quenching Geiger-Müller counters.* Phys. Rev. *63*, 462 (1943).

Spatz, *The factors influencing the plateau characteristics of self-quenching Geiger-Müller counters.* Phys. Rev. *64*, 236 (1943).

Steigman, *A concentration method for certain radioactive metals.* Phys. Rev. *59*, 498 (1941).

Stephens, *A pulsed mass spectrometer with time dispersion.* Phys. Rev. *69*, 691 (1946).

Stevenson and Getting, *A vacuum tube circuit for scaling down counting rates.* Rev. Sci. Instr. *8*, 414 (1937).

Stever, *A directive Geiger counter.* Phys. Rev. *59*, 765 (1941).

Stever, *Circuit independence of charge in fast counter pulses.* Phys. Rev. *60*, 160 (1941).

Stever, *The discharge mechanism of fast Geiger-Müller counters from the deadtime experiment.* Phys. Rev. *61*, 38 (1942).

Stone, Livingston, Sloan and Chaffee, *Radiofrequency high voltage apparatus for X-ray therapy.* Radiology *24*, 153 (1935).

Straus, *A new mass spectrograph and the isotopic constitution of nickel.* Phys. Rev. *59*, 430 (1941).

Swann, *Fluctuations in measurements of ionization per cm of path in proportional counters.* Phys. Rev. *69*, 690 (1946).

Szilard and Chalmers, *Chemical separation of the radioactive element from its bombarded isotope in the Fermi effect.* Nature *134*, 462 (1934).

Taylor, *A modified Aston-type mass spectrometer and some preliminary results.* Phys. Rev. *47*, 666 (1935).

Taylor, *A mass spectrometer and gaseous thermal diffusion isotope separator.* Rev. Sci. Instr. *15*, 1 (1944).

Thode, Graham and Ziegler, *A mass spectrometer and the measurement of isotope exchange factors.* Canad. J. Res. *23B*, 40 (1945).

Thomas, *The paths of ions in the cyclotron. I. Orbits in the magnetic field.* Phys. Rev. *54*, 580 (1938).

Thomas, *The paths of ions in the cyclotron. II. Paths in the combined electric and magnetic fields.* Phys. Rev. *54*, 588 (1938).

Trump and Van de Graaff, *A compact pressure-insulated electrostatic X-ray generator.* Phys. Rev. *55*, 1160 (1939).

Tsien, Chastel, Faraggi and Vigneron, *A new photographic plate for the measurement of the paths of alpha rays.* C. R., Paris *223*, 571 (1946).

Tuve, Dahl and Hafstad, *The production and focusing of intense positive ion beams.* Phys. Rev. *48*, 241 (1935).

Tuve, Hafstad and Dahl, *High voltage technique for nuclear physics studies.* Phys. Rev. *48*, 315 (1935).

Van Atta, Van de Graaff and Barton, *A new design for a high voltage discharge tube.* Phys. Rev. *43*, 158 (1933).

Van Atta, Northrup, Van Atta and Van de Graaff, *The design, operation and performance of the Round Hill electrostatic generator.* Phys. Rev. *49*, 761 (1936).

Van Atta, Northrup, Van de Graaff and Van Atta, *Electrostatic generator for nuclear research at M. I. T.* Rev. Sci. Instr. *12*, 534 (1941).

Van de Graaff, *A 1,500,000 volt electrostatic generator.* Phys. Rev. *38*, 1919 (1931).

Van de Graaff, Compton and Van Atta, *The electrostatic production of high voltage for nuclear investigations.* Phys. Rev. *43*, 149 (1933).

Van de Graaff, Buechner, Woodward, McIntosh, Burrill and Sperduto, *An electrostatic generator for nuclear research.* Phys. Rev. *70*, 797 (1946).

Van Gemert, Den Hartog, and Muller, *Measurements on self-quenching Geiger-Müller counters* (in English). Physica *9*, 556 (1942).

Van Voorhis, Kuper and Harnwell, *Proton source for atomic disintegration experiments.* Phys. Rev. *45*, 492 (1934).

Van Voorhis, *The artificial radioactivity of copper, a branch reaction.* Phys. Rev. *49*, 876 (1936).

Van Voorhis, *Apparatus for the measurement of artificial radioactivity.* Phys. Rev. *49*, 889 (1936).

Veksler, *A new method for acceleration of relativistic particles* (in English). C. R., URSS *43*, 329 (1944).

Veksler, *A new method of acceleration of relativistic particle.* J. Phys. USSR *9*, 153 (1945).

Veksler, *Concerning some new methods of acceleration of relativistic particles.* Phys. Rev. *69*, 244 (1946).

de Vries, *The construction of counter tubes* (in English). Physica *11*, 433 (1946).

Wäffler, *Problems and research methods of nuclear physics.* Bull. Ass. Suisse Elect. *31*, 25, 282 (1940).

Walcher, *Enrichment of rubidium isotopes.* Phys. Z. *38*, 961 (1937).

Walcher, *On the possible uses of the Kunsman anodes for the mass spectrographic separation of isotopes.* Z. Phys. *121*, 604 (1943).

Walcher, *An ion source for mass spectrographic isotope separation.* Z. Phys. *122*, 62 (1944).

Walcher, *Separation of the thallium isotope. I. Mass spectroscopic separation.* Z. Phys. *122*, 401 (1946).

Walchli, *Performance tests on the Penn State type of Geiger-Müller circuits.* Phys. Rev. *65*, 346 (1944).

Wang, Marvin and Stenstrom, *A modification of the Barnes Geiger-Müller counter.* Rev. Sci. Instr. *13*, 81 (1942).

Wang, *On increasing the effectiveness of a betatron.* Phys. Rev. *69*, 42 (1946).

Ward, *A new method of determining half-value periods from observations with a single Geiger counter.* Proc. Roy. Soc. *181*, 183 (1942).

Washburn and Berry, *Effects of high initial energies on mass spectra.* Phys. Rev. *70*, 559 (1946).

Watson, *Concentration of heavy carbon by thermal diffusion.* Phys. Rev. *56,* 703 (1939).

Watson, *Thermal separation of isotopes.* Phys. Rev. *57,* 899 (1940).

Watson, Simon and Woernley, *Thermal diffusion separation of isotopes.* Phys. Rev. *62,* 558 (1942).

Weisz, *The Geiger-Müller tube.* Electronics *14,* 18 (1941).

Weisz, *Self-quenching Geiger-Müller counters.* Phys. Rev. *61,* 392 (1942).

Weisz, *Note on the nature of the gas mixture in self-quenching Geiger-Müller tubes.* Phys. Rev. *62,* 477 (1942).

Weisz, *Particle accelerators as mass analyzers.* Phys. Rev. *70,* 91 (1946).

Welles, *Partial separation of the oxygen isotopes by thermal diffusion.* Phys. Rev. *59,* 920 (1941).

Welles, *Partial separation of the oxygen isotopes by thermal diffusion and the deutron bombardment of O^{17}.* Phys. Rev. *69,* 586 (1946).

Weltin, *Soft radiation Geiger counter.* Rev. Sci. *14,* 278 (1943).

Westendorf, *Biased betatron in operation.* Phys. Rev. *71,* 271 (1947).

Westhover and Brewer, *Efficiency of thermal diffusion process for separating isotopes.* J. Chem. Phys. *8,* 314 (1940).

White, Henderson, Henderson and Ridenour, *The production of high energy alpha particles by the Princeton cyclotron.* Phys. Rev. *51,* 1012 (1937).

Wick, *On diffusion problems.* Z. Phys. *121,* 702 (1943).

Wideröe, *On bringing the beam out of a betatron.* Phys. Rev. *71,* 376 (1947).

Wiebe, *The problem of focusing in a cyclotron.* Z. Phys. *122,* 451 (1944).

Wilkening and Kanne, *Localization of the discharge in Geiger-Müller counters.* Phys. Rev. *62,* 534 (1942).

Wilkening and Kanne, *The localization phenomenon in Geiger-Müller counters.* Phys. Rev. *63,* 63 (1943).

Wilkens and St. Helens. *Grain spacing of alpha ray, proton and deuteron tracks in photographic emulsions.* Phys. Rev. *54,* 783 (1938).

Wilson, *On the comparative efficiency as condensation nuclei of positively and negatively charged ions.* Phil. Trans. Roy. Soc. *193,* 289 (1900).

Wilson, *Magnetic and electrostatic focusing in the cyclotron.* Phys. Rev. *53,* 408 (1938).

Wilson and Kamen, *Internal targets in the cyclotron.* Phys. Rev. *54,* 1031 (1938).

Wilson, *Formation of ions in the cyclotron.* Phys. Rev. *56,* 459 (1939).

Witcher, *An electron lens type of beta ray spectrometer.* Phys. Rev. *60,* 32 (1941).

Woodyard, *A comparison of the high frequency accelerator and betatron as a source of high energy electrons.* Phys. Rev. *69,* 50 (1946).

Wooldridge and Smythe, *The separation of gaseous isotopes by diffusion.* Phys. Rev. *50,* 233 (1936).

Wright and Richardson, *Frequency modulated cyclotron characteristics.* Phys. Rev. *70,* 445 (1946).

Wynn-Williams, *Use of thyratrons for high speed automatic counting of physical phenomena.* Proc. Roy. Soc. *132,* 295 (1931).

Wynn-Williams, *A thyratron "scale-of-two" automatic counter.* Proc. Roy. Soc. *136,* 312 (1932).

Yates, *Separation of isotopes for investigation of nuclear transformations.* Proc. Roy. Soc. *168,* 148 (1938).

Yetter, *Another circuit for use with Geiger-Müller counters.* Phys. Rev. *53,* 612 (1938).

York, Hildebrand, Putnam and Hamilton. *Acceleration of stripped light nuclei.* Phys. Rev. *70,* 446 (1946).

Yukawa and Sakata, *Efficiency of gamma ray counter.* Tokyo Inst. Phys. Chem. Res. *31,* 187 (1937).

Zahn and Spees, *An improved method for the determination of the specific charge of beta particles.* Phys. Rev. *53,* 357 (1938).

Zinn, *Low voltage positive ion source.* Phys. Rev. *52,* 650 (1937).

Zinn and Seely, *A neutron generator utilizing the deuteron-deuteron reaction.* Phys. Rev. *52,* 919 (1937).

LIST OF JOURNALS

Acta Phys. Polonica	Acta Physica Polonica. Warsaw.
Acta Soc. Sci. Fennicae	Acta Societas Scientiarum Fennicae. Helsingfors.
Amer. J. Phys.	American Journal of Physics. New York.
Amer. J. Sci.	American Journal of Science. New Haven, Conn.
Ann. N. Y. Acad. Sci.	Annals of the New York Academy of Science. New York.
Ann. Phys., Leipzig	Annalen der Physik. Leipzig.
Ann. Soc. Sci. Brux.	Annales de la Société Scientifique de Bruxelles. Louvain.
Anz. Akad. Wiss. Wien	Anzeiger der Akademie der Wissenschaften in Wien. Mathematisch-naturwissenschaftliche Klasse. Vienna.
Ark. Kemi. Min. Geol.	Arkiv för Kemi, Mineralogi och Geologi. Stockholm.
Ark. Mat. Astr. Fys.	Arkiv för Matematik Astronomi och Fysik. Stockholm.
Astrophys. J.	Astrophysical Journal. Chicago.
Atti Accad. Lincei.	Atti della reale accademia nazionale dei Lincei. Rendiconti. Classe di scienze fisiche, matematiche e naturali. Rome.
Ber. Deutsch. Chem. Ges.	Berichten der deutschen chemischen Gesellschaft. Berlin.
Bull. Ass. Suisse Elect.	Bulletin de L'Association Suisse des Électriciens. Zürich.
Bull. Soc. Roumaine de Phys.	Bulletin de la société roumaine de physique. Bucharest.
Canad. J. Res.	Canadian Journal of Research. Ottawa.
Chem. Rev.	Chemical Reviews. Baltimore.
C.R., Paris	Compte Rendu Hebdomadaire des Séances de L'Académie des Sciences, Paris.
C.R., URSS	Comptes Rendus de l'Académie des Sciences de l'URSS. Moscow.
Curr. Sci.	Current Science. Bangalore.
Electronics	Electronics. New York and London.
Endeavour	Endeavour. London.
Ergeb. d. exakt. Naturwiss.	Ergebnisse der exakten Naturwissenschaften. Berlin.
Gött. Nachr.	Königliche Gesellschaft der Wissenschaften zu Göttingen. Nachrichten. Göttingen.

Helv. Phys. Acta	Helvetica Physica Acta. Basle.
Indian J. Phys.	Indian Journal of Physics. Calcutta.
Int. Conf. Phys. Lond.	International Conference on Physics. London.
J. Amer. Chem. Soc.	Journal of the American Chemical Society. Easton, Penna.
J. Appl. Phys.	Journal of Applied Physics (formerly Physics). New York.
J. Chem. Phys.	Journal of Chemical Physics. New York.
J. Chem. Soc.	Journal of the Chemical Society. London.
J. Exp. Theor. Phys. USSR	Journal of Experimental and Theoretical Physics of the USSR (Zhurnal Exsperimentalnoi i Teoreticheskoi Fiziki). Moscow.
J. Franklin Inst.	Journal of the Franklin Institute. Lancaster and Philadelphia, Penna.
J. Mysore Univ.	Journal of the Mysore University. Bangalore.
J. Phys. Chem.	Journal of Physical Chemistry. Baltimore, Md., and London.
J. Phys. Rad.	Journal de physique et le radium. Paris.
J. Phys., USSR	Journal of Physics. Moscow.
J. Res. Nat. Bur. Stand.	Journal of Research of the National Bureau of Standards. Washington, D. C.
J. Sci. Instr.	Journal of Scientific Instruments. London.
K. Fys. Sällsk. Lund, Förh.	Kungliga Fysiografiska Sällskapets i Lund Förhandlingar. Lund, Sweden.
K. Ned. Akad. Verh.	Koninklijke Akademie van Wetenschappen, Verhanelingen. Amsterdam.
Kgl Danske Vid. Sels.	Kongelige Danske Videnskabernes Selskab. Copenhagen.
Mem. Coll. Sci. Kyoto Imp. Univ.	Memoirs of the College of Science Kyoto Imperial University. Tokyo.
Metrop-Vick. Gaz.	Metropolitan-Vickers Gazette. Manchester.
Nature	Nature. London.
Naturwiss.	Naturwissenschaften. Berlin.
Ned. T. Natuurk.	Nederlandsch Tijdschrift voor Natuurkunde. The Hague.
Nuovo Cim.	Il Nuovo Cimento. Bologna.
Phil. Mag.	The London, Edinburgh and Dublin Philosophical Magazine and Journal of Science. London.
Phil. Trans. Roy. Soc.	Philosophical Transactions of the Royal Society of London. London.
Phys. in regelmäss Ber.	Die Physik in regelmässigen Berichten. Leipzig.
Phys. Rev.	Physical Review. New York.

Phys. Z.	Physikalische Zeitschrift. Leipzig.
Phys. Z. Sowjet.	Physikalische Zeitschrift der Sowjetunion. Leipzig.
Physica	Physica. The Hague.
Portugaliae Physica	Portugaliae Physica. Lisbon.
Proc. Amer. Phil. Soc.	Proceedings of the American Philosophical Society. Philadelphia, Penna.
Proc. Camb. Phil. Soc.	Proceedings of the Cambridge Philosophical Society. London.
Proc. Indian Acad. Sci.	Proceedings of the Indian Academy of Sciences. Bangalore.
Proc. Iowa Acad. Sci.	Proceedings of the Iowa Academy of Science. Ames, Iowa.
Proc. K. Akad. Amsterdam	Proceedings of the "Koninklijke Akademie van Wetenschappen te Amsterdam." Amsterdam.
Proc. Nat. Acad. Sci.	Proceedings of the National Academy of Sciences. Washington, D. C.
Proc. Nat. Inst. Sci. India.	Proceedings of the National Institute of Sciences of India. Calcutta.
Proc. Phys.-Math. Soc. Jap.	Proceedings of the Physico-Mathematical Society of Japan. Tokyo.
Proc. Phys. Soc., Lond.	Proceedings of the Physical Society. London.
Proc. R. Irish Acad.	Proceedings of the Royal Irish Academy. Dublin and London.
Proc. Roy. Soc.	Proceedings of the Royal Society of London. London.
Publ. Fac. Cienc. Fis.-Mat. La Plata	Publicaciones de la Facultad de Ciencias Físico-Matemáticas de la Universidad Nacional de la Plata. La Plata.
Radiology	Radiology. Syracuse, New York.
Rep. Phys. Soc. Progr. Phys.	Reports of the Physical Society on Progress in Physics. London.
Rev. Bras. Biol.	Revista Brasiliera de Biologia. Rio de Janeiro.
Rev. Gén. Elect.	Revue Génerale d'Electricité. Paris.
Rev. Mod. Phys.	Reviews of Modern Physics. New York.
Rev. Sci. Instr.	Review of Scientific Instruments. New York.
Rev. Sci., Paris	Revue Scientifique. Paris.
Ric. Sci. Ricostruz.	Ricerca Scientifica e Ricostruzione. Rome.
S.B. Akad. Wiss. Wien.	Sitzungsberichte der Akademie der Wissenschaften in Wien. Vienna.
S.B. preuss Akad.	Sitzungsberichte der preussische Akademie der Wissenschaften. Berlin.
Science and Culture	Science and Culture.

Science	Science. Lancaster, Penna.
Sci. Rec. Acad. Sin.	Science Record Academia Sinica. Chungking.
Tohoku Univ. Sci. Rep.	Scientific and Technology reports of the Tohoku Imperial University. Sendai, Japan.
Tokyo Inst. Phys. Chem. Res.	Scientific Papers of the Institute of Physical and Chemical Research. Tokyo.
Trans. Faraday Soc.	Transactions of the Faraday Society. London.
Univ. Lund (Thesis Phys. Inst.)	University of Lund. Thesis of the Physics Institute. Lund, Sweden.
Z. angew. Chem.	Zeitschrift für angewandte Chemie. Berlin.
Z. anorg. allg. Chem.	Zeitschrift für anorganische und allgemeine Chemie. Leipzig.
Z. Naturforsch.	Zeitschrift für Naturforschung. Wiesbaden, Germany.
Z. Phys.	Zeitschrift für Physik. Berlin.
Z. phys. Chem.	Zeitschrift für physikalische Chemie. Leipzig.
Z. techn. Phys.	Zeitschrift für technische Physik. Leipzig.

The language of the article is the same as that used in the name of the journal unless otherwise indicated.

THE THEORY OF ELECTRONS
by H. A. Lorentz

This is an unabridged reproduction of the second (1915) edition of a famous course of lectures delivered at Columbia University by Nobel Laureate H. A. Lorentz. A great classic of science it is also still astonishingly modern. It has been characterized as differing basically from more recent books only in that "not all more modern books contain the same outstanding discussion of general principles and experimental facts."

Beginning with Maxwell's electromagnetic equations, the author discusses the emission and absorption of electro-magnetic radiation, the theory of the Zeeman effect, the propagation of electromagnetic waves in bodies composed of molecules, and optical phenomena in moving bodies. 109 pages of notes give detailed examinations of the mathematics involved.

PARTIAL CONTENTS. Chapter cover: General principles of the theory of electrons, electromagnetic equations, field, potential, polarized particles, theory of free electrons, etc. Emission and absorption of heat, energy of radiation, Boltzmann's, Wien's law, Jean's theory, etc. Theory of the Zeeman-effect, electrons in magnetic fields, series of spectral lines, rotation of particles in magnetic fields, polarization of radiation, etc. Propagation of light in a body composed of molecules, theory of the inverse Zeeman-effect, equations of motion, refractivity, dispersion & absorption of light, transverse Zeeman-effect, rotation of refraction planes, magnetic double refraction, etc. Optical phenomena in moving bodies, Huygen's principle, Stokes's theory of aberration, velocity of light in a moving medium, Fresnel's coefficient, Michelson's experiment, moving electristatic systems, molecular motion, general electromagnetic equations, Einstein, principle of relativity, etc. 109 pages of notes.

Index. 9 figures. 325pp. 5⅜ x 8.

Paperbound **$1.00**

ORDINARY DIFFERENTIAL EQUATIONS
by E. L. Ince

The theory of ordinary differential equations in real and complex domains is here clearly explained and analyzed. The author covers not only classical theory, but also main developments of more recent times.

The pure mathematician will find valuable exhaustive sections on existence and nature of solutions, continuous transformation groups, the algebraic theory of linear differential systems, and the solution of differential equations by contour integration. The engineer and physicist will be interested in an especially fine treatment of the equations of Legendre, Bessel, and Mathieu; the transformations of Laplace and Mellin; the conditions for the oscillatory character of solutions of a differential equation; the relation between a linear differential system and an integral equation; the asymptotic development of characteristic numbers and functions; and many other topics.

PARTIAL CONTENTS: **Real Domain**. Elementary methods of integration. Existence and nature of solutions. Continuous transformation-groups. Linear differential equations — theory of, with constant coefficients, solutions of, algebraic theory of. Sturmian theory, its later developments. Boundary problems. **Complex Domain**. Existence theorems. Equations of first order. Non-linear equations of higher order. Solutions, systems, classifications of linear equations. Oscillation theorems.

"Will be welcomed by mathematicians, engineers, and others," MECH. ENGINEERING. "Highly recommended," ELECTRONICS INDUSTRIES. "Deserves the highest praise," BULLETIN, AM. MATH. SOC.

Historical appendix. Bibliography. Index. 18 figures. viii + 558pp. 5⅜ x 8.

.S349 Paperbound **$2.45**

TREATISE ON THE DYNAMICS OF A SYSTEM OF RIGID BODIES (Advanced Part)

by Edward John Routh

This is an unabridged republication of the 6th revised edition of a standard work on the fundamentals of dynamics. It provides a full coverage of basic theorems, motions and forces, and applications of calculus to dynamics studies. It is especially valuable for its full demonstrations and analyses, and contains much material that has not been duplicated in more recent texts: application of the calculus of finite differences to the dynamics of rigid bodies. It is highly concrete and practical, with hundreds of applied situations and hundreds of full demonstrations. More than 400 problems are provided for the student to work out; in most cases instructions are provided for solution. A brilliant section on the calculus of variations is of special interest.

"Expert handling and masterly presentation give this book its value," AUSTRALIAN ENGINEER. "A profusion of individual problems and methods, such as is seldom treated so extensively and so basically," FACULTE DES SCIENCES, UNIVERSITY OF INSTANBUL. "Much of its material has never been duplicated, of great value," AERO DIGEST.

PARTIAL CONTENTS. Chapters cover Moving axes and relative motion. Oscillations about equilibrium. Oscillations about a state of motion. Motion of a body under no forces. Motion of a body under any forces. Nature of motion given by linear equations, conditions of stability. Free and forced vibrations. Determination of constants of integration in terms of initial conditions. Calculus of finite differences. Calculus of variations. Procession and nutation. Motion of the moon about its center. Motion of a string or chain. Motion of a membrane.

Index. 64 figures. xiv + 484pp. 5⅜ x 8.

S229 Paperbound $2.35

PARTIAL DIFFERENTIAL EQUATIONS OF MATHEMATICAL PHYSICS

by A. G. Webster

Still one of the most important treatises on partial differential equations in any language, this comprehensive work by one of America's greatest mathematical physicists covers the basic method, theory and application of partial differential equations. There are clear and full chapters on

> Fourier series
> integral equations
> elliptic equations
> spherical, cylindrical, ellipsoidal harmonics
> Cauchy's method
> boundary problems
> method of Riemann-Volterra
> and many other topics

This is a book complete in itself, developing fully the needed theory and application of every important field.

> vibration
> elasticity
> potential theory
> theory of sound
> wave propagation
> heat conduction
> and others

Professor Webster's work is a keystone book in the library of every mature physicist, mathematical physicist, mathematician, and research engineer. It can also serve as an introduction and supplementary text for the student.

Edited by Samuel J. Plimpton. Second corrected edition. 97 illustrations. vii + 440pp. 5⅜ x 8.

S263 Paperbound $2.00

SELECTED PAPERS IN QUANTUM ELECTRODYNAMICS
edited by Julian Schwinger

The development of quantum mechanics during the first quarter of this century produced a revolution in physical thought even more profound than that associated with the theory of relativity. Nowhere is this more evident than in the area of the theoretical and experimental investigations centering about the properties and the interactions of the electromagnetic field, or, as it is otherwise known, electrodynamics.

In this volume the history of quantum electrodynamics is dramatically unfolded through the original words of its creators. It ranges from the initial successes, to the first signs of crisis, and then, with the stimulus of experimental discovery, to new triumphs leading to an unparalleled quantitative accord between theory and experiment. It terminates with the present position of quantum electrodynamics as part of the larger subject of theory of elementary particles, faced with fundamental problems and the future prospect of even more revolutionary discoveries.

Physicists, mathematicians, electromagnetic engineers, students of the history and philosophy of science will find much of permanent value here. The techniques of quantum electrodynamics are not likely to be substantially altered by future developments, and the subject presents the simplest physical illustration of the challenge posed by the "basic inadequacy and incompleteness of the present foundations of theoretical physics."

Papers are included by Bethe, Bloch, Dirac, Dyson, Fermi, Feynman, Heisenberg, Kusch, Lamb, Oppenheimer, Pauli, Schwinger, Tomonaga, Weisskopf, Wigner, and others. There are a total of 33 papers, 28 of which are in English, 1 in French, 3 in German, and 1 in Italian.

Preface and historical commentary by the editor. xvii + 423pp. 5⅜ x 8.

S444 Paperbound **$2.45**